Springer Proceedings in Physics

Volume 361

Indexed by Scopus

The series Springer Proceedings in Physics, founded in 1984, is devoted to timely reports of state-of-the-art developments in physics and related sciences. Typically based on material presented at conferences, workshops and similar scientific meetings, volumes published in this series will constitute a comprehensive up to date source of reference on a field or subfield of relevance in contemporary physics. Proposals must include the following:

- Name, place and date of the scientific meeting
- A link to the committees (local organization, international advisors etc.)
- Scientific description of the meeting
- List of invited/plenary speakers
- An estimate of the planned proceedings book parameters (number of pages/articles, requested number of bulk copies, submission deadline).

Please contact:

For Americas and Europe: Dr. Zachary Evenson; zachary.evenson@springer.com
For Asia, Australia and New Zealand: Dr. Loyola DSilva; loyola.dsilva@springer.com

José W. F. Valle · Sandhya Choubey ·
Debasish Borah · Rahul Srivastava ·
Debajyoti Dutta
Editors

Proceedings of the International Conference on Future Prospects in Neutrino and Astroparticle Physics

ICFPNAP 2024; 23–24 January; Assam; India

Springer

Preface

The International Conference on Future Prospects in Neutrino and Astroparticle Physics 2024 (ICFPNAP2024) was sponsored by the Science and Engineering Research Board (SERB) and was held from January 23–24, 2024, at Assam Don Bosco University. The Department of Physics at Assam Don Bosco University, Assam, organized the conference in conjunction with the Physics Academy of North East (PANE). Several participants from various Indian universities, colleges, and institutes presented their research work at the conference. A portion of the submitted papers has been included in this book after undergoing the appropriate peer review procedures.

It should be mentioned that the Physics Department at Assam Don Bosco University, which was established in 2018, has been striving to popularize both basic and advanced physics through a variety of additional means, such as workshops, refresher courses, symposiums, and so on. The department believed that an international conference on Future Prospects on Neutrino and Astroparticle Physics could provide a platform for young scientists to exchange ideas and receive advice and support from esteemed academicians in the nation and abroad.

An international advisory committee was formed to carry out this vision. We would like to take this opportunity to thank all of the esteemed members of the advisory committee who have helped us and provided sage advice. We would like to thank each and every reviewer who has improved the proceeding by reviewing each and every document. We also acknowledge the contributions made by each author to the proceedings. Lastly, we would like to thank the Science and Engineering Research Board (SERB), India, for their financial support, without which the event could not have been successfully organized.

Dr. Debajyoti Dutta
Bhattadev University
Bajali, Assam, India

Contents

Contributors

Dharitree Bezboruah Tezpur University, Sonitpur, Assam, India

Subhaditya Bhattacharya Department of Physics, Indian Institute of Technology Guwahati, Assam, India

Debasish Borah Department of Physics, Indian Institute of Technology Guwahati, Assam, India

Happy Borgohain Department of Physics, Silapathar College, Silapathar, India

Hemanshi Bundeliya Department of Physics, Lovely Professional University, Phagwara, Punjab, India

Pralay Chakraborty Department of Physics, Gauhati University, Guwahati, Assam, India

Salvador Centelles Chuliá Max-Planck-Institut für Kernphysik, Heidelberg, Germany

Mrinal Kumar Das Department of Physics, Tezpur University, Napaam, Tezpur, India

Nayan Das Department of Physics, Indian Institute of Technology Guwahati, Assam, India

Arnab Dasgupta Pittsburgh Particle Physics, Astrophysics, and Cosmology Center, Department of Physics and Astronomy, University of Pittsburgh, Pittsburgh, PA, USA

Moon Moon Devi Tezpur University, Sonitpur, Assam, India

Manash Dey Department of Physics, Gauhati University, Guwahati, Assam, India

Sangeeta Dey Cotton University, Guwahati, India

Zannatun Firdowzy Dey Assam Don Bosco University, Sonapur, Assam, India

Debajyoti Dutta Bhattadev University, Pathsala, Bajali, Assam, India

Manoranjan Dutta Department of Physics, North Lakhimpur University, Lakhimpur, Assam, India

Rupak Dutta National Institute of Technology Silchar, Silchar, India

Nayana Gautam Department of Physics, Assam Down Town University, Guwahati, India

Sagar Tirtha Goswami Department of Physics, Gauhati University, Guwahati, India

Ankita Kakoti Department of Physics, Tezpur University, Napaam, Tezpur, India

Raktima Kalita Department of Physics, Cotton University, Guwahati, Assam, India

Monal Kashav Department of Physics and Astronomical Science, Central University of Himachal Pradesh, Dharamshala, India

Lekhashri Konwar Indian Institute of Technology Jodhpur, Jodhpur, India

Ranjeet Kumar Department of Physics, Indian Institute of Science Education and Research - Bhopal, Bhopal, India

Sanjeev Kumar Department of Physics and Astrophysics, University of Delhi, Delhi, India

Jayita Lahiri Institut für Theoretische Physik, Universität Hamburg, Hamburg, Germany

Devabrat Mahanta Department of Physics, Abhayapuri College, Bongaigaon, Assam, India

Soumen Kumar Manna Department of Physics, Indian Institute of Technology Guwahati, Guwahati, Assam, India

Iffat Ara Mazumder National Institute of Technology Silchar, Silchar, India

Abinash Medhi Tezpur University, Sonitpur, Assam, India;
Indian Institute of Technology Guwahati, Guwahati, Assam, India

Manas Kumar Mohapatra University of Hyderabad, Hyderabad, Telangana, India

Partha Pratim Nath Cotton University, Guwahati, India

Mahadev Patgiri Cotton University, Guwahati, India

S. D. Pathak Department of Physics, Lovely Professional University, Phagwara, Punjab, India

Partha Kumar Paul Department of Physics, Indian Institute of Technology Hyderabad, Kandi, Telangana, India

Oleg Popov Department of Biology, Shenzhen MSU-BIT University, Shenzhen, China

Dipankar Pradhan Department of Physics, Indian Institute of Technology Guwahati, Assam, India

Hemant Prajapati Department of Physics, Indian Institute of Science Education and Research - Bhopal, Bhauri, Bhopal, India

Victoria Puyam Department of Physics, Manipur University, Imphal, India

Samiran Roy School of Physics, University of Hyderabad, Hyderabad, India

Subhankar Roy Department of Physics, Gauhati University, Guwahati, Assam, India

Indrajit Saha Department of Physics, Indian Institute of Technology Guwahati, Assam, India

Suchismita Sahoo Department of Physics, Central University of Karnataka, Kalaburagi, India

Sujit Kumar Sahoo Department of Physics, Indian Institute of Technology Hyderabad, Telangana, India

Arnab Sarker Tezpur University, Sonitpur, Assam, India

Debojit Sarma Cotton University, Guwahati, India

Arunansu Sil Department of Physics, Indian Institute of Technology Guwahati, Guwahati, Assam, India

Labh Singh Department of Physics and Astronomical Science, Central University of Himachal Pradesh, Dharamshala, India

Mayengbam Kishan Singh Department of Physics, Manipur University, Imphal, India

Ngangkham Nimai Singh Department of Physics, Manipur University, Imphal, India;
Research Institute of Science and Technology, Manipur University, Imphal, India

Y. Monitar Singh Department of Physics, Manipur University, Imphal, India

Shivaramakrishna Singirala University of Hyderabad, Hyderabad, Telangana, India

Rahul Srivastava Department of Physics, Indian Institute of Science Education and Research - Bhopal, Bhauri, Bhopal, India

Tapender Department of Physics and Astronomical Science, Central University of Himachal Pradesh, Dharamshala, India

Juhi Vardani Indian Institute of Technology Jodhpur, Jodhpur, India

Surender Verma Department of Physics and Astronomical Science, Central University of Himachal Pradesh, Dharamshala, India

Ph. Wilina Department of Physics, Manipur University, Imphal, India

Ajay Kumar Yadav Department of Physics, Central University of Karnataka, Kalaburagi, India

Bhavna Yadav Indian Institute of Technology Jodhpur, Jodhpur, India

Chapter 1
Probing Scalar Leptoquarks Using Neutrino-Nucleus Coherent Scattering

Samiran Roy

Abstract The recent measurements of the coherent elastic neutrino-nucleus scattering (CEνNS) by the COHERENT collaboration give us a capability to examine the various new physics scenarios at low energy neutrino experiments. One such new physics is the scalar leptoquarks (LQs) that arise in many extensions of the Standard Model (SM). We consider the low-scale LQ models that forbid the rapid proton decay by construction. The hypercharges of the scalar LQs are Y = 1/6 and Y = 7/6, which are part of the two electroweak doublet fields, respectively. We constrain the LQs using the COHERENT data that is consistent with the SM prediction. We find that the COHERENT measurements can put a strong bound on the LQs couplings to SM particles over a wide range of LQ masses, and the bounds are competitive with various other experiments.

1.1 Introduction

More than forty years after Freedman's prediction [1], the COHERENT collaboration [2] eventually succeeded in detecting the coherent elastic neutrino-nucleus scattering (CEνNS). The cross-section of the process scales roughly as the square of the total number of protons (Z) and neutrons (N) of the target nucleus for the low momentum transfer (q^2) to the nucleus. In the Standard Model (SM), CEνNS process is mediated by the Z-boson, and the cross-section approximately scales as the square of N since the Z-boson coupling to the neutron is an order of magnitude larger compared to the proton. The observed data by the COHERENT collaboration shows no significant departure from the SM prediction, for CsI [2, 3] and Ar [4] nuclei. Thus, we can use this experimental observation to constrain various beyond the standard model physics, *e.g.* Leptoquarks (LQs), that modifies the CEνNS cross-section. LQs arise in many extensions of the standard model. It carries both the lepton and baryon numbers and can be a scalar or a vector in nature. To prevent the proton decay, the mass of such LQs is typically assumed to be near to the GUT scale. However, there

---

S. Roy (✉)
School of Physics, University of Hyderabad, Hyderabad 500046, India
e-mail: samiranroy.hri@gmail.com

J. W. F. Valle et al. (eds.), *Proceedings of the International Conference on Future Prospects in Neutrino and Astroparticle Physics*, Springer Proceedings in Physics 361,
https://doi.org/10.1007/978-981-97-7441-8_1

are LQ models which prohibit rapid proton decay by construction. We study the LQs, where proton decay does not occur at the tree level since they do not give rise to tree level interactions of the kind $(qlqq)$. Here, we concentrate on the scalar LQs, which are less restricted than the vector case where there is less freedom as the couplings are fixed by the gauge structure of the model. We employ the latest CsI and Ar data to determine the exclusion limits on scalar LQs by altering the SM event rate in the presence of LQs across a broad mass range, spanning from MeV to TeV scales, and compare the results with other experimental bounds on LQs.

1.2 Models

We consider two scalar LQ doublets under $SU(2)_L$, $\tilde{R}_2 = (\Delta_1^{2/3}, \Delta_1^{-1/3})^T$ and $R_2 = (\Delta_2^{5/3}, \Delta_2^{2/3})^T$ with hypercharges Y = 1/6 and Y = 7/6, respectively. The pertinent portion of the Lagrangian involved in the coherent process of neutrino-nucleon scattering is given by

$$\begin{aligned} \mathcal{L}^{\Delta_1} &\supset -y_{ij}^{(1)} \bar{d}_R^i \tilde{R}_2 L_L^j + h.c. \\ &= -y_{ij}^{(1)} \bar{d}^i P_L \ell^j \Delta_1^{2/3} - y_{ij}^{(1)} \bar{d}^i P_L \nu^j \Delta_1^{-1/3} + h.c. \end{aligned} \tag{1.1}$$

$$\begin{aligned} \mathcal{L}^{\Delta_2} &\supset -y_{ij}^{(2)} \bar{u}_R^i R_2 L_L^j + h.c. \\ &= -y_{ij}^{(2)} \bar{u}^i P_L \ell^j \Delta_2^{5/3} - y_{ij}^{(2)} \bar{u}^i P_L \nu^j \Delta_2^{2/3} + h.c., \end{aligned} \tag{1.2}$$

where $P_{L,R}$, are the usual left and right chiral projection operators, $L_L^j = \left(\nu^j \ \ell^j\right)^T$ are the lepton doublets, d^i and u^i are the down-type and up-type quark fields, respectively, and y's are the Yukawa coupling matrices. We only take into account couplings with the first generation of quarks in order to prevent Flavour Changing Neutral Current (FCNC) problem at the tree level. Our forecast for the CEνNS is unaffected by this choice because the process solely depends on the first generation of valence quarks.

1.3 Signal Prediction in COHERENT

In this section, we show the signal prediction for both the SM and LQs cases at COHERENT experiments. The differential neutrino fluxes from the Spallation Neutron Source (SNS) are as follows:

$$\frac{dN_{\nu_\mu}}{dE} = \eta\, \delta\left(E - \frac{m_\pi^2 - m_\mu^2}{2m_\pi}\right) \tag{1.3}$$

$$\frac{dN_{\bar{\nu}_\mu}}{dE} = \eta \frac{64E^2}{m_\mu^3}\Big(\frac{3}{4} - \frac{E}{m_\mu}\Big) \tag{1.4}$$

$$\frac{dN_{\nu_e}}{dE} = \eta \frac{192E^2}{m_\mu^3}\Big(\frac{1}{2} - \frac{E}{m_\mu}\Big), \tag{1.5}$$

where m_π and m_μ are the masses of pion and muon, respectively, $\eta = r\,\mathrm{N_{POT}}/4\text{ß}\mathrm{L}^2$ is a normalization factor, $\mathrm{N_{POT}}$ corresponds to the total number of protons on target (POT), L is the distance from source to detector, and r represents per flavour neutrino produced for each POT. For the CsI detector $r = 0.08$, L = 19.3 m and $\mathrm{N_{POT}} = 3.198 \times 10^{23}$ [3] while for Ar $r = (9 \pm 0.9) \times 10^{-2}$, L = 27.5 m and $\mathrm{N_{POT}} = 13.7 \times 10^{22}$. As a function of the nuclear-recoil kinetic energy (T_{nr}), the SM contribution to the cross-section for a given neutrino flavour (ν_l) is given by

$$\frac{d\sigma_{\nu_l - N}}{dT_{nr}}(E, T_{nr}) = \frac{G_F^2 M}{\pi}\Big(1 - \frac{M T_{nr}}{2E^2}\Big) Q_{l,SM}^2. \tag{1.6}$$

Here E represents the energy of incoming neutrino, M is the detector material mass, G_F is the Fermi constant, and

$$Q_{l,SM}^2 = [g_V^p(\nu_l) Z F_Z(|\vec{q}|^2) + g_V^n N F_N(|\vec{q}|^2)]^2. \tag{1.7}$$

$$g_V^p(\nu_e) = 0.0401, \;\; g_V^p(\nu_\mu) = 0.0318, \;\; g_V^n = -0.5094. \tag{1.8}$$

The distributions of proton and neutron in the nucleus are represented by the Helm form factor [5]

$$F(|\vec{q}|^2) = 3\frac{j_1(|\vec{q}|R_0)}{|\vec{q}|R_0} e^{-|\vec{q}|^2 s^2/2}, \tag{1.9}$$

where $j_1(x) = \sin(x)/x^2 - \cos(x)/x$ is the spherical Bessel function, the rms radius R is connected to R_0 by $R_{p,n}^2 = 3R_0^2/5 + 3s^2$, and $s = 0.9$ fm.

The number of events in the i-th bin of nuclear-recoil energy is calculated using the following equation:

$$N_i^{\mathrm{CE}\nu\mathrm{NS}} = N(\mathcal{N}) \int_{T_{\mathrm{nr}}^i}^{T_{\mathrm{nr}}^{i+1}} dT_{\mathrm{nr}}\, A(T_{\mathrm{nr}}) \int_{E_{\min}}^{E_{\max}} dE \sum_{\nu=\nu_e,\nu_\mu,\bar{\nu}_\mu} \frac{dN_\nu}{dE}\frac{d\sigma_{\nu\text{-}\mathcal{N}}}{dT_{\mathrm{nr}}}(E, T_{\mathrm{nr}}), \tag{1.10}$$

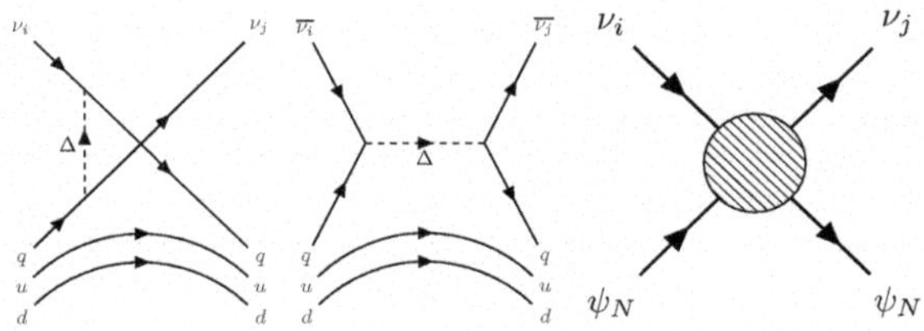

Fig. 1.1 Tree level CEνNS processes mediated by the Leptoquark. The final state neutrinos are experimentally indistinguishable

where $E_{\rm min} = \sqrt{MT_{\rm nr}/2}$, $E_{\rm max} = m_\mu/2 \sim 52.8$ MeV, and $A(T_{\rm nr})$ corresponds to the energy-dependent reconstruction efficiency.

The standard cross-section will be altered due to the extra contribution coming from LQs as shown in Fig. 1.1. As the momentum transfer to the nucleus is low, these diagrams can be represented by the effective four fermion interaction like

$$\mathcal{L}_{\rm eff}^{\Delta} = \frac{y^2}{m_\Delta^2}\,(\bar{\psi}_N P_L \nu)\,(\bar{\nu} P_R \psi_N)\,, \tag{1.11}$$

where ψ_N represents either a u or d quark. Utilizing the Fierz transformations, we can factorize this diagram into a hadronic current and a neutrino current as

$$\mathcal{L}_{\rm eff}^{\Delta} \sim -\frac{y^2}{2\,m_\Delta^2}\,(\bar{\psi}_N \gamma^\mu P_R \psi_N)\,(\bar{\nu}\gamma_\mu P_L \nu). \tag{1.12}$$

The new cross-section combining the effect of both SM and LQs is given by

$$\frac{d\sigma_{\nu_i - N}}{dT_{nr}}(E, T_{nr}) = \frac{G_F^2 M}{\pi}\left(1 - \frac{MT_{nr}}{2E^2}\right) Q_{i,k}^2 \tag{1.13}$$

$$Q_{i,k}^2 = \left((Q_{i,SM} + Q_{ii,\Delta_k})^2 + \sum_{j\neq i} Q_{ij,\Delta_k}^2 \right) \tag{1.14}$$

where

$$\begin{aligned} Q_{ij,\Delta_1} &= \frac{\tilde{y}_{1i}^{(1)}\tilde{y}_{1j}^{(1)}}{4\sqrt{2}\,G_F}\,\frac{ZF_Z(|\vec{q}|^2) + 2\,NF_N(|\vec{q}|^2)}{|\vec{q}|^2 + m^2_{\Delta_1^{-1/3}}}, \\ Q_{ij,\Delta_2} &= \frac{\tilde{y}_{1i}^{(2)}\tilde{y}_{1j}^{(2)}}{4\sqrt{2}\,G_F}\,\frac{2\,ZF_Z(|\vec{q}|^2) + NF_N(|\vec{q}|^2)}{|\vec{q}|^2 + m^2_{\Delta_2^{2/3}}}. \end{aligned} \tag{1.15}$$

Here, we examine three possible cases for simplicity. In case A, Δ_k has the coupling to both ν_e and ν_μ. In cases B and C, Δ_k couples only to one of ν_μ and ν_e type neutrinos, respectively, as mentioned in Table 1.1. For more details, see the reference [6].

Table 1.1 The benchmark cases for LQs

LQCase	A	B	C
$\Delta_1^{-1/3}$	$\tilde{y}^{(1)} = \begin{pmatrix} g_{\Delta_1} & g_{\Delta_1} & 0 \\ 0 & 0 & 0 \\ 0 & 0 & 0 \end{pmatrix}$	$\tilde{y}^{(1)} = \begin{pmatrix} 0 & g_{\Delta_1} & 0 \\ 0 & 0 & 0 \\ 0 & 0 & 0 \end{pmatrix}$	$\tilde{y}^{(1)} = \begin{pmatrix} g_{\Delta_1} & 0 & 0 \\ 0 & 0 & 0 \\ 0 & 0 & 0 \end{pmatrix}$
$\Delta_2^{2/3}$	$\tilde{y}^{(2)} = \begin{pmatrix} g_{\Delta_2} & g_{\Delta_2} & 0 \\ 0 & 0 & 0 \\ 0 & 0 & 0 \end{pmatrix}$	$\tilde{y}^{(2)} = \begin{pmatrix} 0 & g_{\Delta_2} & 0 \\ 0 & 0 & 0 \\ 0 & 0 & 0 \end{pmatrix}$	$\tilde{y}^{(2)} = \begin{pmatrix} g_{\Delta_2} & 0 & 0 \\ 0 & 0 & 0 \\ 0 & 0 & 0 \end{pmatrix}$

1.4 Results

For each scenario listed in Table 1.1, we present the bounds on the Yukawa coupling strength as a function of the LQ masses in Fig. 1.2. The constraints on case A only and cases B and C are shown in the left and right panels, whereas the top and bottom panels correspond to CsI and Ar detectors, respectively. We represent the limits on $\Delta_{1,2}$ as a single solid line since they differ only at the percent level. As anticipated, the strongest bounds come from scenario A, as all of the ν_μ , $\bar{\nu}_\mu$, and ν_e fluxes contribute to the CEνNS process, while for the B and C cases only the $\nu_{\mu,\bar{\mu}}$ and only the ν_e participate in the process, respectively. The constraints become insensitive to very low LQs masses (around 10 MeV), since the cross-section depends mostly on the momentum

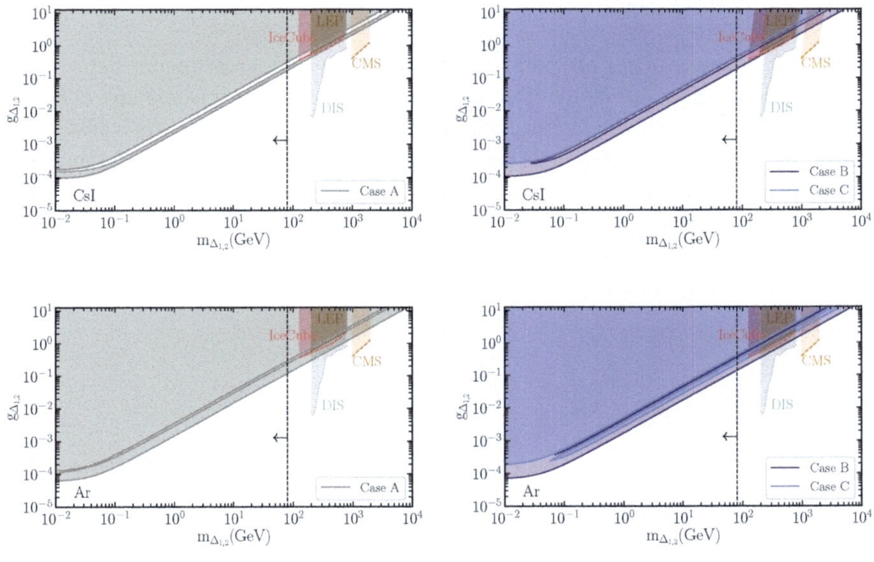

Fig. 1.2 Bounds obtained in the plane $g_\Delta - m_\Delta$ for various cases. In the upper (lower) panel we represent the constraints obtained using CsI (Ar)

transfer (q^2) in this domain. In each figure, we display some of the current limits on LQs for comparison. The LEP searches via the unsuppressed channel ($e^+e^- \to \gamma^* \to \Delta^+\Delta^-$) severely limit the low mass (< 80 GeV) LQs. The constraints coming from COHERENT data are comparable to other experiment bounds such as IceCube, LEP, and CMS and in some region the bounds improved slightly. There are some regions where we cannot put constraint due to the degeneracy with SM prediction. By looking at Eqs. 1.7 and 1.8 and considering the values of the form factors to be unity, we get the value of $Q_{i\,SM} \simeq -N/2$. On the other hand, $Q_{ij,\Delta_K} > 0$. Hence, there is a degenerate point where the total new charge becomes $N/2$. From Eq. 1.14, we get

$$\begin{aligned} \text{for case A} : \left(-\frac{N}{2} + Q_{ii,\Delta_K}\right)^2 + \sum_{i \neq j} Q^2_{ij,\Delta_K} &\simeq \left(\frac{N}{2}\right)^2, \\ \text{for case B and C} : \left(-\frac{N}{2} + Q_{ii,\Delta_K}\right)^2 &\simeq \left(\frac{N}{2}\right)^2, \end{aligned} \tag{1.16}$$

which almost replicates the SM values. As the ratio g_Δ/m_Δ increases from zero, the predicted events with LQs are initially decreasing from the SM prediction. At a certain g_Δ/m_Δ the predicted events become equal to the SM one (see Eq. 1.16), and the $\Delta\chi^2 = 0$. This is the degeneracy which gives the discontinuity in our constraints plots.

1.5 Conclusions

We constrain the scalar LQs parameter space using the most recent data of the COHERENT collaboration for CSI and Ar detectors. Neutrino-nucleus coherent scattering involves just the valence quarks (u and d) of the nuclei. As a result, the specific form of the Yukawa matrix that links to the first generation of quarks and neutrinos can be constrained as listed in Table 1.1. We are able to strongly constrain the scalar LQs Yukawa coupling over a wide mass range using the low energy COHERENT experiment. The bounds are comparable to the constraints coming from other existing high energy experiments.

References

1. D.Z. Freedman, Phys. Rev. D **9**, 1389–1392 (1974). https://doi.org/10.1103/PhysRevD.9.1389
2. D. Akimov et al., Science **357**(6356), 1123–1126 (2017). https://doi.org/10.1126/science.aao0990. arXiv:1708.01294 [nucl-ex]
3. D. Akimov et al., Phys. Rev. Lett. **129**(8), 081801 (2022). https://doi.org/10.1103/PhysRevLett.129.081801. arXiv:2110.07730 [hep-ex]
4. D. Akimov et al., Phys. Rev. Lett. **126**(1), 012002 (2021). https://doi.org/10.1103/PhysRevLett.126.012002. arXiv:2003.10630 [nucl-ex]

5. R.H. Helm, Phys. Rev. **104**, 1466–1475 (1956). https://doi.org/10.1103/PhysRev.104.1466
6. R. Calabrese, J. Gunn, G. Miele, S. Morisi, S. Roy, P. Santorelli, Phys. Rev. D **107**(5), 055039 (2023). https://doi.org/10.1103/PhysRevD.107.055039. arXiv:2212.11210 [hep-ph]

Chapter 2
Relevance of Golden Ratio in Neutrino Physics

Ngangkham Nimai Singh and Y. Monitar Singh

Abstract We study the implication of golden ratio in neutrino physics where the solar mixing angle (θ_{12}) which is one of the three neutrino mixing angles is closely related to this ratio. An exact leptonic mixing pattern which predicts golden ratio can be generated by certain discrete symmetry groups such as A_5. We use these three mixing angles given by the exact golden ratio neutrino mixing pattern as input values at a high energy scale, to obtain the low energy neutrino oscillation parameters through the numerical analysis of the relevant renormalization group equations (RGEs) of neutrino masses and mixing angles. Such radiative correction establishes the validity of golden ratio neutrino mixings defined at high energy scale from certain discrete symmetry, consistent with latest Planck cosmological data $\sum |m_i| < 0.12$ eV, for normal hierarchical mass model at a larger value of $\tan\beta > 60$ and SUSY breaking scale m_s=1TeV. The sensitivity on the values of $\sum |m_i|$ on m_s is also discussed.

2.1 Introduction

The relevance of golden ratio is naturally found in classic architecture(The Great Pyramid of Giza, Parthenon, Taj Mahal, etc.), artwork (Famous Artworks by Leonardo da Vinci: Mona Lisa, Vitruvian Man, etc.), nature(Plants, Animals), and even music. It is mathematically defined as the ratio of a line segment cut into two pieces of different lengths such that the ratio of the whole segment (**a+b**) to that of the longer segment (**a**) is equal to the longer segment (**a**) to shorter segment (**b**), i.e. $\frac{\mathbf{a+b}}{\mathbf{a}} = \frac{\mathbf{a}}{\mathbf{b}}$. This ratio numerically comes out as $\phi = \frac{1+\sqrt{5}}{2} \approx 1.618$. This ratio is also related to the neutrino physics in the sense that it can give the value of solar neutrino mixing angle as $\theta_{12} = \tan^{-1}(\frac{1}{\phi}) \approx 31.7^0$ [1, 2]. The value of solar neutrino mixing angle (θ_{12}) is one of the very special parameters for characterizing various

N. N. Singh · Y. M. Singh
Department of Physics, Manipur University, Imphal 795003, India

N. N. Singh (✉)
Research Institute of Science and Technology (RIST), Manipur 795003, India
e-mail: nimai03@yahoo.com

J. W. F. Valle et al. (eds.), *Proceedings of the International Conference on Future Prospects in Neutrino and Astroparticle Physics*, Springer Proceedings in Physics 361,
https://doi.org/10.1007/978-981-97-7441-8_2

leptonic mixing patterns. In neutrino physics, some discrete symmetries like A_5 can lead to the generation of exact golden ratio mixing pattern as follows [3, 4]:

$$U_{GR} = U_{PMNS} = \begin{pmatrix} \frac{\phi}{\sqrt{2+\phi}} & \frac{1}{\sqrt{2+\phi}} & 0 \\ -\frac{1}{\sqrt{4+2\phi}} & \frac{\phi}{\sqrt{4+2\phi}} & \frac{1}{\sqrt{2}} \\ \frac{1}{\sqrt{4+2\phi}} & -\frac{\phi}{\sqrt{4+2\phi}} & \frac{1}{\sqrt{2}} \end{pmatrix} \tag{2.1}$$

where ϕ has following properties:

$$\phi = \phi^2 - 1 = 1 + \frac{1}{\phi} = \frac{1+\sqrt{5}}{2} \approx 1.62$$

The three neutrino mixing angles predicted by the above exact golden ratio mixing matrix in Eq. (2.1) are $\theta_{12} = 31.7^0$, $\theta_{23} = 45^0$, and $\theta_{13} = 0$. Although the two mixing angles θ_{12} and θ_{23} are consistent within 3σ bound of experimental data, the third mixing angle, θ_{13}, is not consistent with the experimental non-zero value defined at low energy scale. This inconsistency in the mixing angle θ_{13} can be removed by giving a suitable perturbation to the exact golden ratio mixing pattern in some discrete symmetry models, which can be originated from any one of the followings: (i) contribution from charged lepton sector; (ii) corrections to vacuum alignment of A_5 triplets and quintuplets; and (iii)contribution from corrections in neutrino sector [5]. We can have another approach for studying the effect of perturbation to neutrino masses and mixings given by the exact mixing pattern in Eq. (2.1) through the radiative corrections using renormalization group equations (RGEs). This method of taking the radiative correction is model independent in the sense that all the neutrino masses at high energy scale are taken as free input parameters so that all generated neutrino oscillation parameters at low energy scale are consistent with experimental data within 3σ bound. We also take the three golden ratio mixing angles as input parameters at a high energy scale in running the RGEs to study the validity of the golden ratio in neutrino physics. We impose the latest updated Planck upper bound on the sum of neutrino masses $\sum |m_i| < 0.12$ eV as the most stringent constraint in the generation of neutrino oscillation parameters consistent with experimental data [6].

2.2 Numerical Analysis Through Renormalization Group Equations

The perturbation of the exact golden ratio neutrino mixing pattern Eq. (2.1) is studied through RGEs in running from high to low energy scale within SUSY framework. In such radiative analysis, the exact golden ratio mixing pattern with discrete symmetry group A_5 is assumed to be defined at the high energy scale M_R near the unification scale. The radiative generation of the non-zero value of the reactor neutrino mixing angle (θ_{13}) at low energy scale is consistent with the other two mixing angles, within the 3σ bound of the latest experimental data [6]. One of the most stringent constraints is the sum of three absolute neutrino mass eigenvalues given by the latest Planck data, $\sum |m_i| < 0.12$ eV [7, 8]. The neutrino oscillation parameters include three mixing angles, i.e. the solar mixing angle (θ_{12}), the atmospheric mixing angle θ_{23}, and the reactor mixing angle(θ_{13}) and two mass squared differences, i.e. solar mass squared difference (Δm^2_{21}) and atmospheric mass squared difference ($\Delta m^2_{32} \approx \Delta m^2_{31}$). The radiative correction is numerically evaluated by running renormalization group equations (RGEs). We follow a two-step procedure: (i) bottom-up approach for evaluating the values of three gauge and third generation Yukawa coupling constants from low energy scale to high energy scale, and (ii) top-down approach for generating neutrino oscillation parameters at a low energy scale evolved from high energy scale. The RGEs of neutrino mass eigenvalues are given by the following equations [9, 10]

$$\frac{d}{dt} m_i = -2F_\tau (P_i + Q_i) m_i - F_u m_i, \; (i = 1, 2, 3). \tag{2.2}$$

The RGEs of neutrino mixing angles are given by the following equations [9, 10]

$$\frac{ds_{12}}{dt} = \frac{F_\tau c_{12} \sin 2\theta_{12} s^2_{23}}{2(m^2_2 - m^2_1)} [m^2_1 + m^2_2 + 2m_1 m_2 \cos(2\alpha_2 - 2\alpha_1)] \tag{2.3}$$

$$\begin{aligned} \frac{ds_{23}}{dt} &= \frac{F_\tau c_{23} \sin 2\theta_{23}}{2(m^2_3 - m^2_2)} [c^2_{12}(m^2_3 + m^2_2 + 2m_3 m_2 \cos 2\alpha_2) \\ &+ \frac{s^2_{12}(m^2_3 + m^2_1 + 2m_3 m_1 \cos 2\alpha_1)}{R+1} \end{aligned} \tag{2.4}$$

$$\begin{aligned} \frac{ds_{13}}{dt} &= -\frac{F_\tau c_{13} \sin 2\theta_{12} \sin 2\theta_{23} m_3}{2(m^2_3 - m^2_1)} \\ & [m_1 \cos(2\alpha_1 - \delta) - (1+R) m_2 \cos(2\alpha_2 - \delta) - R m_3 \cos\delta] \end{aligned} \tag{2.5}$$

where $P_1 = (s_{12}s_{23})^2$; $P_2 = (c_{12}s_{23})^2$; $P_3 = (c_{13}c_{23})^2$; $Q_1 = -\frac{1}{2} s_{12} \sin 2\theta_{12} \sin 2\theta_{23} \cos\delta + (c_{12}c_{23}c_{13})^2$; $Q_2 = \frac{1}{2} s_{12} \sin 2\theta_{12} \sin 2\theta_{23} \cos\delta + (s_{12}c_{23}s_{13})^2$; $Q_3 = 0$ and

$$R = \frac{m^2_2 - m^2_1}{m^2_3 - m^2_2}.$$

For scale-dependent VEV in the case of MSSM with $\mu \geq m_s$ [10, 11], we have taken as

$$F_\tau = -\frac{h^2_\tau}{16\pi^2 \cos^2 2\beta}$$

$$F_u = \frac{1}{16\pi^2}\left(\frac{9}{10}g_1^2 + \frac{9}{2}g_2^2\right)$$

and for SM case with $\mu \leq m_s$, we have

$$F_\tau = \frac{3h_\tau^2}{32\pi^2}$$

$$F_u = \frac{1}{16\pi^2}\left(-\frac{9}{10}g_1^2 - \frac{3}{2}g_2^2 + 6h_b^2 - 2\lambda\right)$$

where g_1, g_2, g_3 are gauge couplings, and h_t, h_b, h_τ, and λ are top-quark, bottom-quark, tau-lepton Yukawa couplings, and SM quartic Higgs coupling, respectively. The RGEs for phases and other coupling constants given above are also given in Refs. [10–14]. The RGEs of non-SUSY (SM) are used for energy scale,$m_t \leq \mu \leq m_s$ and the RGEs of SUSY (MSSM) are used for energy scale, $m_s \leq \mu \leq M_R$. The values of coupling constants at a very high energy scale are significant in giving the low energy neutrino oscillation parameters consistent with the latest Planck mass bound, $\sum |m_i| < 0.12$ eV. These coupling constants are affected by the variation of some free parameters such as tan β, the SUSY breaking scale (m_s), the threshold parameter ($\bar{\eta_b}$), and the value of the high energy seesaw scale (M_R). The unification of three gauge couplings (g_1, g_2, g_3) at high energy scale is maintained with the input values of the above parameters.

2.3 Results and Discussion

The observed numerical values of oscillation parameters at low energy scale with suitable input parameters at high energy scale are given in Table 2.2. The values of coupling constants evaluated at high energy scale with various input values of free parameters of tan β, $\bar{\eta_b}$, m_s, and M_R are very significant in the generation of neutrino oscillation parameters at low energy scale consistent with the latest Planck

Table 2.1 Observational data of neutrino oscillation parameters for normal hierarchy (NH) and inverted hierarchy (IH)in 3σ range

Parameter	Normal Hierarchy (best-fit$\pm 1\sigma$)	Inverted Hierarchy (best-fit$\pm 1\sigma$)
$\lvert\Delta m_{21}^2\rvert[10^{-5}eV^2]$	6.82–8.03 ($7.41^{+0.21}_{-0.20}$)	6.82–8.03 ($7.41^{+0.21}_{-0.20}$)
$\lvert\Delta m_{31}^2\rvert[10^{-3}eV^2]$	2.428–2.597 ($2.511^{+0.028}_{-0.027}$)	2.408–2.581 ($2.498^{+0.032}_{-0.025}$)
$\sin\theta_{12}$	0.519–0.585 ($0.303^{+0.012}_{-0.011}$)	0.519–0.585 ($0.303^{+0.012}_{-0.011}$)
$\sin\theta_{23}$	0.636–0.788 ($0.572^{+0.018}_{-0.023}$)	0.640–789 ($0.578^{+0.016}_{-0.021}$)
$\sin\theta_{13}/10^{-2}$	0.142–0.155 ($2.203^{+0.056}_{-0.059}$)	0.143–0.156 ($2.219^{+0.060}_{-0.057}$)
$\delta_{CP}/^o$	105–405 ($197^{+0.42}_{-0.25}$)	192–361 ($286^{+0.27}_{-0.32}$)

Table 2.2 Low energy neutrino oscillation parameters obtained from radiative corrections in MSSM while running from $M_R = 1.84 \times 10^{15}$GeV to low energy scale, $m_t = 172.76$GeV for $\tan\beta = 68$, $m_s = 1TeV$, and $\bar{\eta_b} = 0.02$ for NH model

Parameter	High energy scale input values (NH)	Low energy scale output values (NH)
$m_1[eV]$	0.0224	0.0303
$m_2[eV]$	−0.0255	−0.0316
$m_3[eV]$	0.0482	0.0579
$\|\Delta m^2_{21}\|[10^{-5}eV^2]$	14.84	7.54
$\|\Delta m^2_{31}\|[10^{-3}eV^2]$	1.82	2.43
$\sin\theta_{12}$	0.5257	0.5355
$\sin\theta_{23}$	0.7071	0.789
$\sin\theta_{13}$	0.0	0.1428
$\delta_{CP}/^o$	175	200.63
α_1	2	12.35
α_2	0.5	4.55
$\sum\|m_i\|[eV]$	0.0962	0.119

upper bound. The effect of the variation of free parameters $\tan\beta$, $\bar{\eta_b}$, m_s and M_R with $\sum|m_i|$ is also studied and their graphical representations are shown in Fig. 2.1. At particular specific values of these parameters, $\tan\beta = 68$, $m_s = 1$TeV, $\bar{\eta_b} = 0.02$, and $M_R = 1.84 \times 10^{15}$ GeV, the golden ratio mixing pattern is valid in giving the low energy oscillation parameters consistent with the latest Planck upper bound. There is a wide range of allowed values of the free parameters if the cosmological upper bound on the sum of three neutrino masses is relaxed upto $\sum|m_i| < 0.23$ eV [15, 16] and the graphical representation is shown in Fig. 2.2. The validity of the golden ratio mixings at high energy scale lies in the values that are chosen for the above three unknown parameters. The SUSY breaking scale (m_s) is one of the very important unknown parameters in the minimal super-symmetric model (MSSM). As there is no evidence for the existence of SUSY particles in the lower TeV range, the validity of golden ratio neutrino mixings in the higher TeV range is also discussed. The use of the latest updated Planck cosmological upper bound on the sum of neutrino masses as a constraint is an important factor for the validity of various leptonic mixing patterns in the radiative corrections of neutrino masses and mixings at low energy scale from high energy scale through their respective renormalization group equations. From the analysis of Fig. 2.2, it is seen that the golden ratio mixing pattern is found to be valid for Planck upper bound, $\sum|m_i| < 0.12$ at only around $m_s = 1$TeV, and for $\sum|m_i| < 0.23$ eV, it is valid up to $m_s = 14$ TeV. This higher Planck bound also allows lower values of $\tan\beta \sim 55$ as shown in Fig. 2.1a. This is consistent with the analysis of combined Planck plus James Webb Space Telescope (JWST)[16] and also gives the validity of golden ratio neutrino mixings in inverted hierarchy (IH) in addition to NH (Table 2.2).

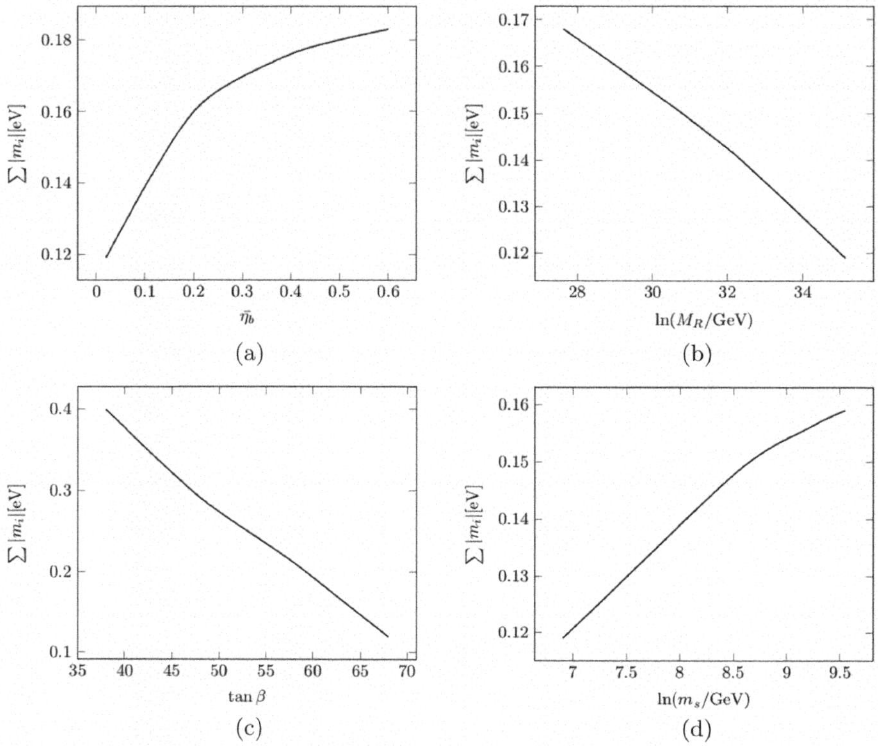

Fig. 2.1 **a** Variation of $\sum |m_i|$[eV] with $\bar{\eta_b}$ for $M_R = 1.84 \times 10^{15}$ GeV,$\tan\beta = 68$,$m_s = 1TeV$; **b** Variation of $\sum |m_i|$[eV] with M_R for $\bar{\eta_b} = 0.02$ GeV, $\tan\beta = 68$,$m_s = 1TeV$; **c** Variation of $\sum |m_i|$[eV] with $\tan\beta$ for $M_R = 1.84 \times 10^{15}$ GeV,$\bar{\eta_b} = 0.02$,$m_s = 1TeV$; **d** Variation of $\sum |m_i|$[eV] with m_s for $M_R = 1.84 \times 10^{15}$ GeV, $\tan\beta = 68$,$\bar{\eta_b} = 0.02$

2.4 Summary and Conclusion

The three neutrino mixing angles after the radiative correction at low energy scale are found to be $\theta_{12} = 32.37^0$, $\theta_{13} = 8.2^0$, and $\theta_{23} = 52.0^0$ which are consistent with the recent oscillation data 3σ range. Thus, all the neutrino oscillation parameters at low energy scale can be generated from high energy scale using the exact golden ratio mixing pattern. The latest Planck cosmological upper bound $\sum |m_i| < 0.12$eV is found to be satisfied in the normal hierarchical mass model. The larger value of $\tan\beta > 60$ is found to be consistent with the Planck bound in the numerical analysis. Our analysis does not favour the inverted hierarchical model within this latest Planck cosmological upper bound. Similar analysis shows that other symmetries based on tri-bimaximal(TBM), bimaximal (BM), and hexagonal mixing (HM) are found to be invalid within this latest Planck cosmological upper bound at low energy scale of

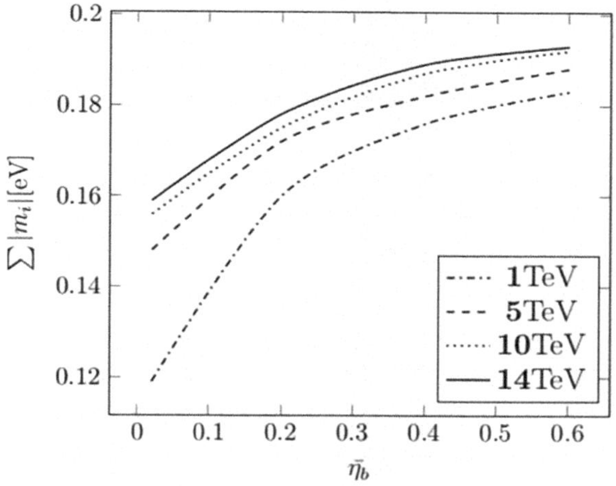

Fig. 2.2 Allowed range of graphical representations for the variation of $\sum |m_i|$[eV] with the different values of SUSY threshold parameter ($\bar{\eta_b} = 0.02, 0.2, 0.4, 0.6$) for different cases of SUSY breaking scale m_s=1TeV, 5TeV, 10TeV,14TeV. Higher values of $\sum |m_i|$ <0.23 eV can accommodate wide range of parameters, ($\bar{\eta_b} = 0.02 - 0.6$) and ($m_s$ =1TeV-14TeV) for particular values of $M_R = 1.84 \times 10^{15}$GeV and $\tan\beta = 68$

neutrino oscillation parameters. The present numerical analysis of radiative correction shows the validity of golden ratio in neutrino physics at around m_s = 1TeV for $\sum |m_i| < 0.12$ eV and it is valid up to m_s = 14 TeV for $\sum |m_i| < 0.23$ eV.

References

1. Y. Kajiyama, M. Raidal, A. Strumia, Golden ratio prediction for solar neutrino mixing. Phys. Rev. D **76**(11), 117301 (2007)
2. A. Adulpravitchai, A. Blum, W. Rodejohann, Golden ratio prediction for solar neutrino mixing. New J. Phys. **11**(6), 063026 (2009)
3. L.L. Everett, A.J. Stuart, Icosahedral (a 5) family symmetry and the golden ratio prediction for solar neutrino mixing. Phys. Rev. D **79**(8), 085005 (2009)
4. S.F. King, Unified models of neutrinos, flavour and cp violation. Progress Particle Nucl. Phys. **94**, 217–256 (2017)
5. V. Puyam, N. Nimai Singh, a_5 symmetry and deviation from golden ratio mixing (2023). arXiv preprint arXiv:2308.05944
6. M.C. Gonzalez-Garcia, M. Maltoni, T. Schwetz, Nufit: three-flavour global analyses of neutrino oscillation experiments. Universe **7**(12), 459 (2021)
7. N. Aghanim, Y. Akrami, M. Ashdown, J. Aumont, C. Baccigalupi, M. Ballardini, A.J. Banday, R.B. Barreiro, N. Bartolo, S. Basak, et al., Planck 2018 results-vi. cosmological parameters. Astron. Astrophys. **641**, A6 (2020)
8. I. Tanseri, S. Hagstotz, S. Vagnozzi, E. Giusarma, K. Freese, Updated neutrino mass constraints from galaxy clustering and CMB lensing-galaxy cross-correlation measurements. J. High Energy Astrophys. **36**, 1–26 (2022)
9. S. Antusch, J. Kersten, M. Lindner, M. Ratz, Running neutrino masses, mixings and CP phases: analytical results and phenomenological consequences. Nucl. Phys. B **674**(1–2), 401–433 (2003)
10. S.K. Agarwalla, M.K. Parida, R.N. Mohapatra, G. Rajasekaran, Neutrino mixings and leptonic cp violation from CKM matrix and Majorana phases. Phys. Rev. D **75**(3), 033007 (2007)

11. N. Nimai Singh, Effects of the scale-dependent vacuum expectation values in the renormalisation group analysis of neutrino masses. Eur. Phys. J. C-Particles Fields **19**(1), 137–141 (2001)
12. K. Helensana Devi, K. Sashikanta Singh, N. Nimai Singh, et al., Stability of the next-to-tribimaximal mixings under radiative corrections with the variation of the Susy breaking scale in MSSM. Adv. High Energy Phys. (2023)
13. K. Helensana Devi, K. Sashikanta Singh, N. Nimai Singh, et al., Effects of variations of susy breaking scale on neutrino parameters at low energy scale under radiative corrections. Adv. High Energy Phys. (2022)
14. K. Sashikanta Singh, S. Roy, N. Nimai Singh, Stability of neutrino parameters and self-complementarity relation with varying Susy breaking scale. Phys. Rev. D **97**(5), 055038 (2018)
15. P.A.R. Ade, N. Aghanim, M. Arnaud, M. Ashdown, J. Aumont, C. Baccigalupi, A.J. Banday, R.B. Barreiro, J.G. Bartlett, N. Bartolo, et al.: Planck 2015 results-xiii. cosmological parameters. Astron. Astrophys. **594**, A13 (2016)
16. J. Liu, Z. Huang, Y. Su, Cosmological constraints on neutrino masses in light of JWST red and massive candidate galaxies (2023). arXiv preprint arXiv:2311.09703

Chapter 3
Randomly Generated Majorana Neutrino Mass Matrix for CP-Conserving Case

Y. Monitar Singh and Ngangkham Nimai Singh

Abstract The present work aims to discriminate among the theoretically predicted different forms of Majorana neutrino mass matrix including texture zeros. The neutrino oscillation parameters are numerically extracted by diagonalizing a general charge-parity (CP)-conserving Majorana neutrino mass matrix whose elements are randomly generated within a certain range of allowed values using adaptive Monte Carlo method. The latest neutrino oscillation experimental data within 3σ determines allowed values of the elements of the neutrino mass matrix. The latest Planck upper bound on the sum of three absolute masses $\sum |m_i| < 0.12$ eV is imposed in the numerical analysis. Both normal hierarchy (NH) and inverted hierarchy (IH) mass models are allowed, thus showing the possibility of both mass hierarchies within 3σ. Further, the detailed numerical analysis confirms that the normal hierarchical mass model is valid up to mass bound, $\sum |m_i| \geq 0.06$ eV while the inverted hierarchical mass model is valid up to mass bound, $\sum |m_i| \geq 0.1$eV. In both models, the value of θ_{23} is allowed in both below and above 45^0. However, $\theta_{23} > 45^0$ is found to be more favourable for NH whereas $\theta_{23} < 45^0$ is more favourable for IH.

3.1 Introduction

The present neutrino oscillation data confirms that neutrinos have very tiny but non-zero masses. These tiny masses of neutrinos are elegantly explained by the celebrated seesaw mechanism. A neutrino mass matrix can be theoretically generated from various discrete symmetries. All neutrino oscillation parameters can be extracted from the mass matrix which is possible for both normal and inverted hierarchical mass models.

Y. M. Singh (✉) · N. N. Singh
Department of Physics, Manipur University, Imphal 795003, India
e-mail: monitar.phd.phy@manipuruniv.ac.in

N. N. Singh
Research Institute of Science and Technology (RIST), Manipur 795003, India

J. W. F. Valle et al. (eds.), *Proceedings of the International Conference on Future Prospects in Neutrino and Astroparticle Physics*, Springer Proceedings in Physics 361,
https://doi.org/10.1007/978-981-97-7441-8_3

There are at present several theoretical models which predict different forms of Majorana mass matrix including various texture zero models, and this ambiguity in the choice of the correct form of this texture is not yet resolved. With the hope to discriminate among these theoretically predicted possible textures, we adopt a different approach that may give some hints on the correct structure of the Majorana neutrino mass matrix consistent with the presently available observational data. One such approach is the adaptive Monte Carlo method for randomly generating Majorana mass matrix.

Another equally important issue in neutrino physics is the problem of neutrino mass hierarchy. This issue is also addressed in the present work by considering a general form of Majorana neutrino mass matrix in a charge-parity (CP)conserving case for which Dirac phase, $\delta_{CP} = 0$. The latest Planck cosmological upper bound $\sum |m_i| < 0.12$ eV [1, 2] is imposed in the numerical analysis of mass matrix. The present study is purely based on numerical analysis which is model independent. This study is also capable of analysing a variety of possible neutrino mass textures predicted by different theoretical models.

A Majorana neutrino mass matrix M_ν is diagonalized by the use of the standard mixing matrix U as follows:

$$M_{diag} = U^T M_\nu U \tag{3.1}$$

and U is given by PMNS mixing matrix,

$$U = U_{PMNS} = \begin{pmatrix} c_{13}c_{12} & c_{13}s_{12} & s_{13} \\ -c_{23}s_{12} - c_{12}s_{13}s_{23} & c_{12}c_{23} - s_{12}s_{13}s_{23} & c_{13}s_{23} \\ s_{12}s_{23} - c_{12}s_{13}c_{23} & -c_{12}s_{23} - c_{23}s_{13}s_{12} & c_{13}c_{23} \end{pmatrix} \tag{3.2}$$

where $s_{ij} = \sin\theta_{ij}$ and $c_{ij} = \cos\theta_{ij}$. As of now, experiments can measure three neutrino mixing angles ($\theta_{12}, \theta_{23}, \theta_{13}$) and two mass-squared differences (Δm^2_{21} and Δm^2_{32}), and also the sum of three absolute neutrino mass eigenvalues, $\sum |m_i| < 0.12$ eV, given by Planck cosmological data. Both normal ordering ($m_3 > m_2 > m_1$) and inverted ordering ($m_2 > m_1 > m_3$) neutrino mass matrix can explain all oscillation data within 3σ. One of the possible ways to discriminate the neutrino mass ordering is to impose the sum of neutrino masses, $\sum |m_i| < 0.12$ eV in the oscillation data predicted by a given neutrino mass matrix. In this present study, a general Majorana mass matrix constructed with random input parameters is diagonalized to generate an eigen-system which in turn predicts neutrino oscillation parameters consistent with the presently available oscillation data.

The paper is organized as follows. In Sect. 3.2, we give a brief outline of adaptive Monte Carlo method for randomly generating CP-conserving Majorana mass matrix. Section 3.3 is devoted to numerical analysis and results. In Sect. 3.4, we give a summary and discussion.

3.2 Adaptive Monte Carlo Method for Randomly Generated Majorana Mass Matrix

A general form of the Majorana neutrino mass matrix is taken as follows:

$$M_\nu = \begin{pmatrix} a & b & c \\ b & d & e \\ c & e & f \end{pmatrix} \tag{3.3}$$

For the present study, we are only considering the mass matrix which is the CP-conserving case as all the elements are real. Only six real numbers are required to construct the mass matrix. These six real numbers are randomly generated within a given range, which is allowed by the latest Planck cosmological upper bound, $\sum |m_i| < 0.12$ eV. For a particular Majorana neutrino mass matrix, the extracted two mass-squared difference (Δm^2_{21} and Δm^2_{32}) and all the elements of neutrino mixing matrix, U_{PMNS}, are compared with the observed experimental values given in Table 3.1. The general structure of neutrino mixing matrix (U_{PMNS}) is given below:

$$U = U_{PMNS} = \begin{pmatrix} U_{e1} & U_{e2} & U_{e3} \\ U_{\mu 1} & U_{\mu 2} & U_{\mu 3} \\ U_{\tau 1} & U_{\tau 2} & U_{\tau 3} \end{pmatrix} \tag{3.4}$$

where $\tan\theta_{12} = \frac{|U_{e2}|}{|U_{e1}|}$, $\tan\theta_{23} = \frac{|U_{\mu 3}|}{|U_{\tau 3}|}$, and $\sin\theta_{13} = |U_{e3}|$.

The rejection rule for a considered mass texture to be a possible solution of Eq. (3.3) is defined as

Table 3.1 The best-fit absolute values of the neutrino mass-squared differences and the elements of U_{PMNS} with the 3σ uncertainties

i	x_i	x_i^{cent}	σ_i
1	$\Delta m^2_{32}[eV^2]$	2.51×10^{-3}	10^{-3}
2	$\Delta m^2_{21}[eV^2]$	7.42×10^{-5}	10^{-5}
3	$\lvert U_{e1}\rvert$	0.835	0.045
4	$\lvert U_{e2}\rvert$	0.54	0.07
5	$\lvert U_{e3}\rvert$	0.148	0.1
6	$\lvert U_{\mu 1}\rvert$	0.355	0.165
7	$\lvert U_{\mu 2}\rvert$	0.575	0.155
8	$\lvert U_{\mu 3}\rvert$	0.7	0.12
9	$\lvert U_{\tau 1}\rvert$	0.365	0.165
10	$\lvert U_{\tau 2}\rvert$	0.59	0.15
11	$\lvert U_{\tau 3}\rvert$	0.685	0.125

$$\frac{(x_i^{cent} - x_i)^2}{\sigma_i^2} \leq \frac{1}{\alpha^2} \tag{3.5}$$

where x_i^{cent} and x_i represent the experimentally observed best-fit values and the numerically observed values of neutrino oscillation parameters, respectively, while σ_i and α are, respectively, the 3σ uncertainties and a parameter which is used for fixing the assumed error σ_i. In the present analysis, we put $\alpha = 1$ so that the results are obtained within 3σ bound. If the extracted values of oscillation parameters from the mass matrix are found to lie within the range of 3σ bound, then it is one of the possible solutions to the general mass matrix of Eq. (3.3). To perform the numerical fitting of the mass matrix, we are using the adaptive Monte Carlo method [3, 4]. This method enables us to analyse a lot of possible solutions to the Majorana mass matrix including some texture zero structures.

3.3 Numerical Analysis and Results

We first find several possible solutions for the general Majorana neutrino mass matrix. From all possible solutions, we estimate all the elements in the mass matrix for both normal hierarchy (NH) and inverted hierarchy (IH), and we then plot their frequency spectrum in Figs. 3.1, 3.2, 3.3, and 3.4. It is observed that out of six input parameters, the numerical values of **a**, **b**, and **c** are found to be concentrated around the zero, indicating that extremely small values are favourable whereas the numerical values of **d**, **e**, and **f** are a little far away from zero, indicating that non-zero values are favourable for NH and the same is opposite for IH. In both NH and IH models, the two mass-squared differences and the three neutrino mixing angles are consistent with experimental data within 3σ [5]. The latest Planck cosmological bound, $\sum |m_i| < 0.12$ eV is also satisfied by the mass eigenvalues obtained in both NH and IH mass models. So, these two hierarchical neutrino mass models are still possible within this cosmological mass bound. Recently, some studies have continuously updated these values on the sum of three neutrino masses, even with $\sum |m_i| < 0.09$ eV, putting under severe tension. Further, detailed numerical analysis is performed with the imposition of $\sum |m_i| < 0.1$ eV in getting the neutrino oscillation parameters. This smaller mass bound is only allowed for NH but not for IH mass model. The numerical values of oscillation parameters obtained in both NH and IH, along with the sum of the mass eigenvalue, are presented in Tables 3.2 and 3.3, respectively. The three neutrino mixing angles are then calculated from the observed elements of U_{PMNS} by using the relations as $\sin\theta_{13} = |U_{e3}|$; $\tan\theta_{12} = \frac{|U_{e1}|}{|U_{e2}|}$; and $\tan\theta_{23} = \frac{|U_{\mu 3}|}{|U_{\tau 3}|}$.

The following graphical representations can be interpreted stepwise. In the first step, possible rough solutions for the general Majorana mass matrix are searched by giving a suitable demanding number on the right side of the rejection rule in Eq. (3.5). A certain number of possible solutions are taken for consideration of numerical analysis and each of the possible solutions is represented by a point in 3-D space,

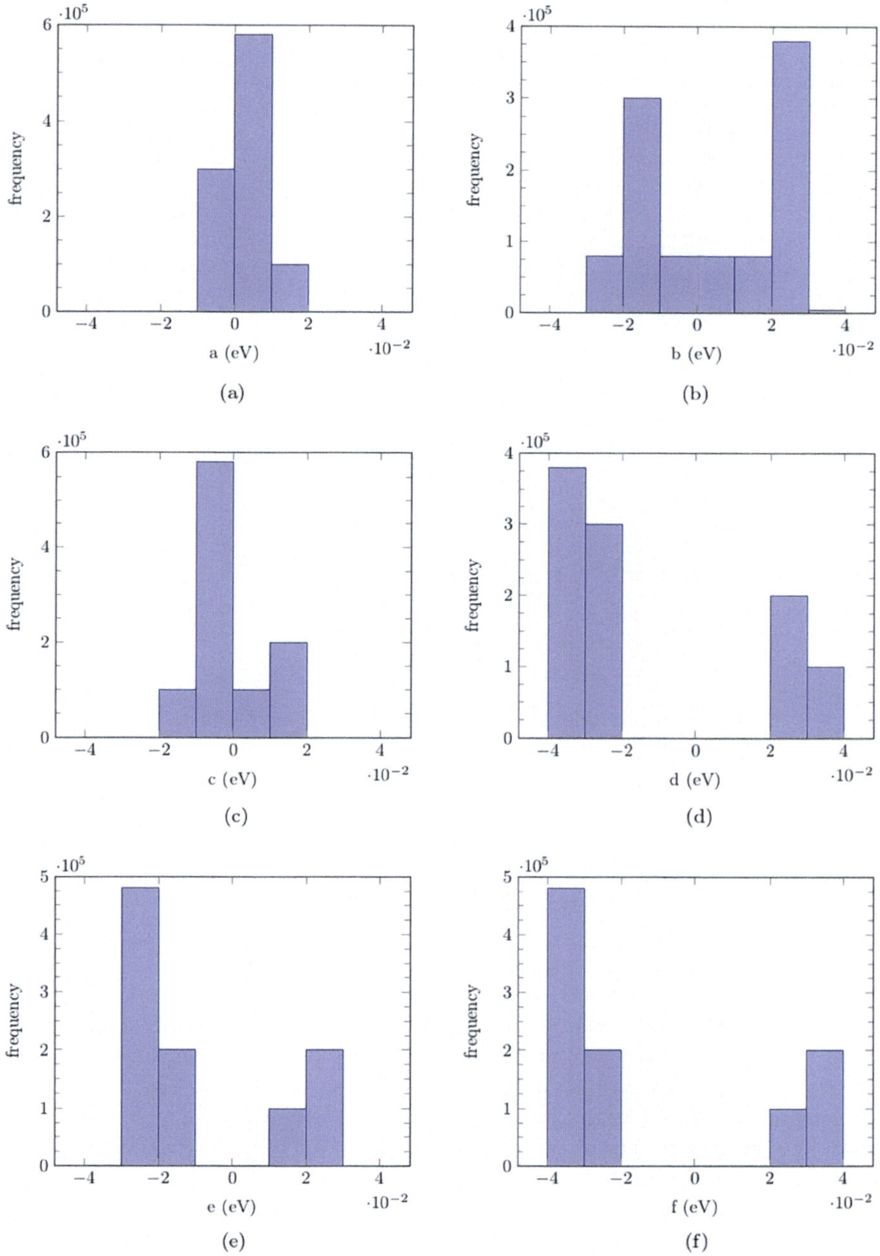

Fig. 3.1 Frequency spectrum for the elements of the neutrino mass matrix in normal mass hierarchy (NH). The numerical values in region of higher peak of frequency spectrum show more possible solutions as compared to lower peak region of the frequency spectrum

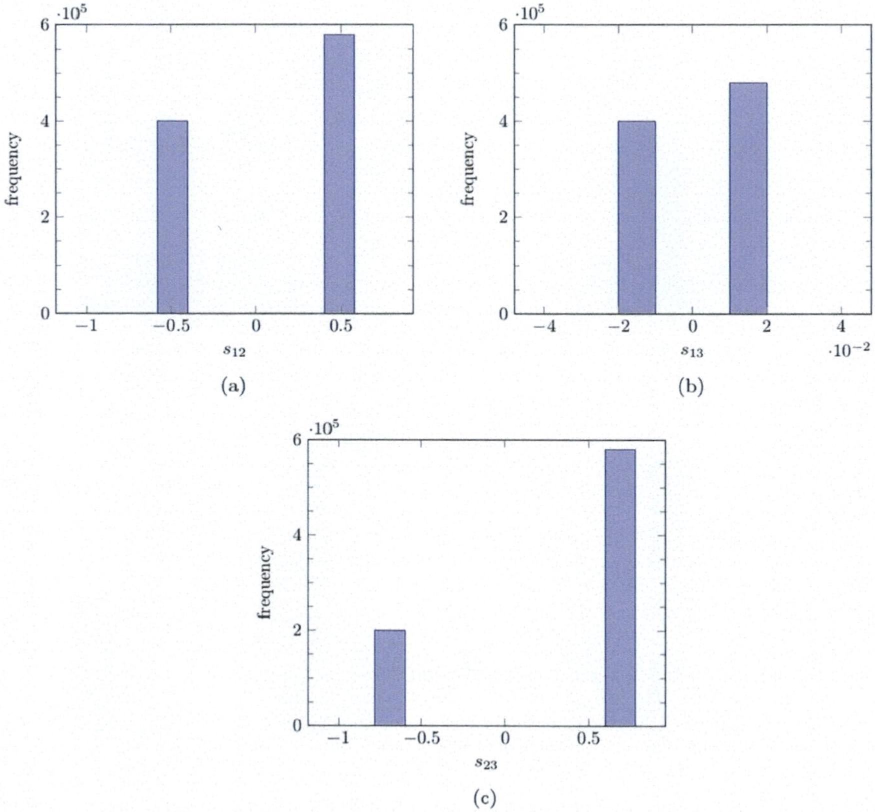

Fig. 3.2 Frequency spectrum for sine of the three mixing angles in normal mass hierarchy (NH). The numerical values in region of higher peak of frequency spectrum show more possible solutions as compared to lower peak region of the frequency spectrum

which is represented as scattered plots in Fig. 3.5a. If the demanding number is equal to one, then all possible solutions are found within 3σ bound of experimental data. In the second step, the algorithm finds a large number of possible solutions about the surroundings of each of the points in the first step, and then selects the best one by using the adaptive Monte Carlo method [3]. The enlarged regions of each point are shown in Fig. 3.5b. The physical interpretation of the frequency spectrum for the elements of the mass matrix is given for both normal hierarchy (NH) in Figs. 3.1 and 3.2 and inverted hierarchy (IH) in Figs. 3.3 and 3.4. In the case of NH, the frequency spectrum for the elements **a**, **b**, and **c** shows preferable small values nearer to centre from both sides on the x-axis. The frequency spectrum for the elements **d**, **e**, and **f** shows preferable small non-zero values farther away from the centre of both sides

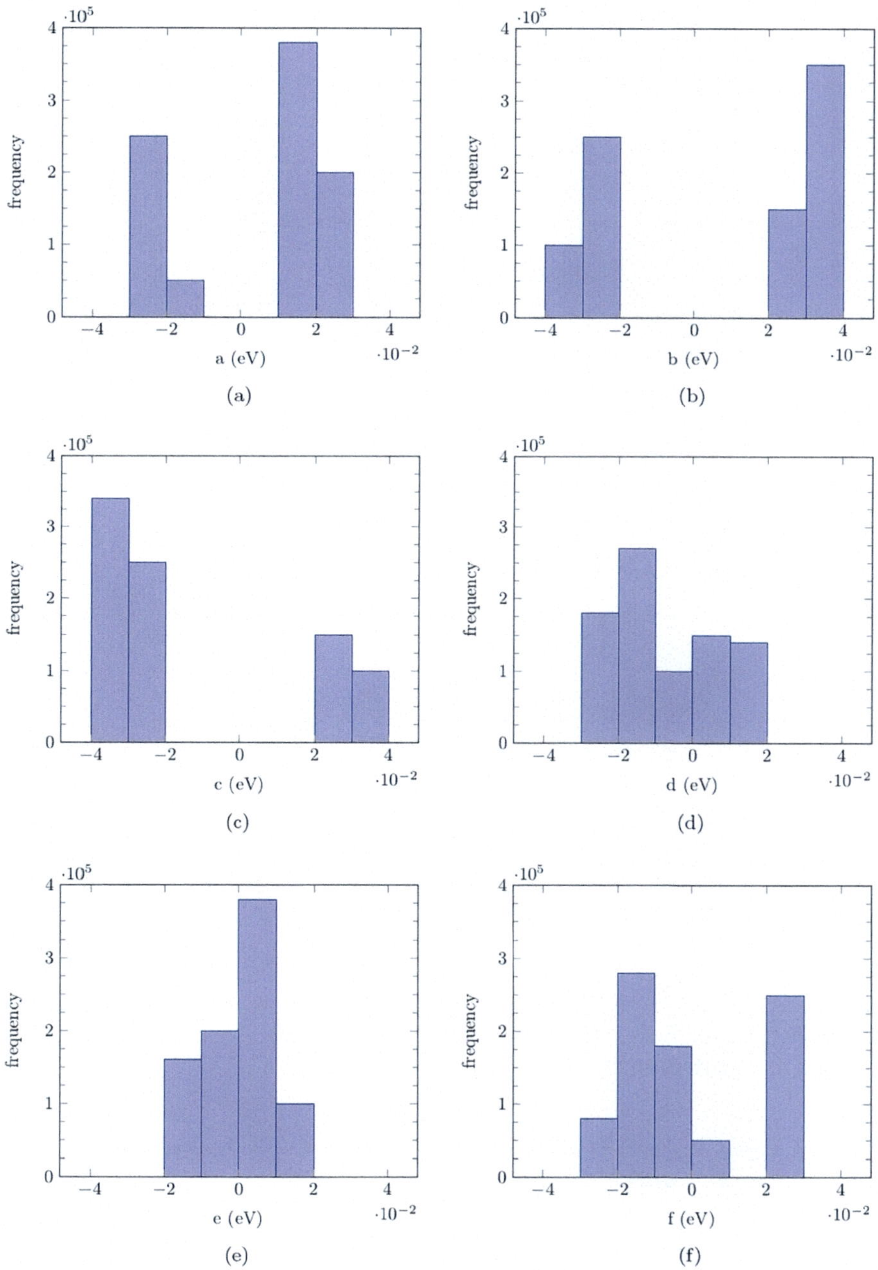

Fig. 3.3 Frequency spectrum for the elements of the neutrino mass matrix in inverted mass hierarchy (IH). The numerical values in region of higher peak of frequency spectrum show more possible solutions as compared to lower peak region of the frequency spectrum

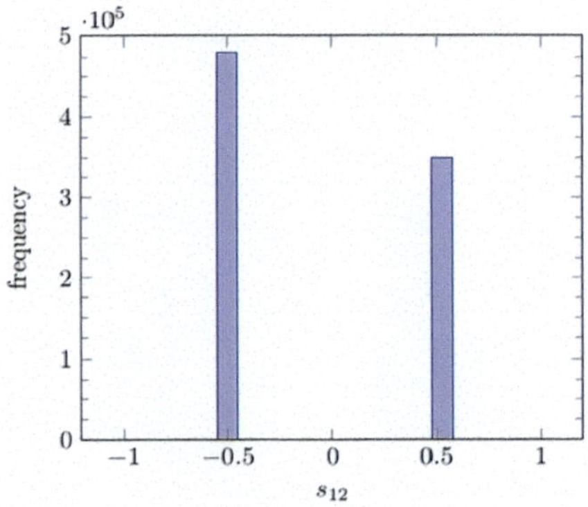

(a)

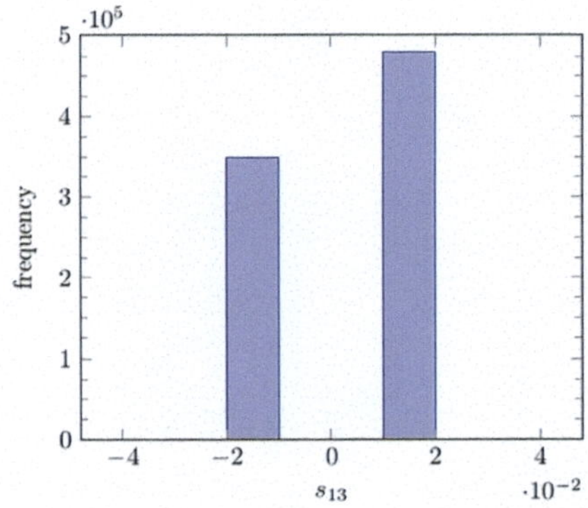

(b)

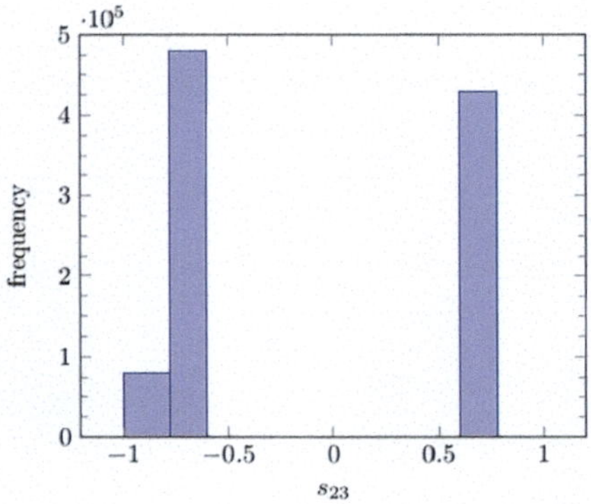

(c)

Fig. 3.4 Frequency spectrum for sine of the three mixing angles in inverted mass hierarchy (IH). The numerical values in region of higher peak of frequency spectrum show more possible solutions as compared to lower peak region of the frequency spectrum

on the x-axis. The sine of three mixing angles in NH shows two possible allowed frequency peaks on both sides of the x-axis. The same case is opposite in IH for the elements **a**, **b**, **c**, **d**, **e**, and **f**.

3.4 Summary and Conclusion

To summarize, we use the adaptive Monte Carlo method for numerical analysis where elements of the Majorana neutrino mass matrix are randomly generated. The main result of the numerical analysis shows that the present neutrino oscillation parameters are still valid for both normal and inverted hierarchical mass models within 3σ bound, consistent with the latest Planck upper mass bound, $\sum |m_i| < 0.12$ eV. Some studies

Table 3.2 The best central values of the neutrino oscillation parameters within 3σ bound of experimental data calculated by using adaptive Monte Carlo method in NH. All the neutrino oscillation parameters are found within 3σ bound of experimental data consistent with the latest Planck cosmological upper bound $\sum |m_i| < 0.12$ eV and still valid upto $\sum |m_i| > 0.06$ eV

Section	I	II	III	IV
$\Delta m^2_{32}[eV^2]$	2.47×10^{-3}	2.43×10^{-3}	2.54×10^{-3}	2.57×10^{-3}
$\Delta m^2_{21}[eV^2]$	7.40×10^{-5}	7.03×10^{-5}	7.33×10^{-5}	7.51×10^{-5}
$\|U_{e1}\|$	0.826	0.820	0.804	0.826
$\|U_{e2}\|$	0.544	0.550	0.575	0.543
$\|U_{e3}\|$	0.145	0.153	0.146	0.148
$\|U_{\mu 1}\|$	0.468	0.289	0.497	0.466
$\|U_{\mu 2}\|$	0.519	0.632	0.517	0.512
$\|U_{\mu 3}\|$	0.714	0.718	0.696	0.721
$\|U_{\tau 1}\|$	0.313	0.491	0.324	0.316
$\|U_{\tau 2}\|$	0.658	0.545	0.633	0.664
$\|U_{\tau 3}\|$	0.684	0.678	0.702	0.676
$\theta_{12}/^0$	33.34	33.82	35.56	33.31
$\theta_{13}/^0$	8.33	8.80	8.39	8.51
$\theta_{23}/^0$	46.20	46.61	44.74	46.82
$\sum \|m_i\|[eV]$	0.0604	0.1	0.064	0.0595

Table 3.3 The best central values of the neutrino oscillation parameters within 3σ bound of experimental data calculated by using adaptive Monte Carlo method in IH. All the neutrino oscillation parameters are found within 3σ bound of experimental data consistent with the latest Planck cosmological upper bound $\sum |m_i| < 0.12$ eV and still valid upto $\sum |m_i| > 0.1$ eV

Section	I	II	III	IV
$\Delta m^2_{32}[eV^2]$	2.59×10^{-3}	2.46×10^{-3}	2.37×10^{-3}	2.5×10^{-3}
$\Delta m^2_{21}[eV^2]$	7.44×10^{-5}	7.38×10^{-5}	7.29×10^{-5}	7.06×10^{-5}
$\|U_{e1}\|$	0.819	0.825	0.837	0.805
$\|U_{e2}\|$	0.554	0.544	0.527	0.572
$\|U_{e3}\|$	0.148	0.149	0.144	0.153
$\|U_{\mu 1}\|$	0.305	0.299	0.469	0.346
$\|U_{\mu 2}\|$	0.640	0.646	0.557	0.665
$\|U_{\mu 3}\|$	0.704	0.701	0.684	0.660
$\|U_{\tau 1}\|$	0.485	0.478	0.280	0.480
$\|U_{\tau 2}\|$	0.531	0.534	0.641	0.479
$\|U_{\tau 3}\|$	0.694	0.696	0.714	0.734
$\theta_{12}/^0$	34.07	33.39	32.19	35.40
$\theta_{13}/^0$	8.51	8.56	8.27	8.80
$\theta_{23}/^0$	45.39	45	43.76	41.95
$\sum \|m_i\|[eV]$	0.113	0.108	0.1	0.1039

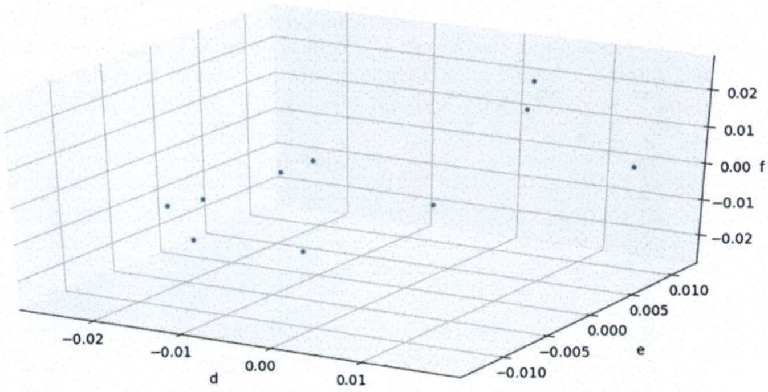

(a)

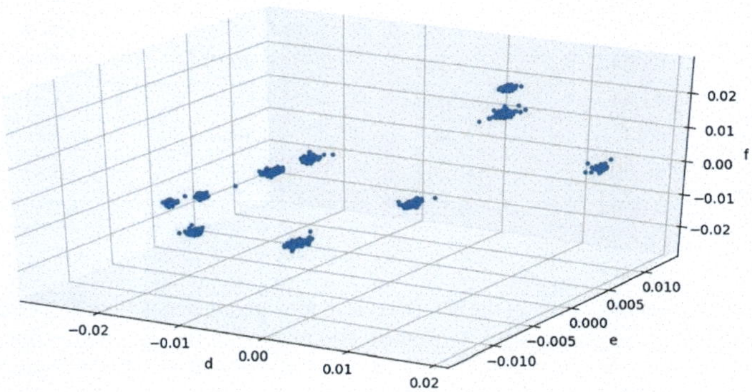

(b)

Fig. 3.5 Three-dimensional plots of allowed input parameters of a general Majorana mass texture generated by using Adaptive Monte Carlo simulation. Thin and thick plots of the above figure represent scatter plots for rough solutions and enlarged solutions for the general Majorana mass matrix in inverted hierarchy (IH)

[6, 7] have also indicated the most constraining bound to date, $\sum |m_i| < 0.09$ eV. When we impose this bound in our numerical analysis, there is no solution for the mass matrix in IH. Our numerical analysis confirms the mass bound of both actual normal and inverted hierarchy ($m_1 = 0$ and $m_3 = 0$) calculated by using the two mass-squared differences of experimental best-fit values. The present numerical analysis is focused on the CP-conserving Majorana neutrino mass matrix. In this case, the Dirac phase is taken exactly equal to zero. The recent experimental data observed the non-zero value of Dirac phase which is responsible for CP violation in neutrino sector, i.e. CP non-conserving case. The non-zero value of Dirac phase can only be extracted from the U_{UPMNS} of complex Majorana neutrino mass matrix. The full detailed analysis of complex Majorana neutrino mass matrix can be studied by adaptive Monte Carlo simulation method and shall be reported in our future communication.

References

1. N. Aghanim, Y. Akrami, M. Ashdown, J. Aumont, C. Baccigalupi, M. Ballardini, A.J. Banday, R.B. Barreiro, N. Bartolo, S. Basak, et al., Planck 2018 results-vi. cosmological parameters. Astron. Astrophys. **641**, A6 (2020)
2. I. Tanseri, S. Hagstotz, S. Vagnozzi, E. Giusarma, K. Freese, Updated neutrino mass constraints from galaxy clustering and cmb lensing-galaxy cross-correlation measurements. J. High Energy Astrophys. **36**, 1–26 (2022)
3. B. Dziewit, K. Kajda, J. Gluza, M. Zrałek, Majorana neutrino textures from numerical considerations: the cp conserving case. Phys. Rev. D **74**(3), 033003 (2006)
4. K. Kiers, J. Kolb, J. Lee, A. Soni, W. Guo-Hong, Ubiquitous cp violation in a top-inspired left-right model. Phys. Rev. D **66**(9), 095002 (2002)
5. M.C. Gonzalez-Garcia, M. Maltoni, T. Schwetz, Nufit: three-flavour global analyses of neutrino oscillation experiments. Universe **7**(12), 459 (2021)
6. E. Di Valentino, S. Gariazzo, O. Mena, Most constraining cosmological neutrino mass bounds. Phys. Rev. D **104**(8), 083504 (2021)
7. E. Di Valentino, A. Melchiorri, Neutrino mass bounds in the era of tension cosmology. Astrophys. J. Lett. **931**(2), L18 (2022)

Chapter 4
Scalar NSI: A Unique Probe to Explore New Physics in Neutrinos

Abinash Medhi, Arnab Sarker, Dharitree Bezboruah, Moon Moon Devi, and Debajyoti Dutta

Abstract The discovery of neutrino oscillations marked a groundbreaking moment, presenting the first experimental evidence of physics beyond the Standard Model (BSM). The BSM physics exploration often involves Non-Standard Interactions (NSIs), introducing unknown neutrino couplings. Scalar NSI, a unique interaction between neutrinos and matter fermions through a scalar, manifests as a medium-dependent correction to the neutrino mass term, introducing unique phenomenology in neutrino oscillations. This study delves into the impact of scalar NSI on the measurement sensitivities of oscillation parameters in upcoming long-baseline experiments (DUNE, T2HK, T2HKK). Scalar NSI introduces medium-dependent corrections to neutrino mass terms, influencing oscillation probabilities and detector event rates. Through a synergy study among experiments (DUNE+T2HK, DUNE+T2HKK), enhanced capabilities are demonstrated in constraining scalar NSI parameters, improving sensitivity to CP-violation and mass hierarchy. The exploration of scalar NSI also emerges as a valuable avenue for constraining absolute neutrino masses in the realm of neutrino oscillation experiments.

A. Medhi · A. Sarker · D. Bezboruah · M. M. Devi
Tezpur University, Sonitpur, Assam 784028, India
e-mail: arnabs@tezu.ernet.in

D. Bezboruah
e-mail: dbbphy1@tezu.ernet.in

M. M. Devi
e-mail: devimm@tezu.ernet.in

A. Medhi (✉)
Indian Institute of Technology Guwahati, Guwahati 781039, Assam, India
e-mail: abinashmedhi0@gmail.com

D. Dutta
Bhattadev University, Pathsala, Bajali, Assam 781325, India

J. W. F. Valle et al. (eds.), *Proceedings of the International Conference on Future Prospects in Neutrino and Astroparticle Physics*, Springer Proceedings in Physics 361,
https://doi.org/10.1007/978-981-97-7441-8_4

4.1 Introduction

The joint discovery of neutrino oscillations by Super-Kamiokande (SK) [1] and Sudbury Neutrino Observatory (SNO) [2] has opened new avenues for probing physics beyond the Standard Model (BSM). Neutrino oscillations confirm the existence of neutrino mass and offer the first experimental glimpse into BSM physics. Parameters related to neutrino oscillations are extensively studied in various experiments [3–6], with neutrinos serving as promising probes of BSM physics in the leptonic sector. Non-Standard Interactions (NSIs), involving unknown neutrino couplings, are explored in BSM models describing neutrino masses and mixing.

This work investigates the impact of scalar-mediated NSI on the measurement of the leptonic phase δ_{CP} in three long-baseline (LBL) neutrino experiments: DUNE [7], T2HK [8], and T2HKK [9]. A synergy analysis is conducted to probe scalar NSI effects in a model-independent manner. Current and upcoming neutrino experiments strive for precise measurements of oscillation parameters. However, parameter degeneracy among mixing parameters poses challenges. Combining data from various experiments helps overcome these challenges, as the degenerate parameter space differs among experiments. The synergy between experiments not only improves sensitivity but also reveals potential synergies. Previous studies have demonstrated enhanced CP-violation, mass hierarchy, and octant discovery sensitivities through the combination of different LBL experiments.

In the era of precision neutrino physics, ongoing experiments aim to address key unknowns: the hierarchy of neutrino masses [10], the octant of mixing angle θ_{23} [11], and the determination of the CP phase (δ_{CP}) [12]. Ongoing and future experiments, due to their robust nature, are sensitive to subdominant effects such as NSIs. This study focuses on scalar NSIs, which affect the neutrino mass in the oscillation Hamiltonian. Unlike vector NSIs, scalar NSI effects scale linearly with environmental matter density, making LBL experiments ideal for their exploration. The exploration of scalar NSI is motivated by its potential impact on CP-violation sensitivities, particularly in the DUNE experiment. This study, for the first time, conducts a model-independent synergy analysis of scalar NSI effects on DUNE, T2HK, and T2HKK. The results demonstrate significant impacts on CP-violation sensitivities, with the combination of experiments improving overall precision and helping alleviate degeneracies in CP measurements.

In conclusion, the paper emphasizes the importance of considering scalar NSIs in neutrino experiments, providing valuable insights into their effects on CP-violation measurements. The model-independent synergy analysis enhances our understanding of the scalar NSI impact on different LBL experiments, highlighting their crucial role in refining our knowledge of neutrino oscillations and BSM physics.

4.2 Formalism and Methodology

Neutrinos interact with matter through $W^{\pm}$ and Z^0 mediators, influencing ν-oscillations [13–16]. These interactions introduce an additional potential term in the ν-oscillation Hamiltonian [17–20]. Exploring scalar interactions of neutrinos is intriguing, where neutrinos couple with a scalar (Higgs boson) possessing a non-zero vacuum expectation, contributing to their mass generation. The Lagrangian for such non-standard ν coupling with environmental fermions via a scalar is given by [21–24],

$$\mathcal{L}_{\text{eff}}^{\text{S}} = \frac{y_f y_{\alpha\beta}}{m_\phi^2}(\bar{\nu}_\alpha(p_3)\nu_\beta(p_2))(\bar{f}(p_1)f(p_4)), \tag{4.1}$$

where y_f denotes the Yukawa coupling between the scalar mediator and environmental fermions f, while $Y_{\alpha\beta}$ represents the coupling between the scalar mediator and neutrinos. The alteration in $\mathcal{L}$ results in a correction to the Dirac equation, as illustrated below:

$$\overline{\nu}_\beta \left[i\partial_\mu \gamma^\mu + \left(M_{\beta\alpha} + \frac{\sum_f n_f y_f Y_{\alpha\beta}}{m_\phi^2} \right) \right] \nu_\alpha = 0, \tag{4.2}$$

where m_ϕ signifies the mass of the scalar mediator. As demonstrated in reference [21, 22], scalar Non-Standard Interactions (SNSI) can manifest as corrections to the neutrino mass matrix. The effective Hamiltonian in the presence of SNSI can be formulated from the modified Lagrangian, as indicated in Eq. 4.1.

$$\mathcal{H} \approx E_\nu + \frac{M_{eff} M_{eff}^\dagger}{2E_\nu} \pm V_{SI}, \tag{4.3}$$

where M_{eff}=M+M_{SNSI} with $M_{SNSI} \equiv \sum_f n_f y_f Y/m_\phi^2$. The neutrino mass matrix can undergo diagonalization using a modified mixing matrix, expressed as $\mathcal{U}' = P\mathcal{U}Q^\dagger$. The Majorana rephasing matrix Q is absorbed, resulting in $QD_\nu Q^\dagger = D_\nu = diag(m_1, m_2, m_3)$. The unphysical diagonal rephasing matrix, P, can be rotated away into the Scalar Non-Standard Interaction (SNSI) contribution:

$$M_{eff} = \mathcal{U}D_\nu\mathcal{U}^\dagger + P^\dagger M_{SNSI} P = M + \delta M. \tag{4.4}$$

We parameterize the SNSI contribution δM in a model-independent manner as follows.

$$\delta M \equiv \sqrt{|\Delta m_{31}^2|} \begin{pmatrix} \eta_{ee} & \eta_{e\mu} & \eta_{e\tau} \\ \eta_{e\mu}^* & \eta_{\mu\mu} & \eta_{\mu\tau} \\ \eta_{e\tau}^* & \eta_{\mu\tau}^* & \eta_{\tau\tau} \end{pmatrix}, \tag{4.5}$$

Table 4.1 The benchmark values of oscillation parameters used [31]

Parameters	True values
θ_{12}	34.51°
θ_{13}	8.44°
θ_{23}	47°
δ_{CP}	$-\pi/2$
Δm^2_{21}	$7.56 \times 10^{-5}\ eV^2$
Δm^2_{31}	$2.43 \times 10^{-3}\ eV^2$

where the dimensionless parameter $\eta_{\alpha\beta}$ gauges the strength of Scalar Non-Standard Interaction (SNSI), with $\sqrt{|\Delta m^2_{31}|}$ serving as a scaling factor. In the standard scenario, subtracting a common m_i^2 term yields a dependence of neutrino oscillations on mass-squared splittings, i.e., Δm^2_{21} and Δm^2_{31}. However, in the presence of SNSI, the cross terms, $M\delta M^\dagger$ and $M^\dagger \delta M$, introduce a direct dependence on the absolute masses of neutrinos, as there is no common term that can be subtracted from the mass matrix. Consequently, oscillation probabilities will directly rely on the absolute mass of neutrinos in the presence of SNSI.

To investigate the influence of NSI on various neutrino experiments, we utilized GLoBES (Global Long-Baseline Experiment Simulator) [27–29], a widely employed simulator for long-baseline experiments. The mixing parameters used in our simulations are detailed in Table 4.1. In our analysis, we assumed the normal hierarchy and higher octant as the true hierarchy and octant, respectively. We focused on three proposed super-beam experiments: DUNE, T2HK, and T2HKK, examining the impact of scalar NSI. Systematics and background details were incorporated from the corresponding Technical Design Reports (TDR) of the experiments, with uncertainties on signal and background summarized in Table 4.2. The study considered individual diagonal scalar NSI elements, exploring their effects at both the probability and event levels at the detector.

Table 4.2 The baselines, L/E, and fiducial volumes of each detector for T2HK, T2HKK, and DUNE

Experiment	Baseline (L in km)	L/E (in km/GeV)	Fiducial volume (in kton)
T2HK	295 km	527	187 × 2
T2HKK	295 km; 1100 km	527 (295 km); 1964 (1100 km)	187 (295 km) + 187 (1100 km)
DUNE	1300 km	1543	40

4.3 Results and Discussion

In this section, we discuss the impact of scalar NSI on the neutrino oscillation probabilities as well as on the physics sensitivities of the experiments.

4.3.1 *Effects on Oscillation Probabilities*

In this section, we examine the impact of scalar Non-Standard Interactions (NSI) on neutrino oscillation probabilities, specifically the three diagonal cases. Our analysis utilizes the NuOscProbExact package [30], a versatile Python-based numerical oscillation probability calculator for both two and three-flavor scenarios. The package employs SU(2) and SU(3) expansions of the evolution operators to compute numerical probabilities for time-independent Hamiltonians. We have adapted the neutrino Hamiltonian, as specified in Eq. 4.3, to include the three scalar NSI cases. The oscillation parameter values from Table 4.1 are used, assuming NH as the true mass hierarchy and HO as the true octant, unless stated otherwise.

Figure 4.1 shows the impact of diagonal scalar Non-Standard Interaction (NSI) elements, η_{ee} (left column), $\eta_{\mu\mu}$ (middle column), and $\eta_{\tau\tau}$ (right column), on $P_{\mu e}$ as a function of neutrino energy. The plots represent DUNE (top row), T2HK (middle row), and T2HKK (bottom row) baselines. Calculations assume $\delta_{CP} = -90°$ and $\theta_{23} = 47°$. In all plots, the solid-red line corresponds to the case without scalar NSI ($\eta_{\alpha\beta} = 0$). The solid (dashed) lines in black, blue, and magenta represent positive (negative) values of η_{ee}, $\eta_{\mu\mu}$, and $\eta_{\tau\tau}$, respectively. The scalar NSI parameters significantly affect oscillation probabilities at all three baselines, particularly around oscillation maxima. Positive (negative) η_{ee} enhances (suppresses) probabilities around oscillation maxima, while positive (negative) $\eta_{\tau\tau}$ shows complementary variations. Positive (negative) $\eta_{\mu\mu}$ shifts oscillation maxima toward higher (lower) energies with minor amplitude suppression.

4.3.2 *Effects on CP-Violation Sensitivity*

We examine the potential impact of $\eta_{\alpha\beta}$ on the measurement capability of δ_{CP} in three experiments. We assess their sensitivity to CP-conserving and CP-violating phases using the statistical χ^2 defined as follows:

$$\chi^2 \equiv \min_{\eta} \sum_i \sum_j \frac{\left[N^{i,j}_{true} - N^{i,j}_{test}\right]^2}{N^{i,j}_{true}}, \tag{4.6}$$

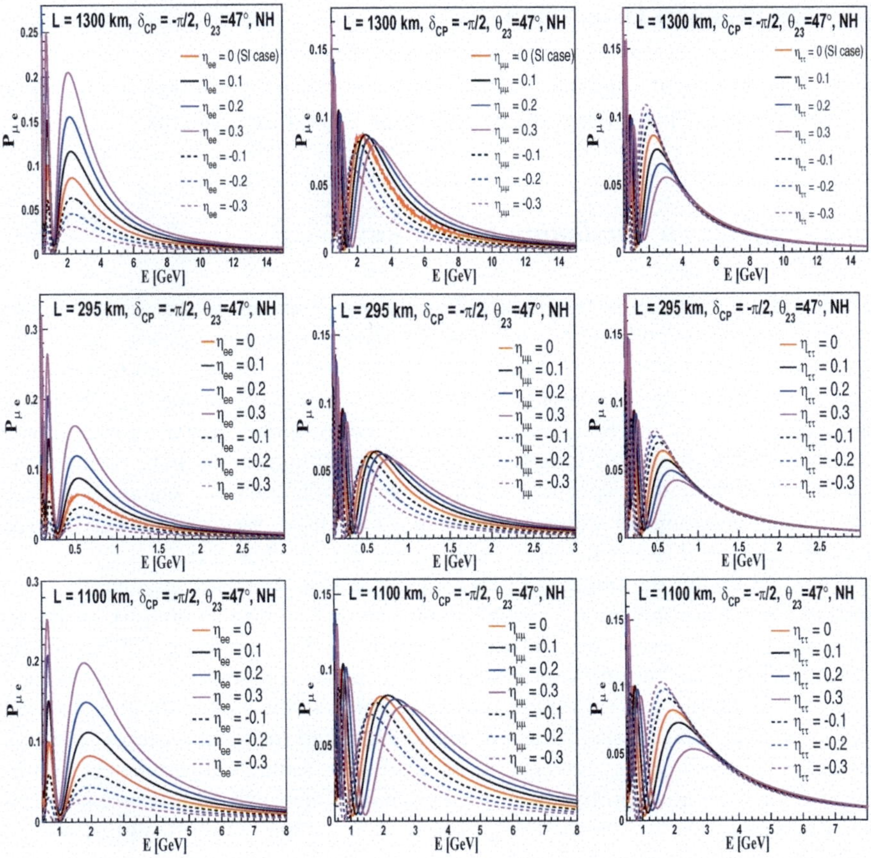

Fig. 4.1 The effects of η_{ee} (left column), $\eta_{\mu\mu}$ (middle column), and $\eta_{\tau\tau}$ (right column) on $P_{\mu e}$ at the baselines corresponding to DUNE (top row), T2HK (middle row), and T2HKK (bottom row). Here, $\delta_{CP} = -\pi/2$, $\theta_{23} = 47°$, and true mass Hierarchy = NH. In all the plots, the red solid-curve is for no-NSI case while other solid (dashed) curves are for positive (negative) NSI parameters.

where $N^{i,j}_{true}$ and $N^{i,j}_{test}$ represent the true and test events in the i, j-th bin, respectively. We conduct a sensitivity analysis to gauge the experiments' ability to constrain $\eta_{\alpha\beta}$. Additionally, we explore how $\eta_{\alpha\beta}$ influences CP-violation measurements, defined as the experiments' capacity to distinguish CP-conserving from CP-violating values. Sensitivities are initially determined for each experiment individually, followed by an investigation of combining DUNE+T2HK and DUNE+T2HKK configurations to uncover potential synergies.

Here, we examine the impact of $\eta_{\alpha\beta}$ on CPV sensitivities in DUNE, T2HK, and T2HKK. Sensitivities are derived by varying the true δ_{CP} within the range $[-\pi, \pi]$. Other mixing parameters, detailed in Table 4.1, maintain their true values. The test spectrum for δ_{CP} includes only CP-conserving values (0 and $\pm\,\pi$). We marginalize

θ_{23} and Δm^2_{31} over their allowed 3σ ranges [31] and minimize the χ^2 across all marginalization ranges. CPV sensitivity is calculated as follows:

$$\Delta\chi^2_{\rm CPV}\,(\delta^{\rm true}_{\rm CP}) = \min\,\left[\chi^2\,(\delta^{\rm true}_{CP}, \delta^{\rm test}_{CP} = 0),\ \chi^2(\delta^{\rm true}_{CP}, \delta^{\rm test}_{CP} = \pm\pi)\right]. \tag{4.7}$$

In Fig. 4.2, we illustrate the impact of scalar Non-Standard Interactions (NSI) on CP-Violation (CPV) sensitivity for DUNE (left column), T2HK (middle column), and their combination, DUNE + T2HK (right column). The inclusion of scalar NSI significantly affects the CPV sensitivity of both experiments. The plots depict the

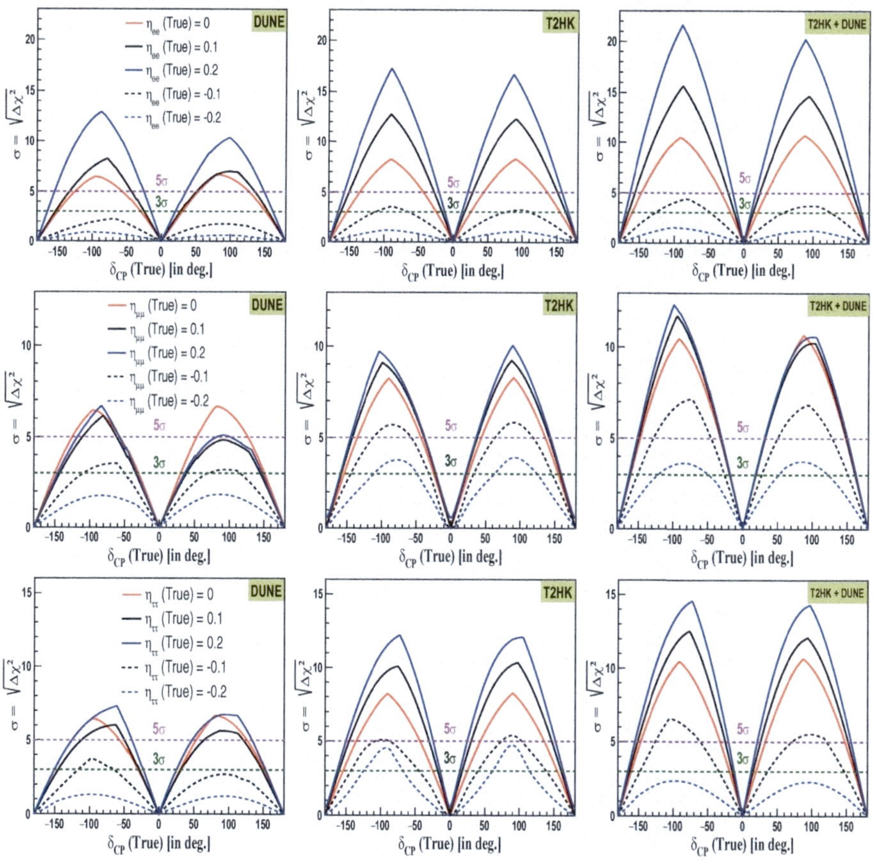

Fig. 4.2 The CPV sensitivity of DUNE (left column), T2HK (middle column), and DUNE + T2HK (right column) in presence of scalar NSI. The plots for η_{ee}, $\eta_{\mu\mu}$, and $\eta_{\tau\tau}$ are included in the top row, middle row, and bottom row, respectively. The solid-red curve is for the no scalar NSI case whereas solid (dashed) black and blue curves are for positive (negative) $\eta_{\tau\tau}$.

statistical significance ($\sigma = \sqrt{\Delta\chi^2_{CPV}}$) in relation to the true δ_{CP}. The top, middle, and bottom rows represent the results for η_{ee}, $\eta_{\mu\mu}$, and $\eta_{\tau\tau}$, respectively. The solid-red curve indicates the absence of scalar NSI ($\eta_{\alpha\beta} = 0$), while solid (dashed) black and blue curves represent positive (negative) values of $\eta_{\alpha\beta}$. The observations from Fig. 4.2 are summarized as follows. Positive (negative) η_{ee} enhances (suppresses) CPV sensitivities at DUNE and T2HK. Positive $\eta_{\mu\mu}$ deteriorates sensitivities in the upper half of δ_{CP} at DUNE but enhances them at T2HK. Negative $\eta_{\mu\mu}$ leads to significant suppression. Positive $\eta_{\tau\tau}$ induces marginal fluctuations at DUNE and enhances sensitivity at T2HK. The combined study of DUNE + T2HK improves sensitivities for all cases, particularly in overlapped regions, due to data collection in a broader range of degenerate spaces.

In Fig. 4.3, we present the impact of scalar Non-Standard Interactions (NSI) on CP-violation (CPV) sensitivities at DUNE (left panel), T2HKK (middle panel), and the combined analysis of DUNE + T2HKK (right panel). The influence of the NSI parameter $\eta_{\alpha\beta}$ on CPV sensitivities is noteworthy for both DUNE and T2HKK individually. The joint analysis reveals an enhanced sensitivity by combining the two experiments, enabling the collection of larger statistics across a broader range of degenerate parameter spaces. Marginalization over NSI parameters and θ_{23} within the range [40°, 50°] has been performed. The solid-red line represents the standard case, while other colored solid (dashed) lines correspond to positive (negative) values of $\eta_{\alpha\beta}$. The effects of η_{ee}, $\eta_{\mu\mu}$, and $\eta_{\tau\tau}$ are depicted in the top, middle, and bottom panels, respectively. The dashed green and magenta lines represent the 3σ and 5σ confidence levels, respectively. Notably, positive (negative) η_{ee} enhances (suppresses) CPV sensitivities, predominantly in DUNE and T2HKK. The joint analysis resolves degeneracies, improving overall sensitivities. A negative $\eta_{\mu\mu}$ degrades CPV sensitivity, while a positive $\eta_{\mu\mu}$ introduces degeneracies, which can be resolved through the joint analysis of DUNE + T2HKK. Similarly, a negative $\eta_{\tau\tau}$ suppresses CPV sensitivities, while a positive $\eta_{\tau\tau}$ improves them. The joint sensitivity of DUNE + T2HKK overcomes degeneracies, leading to enhanced CPV sensitivities in the presence of NSI. A detailed analysis of this study can be found in [23, 24].

4.3.3 Effects on Neutrino Mass Ordering

We investigate how scalar non-standard interactions (NSI) affect the sensitivity of the mass ordering (MO) at the Deep Underground Neutrino Experiment (DUNE). Figure 4.4 illustrates the impact of diagonal scalar NSI components (η_{ee}, $\eta_{\mu\mu}$, $\eta_{\tau\tau}$) on the neutrino oscillation probability ($P_{\mu e}$) across varying neutrino energies (E) at DUNE. It is observed that, in the absence of scalar NSI, DUNE's baseline capability to differentiate between the normal ordering (NO) and inverted ordering (IO) is evident within the energy range of approximately [1.5, 4] GeV. Introduction of η_{ee} extends both the energy range and the discernibility between the mass orderings. Conversely, the presence of $\eta_{\mu\mu}$ slightly diminishes the separation between the mass orderings and

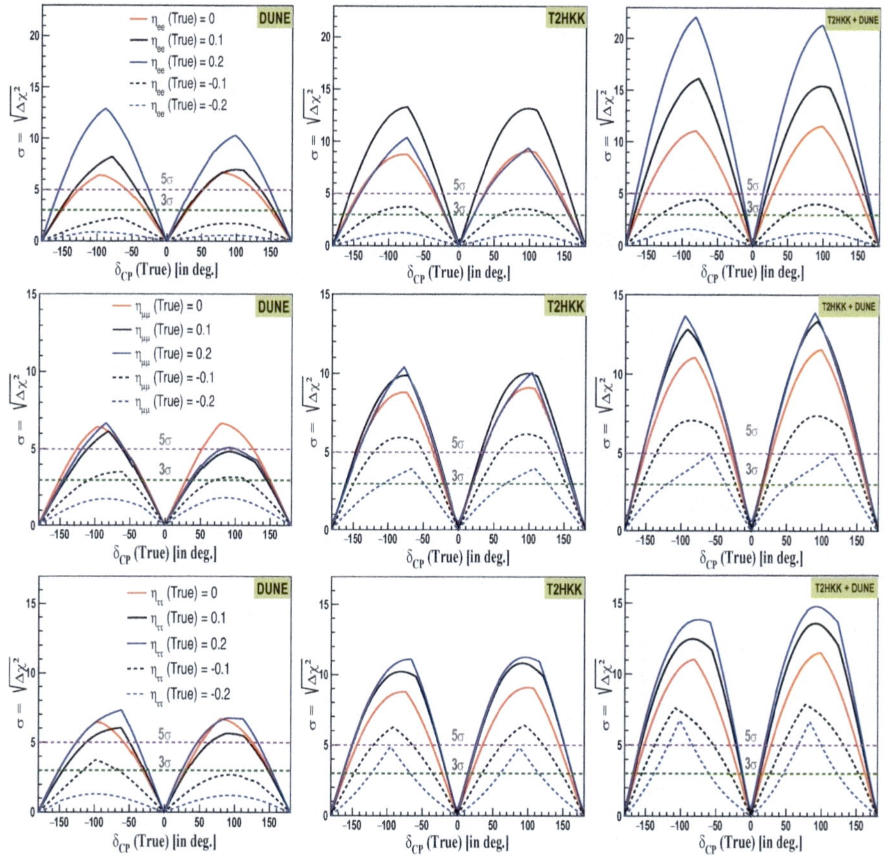

Fig. 4.3 The CPV sensitivity of DUNE (left column), T2HKK (middle column), and DUNE + T2HKK (right column) in presence of scalar NSI. The plots for η_{ee}, $\eta_{\mu\mu}$, and $\eta_{\tau\tau}$ are included in the top row, middle row, and bottom row, respectively. The solid-red curve is for the no scalar NSI case whereas solid (dashed) black and blue curves are for positive (negative) $\eta_{\tau\tau}$.

narrows the energy window of differentiation. Notably, the inclusion of $\eta_{\tau\tau}$ results in a significant overlap between the observed bands. For a more comprehensive analysis of the findings from this research, please refer to [26].

4.3.4 *Effects on Varying Neutrino Mass*

In this study, we've explored the potential to constrain the absolute masses of neutrinos by examining the impacts of scalar non-standard interactions. In Fig. 4.5, we investigate the influence of various permissible values of the lightest neutrino mass on

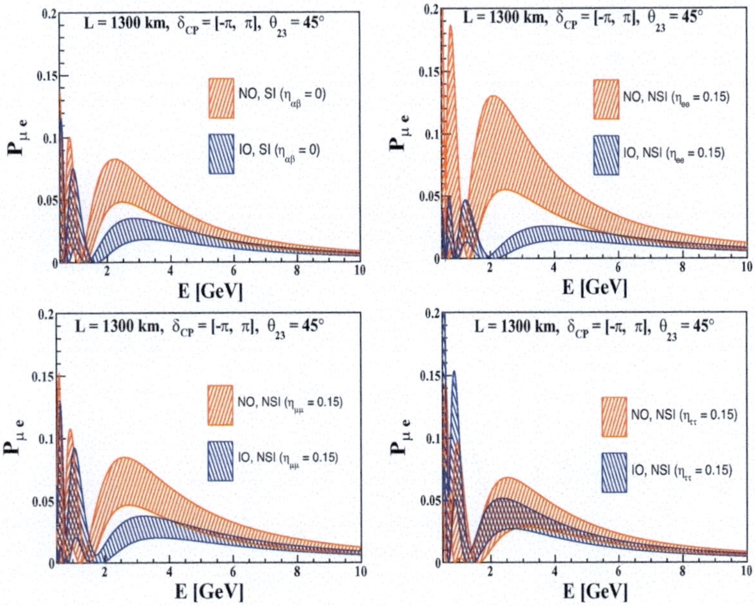

Fig. 4.4 Effects on $P_{\mu e}$ at the DUNE in the presence of scalar NSI, for NO (red band) and IO (blue band). Both the bands correspond to δ_{CP} values $\in [-\pi, \pi]$. The panels correspond to no scalar NSI (top-left), η_{ee} (top-right), $\eta_{\mu\mu}$ (bottom-left), and $\eta_{\tau\tau}$ (bottom-right) for $\delta_{CP} = -\pi/2$ and $\theta_{23} = 45°$.

the μ to e neutrino oscillation probability ($P_{\mu e}$) for η_{ee} (left panel), $\eta_{\mu\mu}$ (middle panel), and $\eta_{\tau\tau}$ (right panel) at DUNE. For the NH case (top panel), we note a slight deviation as the absolute neutrino mass scale shifts to higher values. The impact on $P_{\mu e}$ varies for different values of m_1, with distinct effects observed for positive and negative η_{ee}. Positive (negative) $\eta_{\mu\mu}$ leads to a suppression (enhancement) of probabilities for higher neutrino mass values. Similarly, a significant suppression (enhancement) of $P_{\mu e}$ is observed for positive (negative) $\eta_{\tau\tau}$, as depicted in the top-right panel. For the IH scenario (bottom panel), a positive η_{ee} results in suppression with increasing neutrino mass (m_3), while a negative η_{ee} causes enhancement. Positive (negative) $\eta_{\mu\mu}$ induces a marginal shift of the second peak toward lower (higher) energies with increasing neutrino mass. Additionally, we observe enhancement (suppression) of $P_{\mu e}$ with increasing neutrino mass for positive (negative) $\eta_{\tau\tau}$. For further elaboration on the findings of this study, readers can refer to [25].

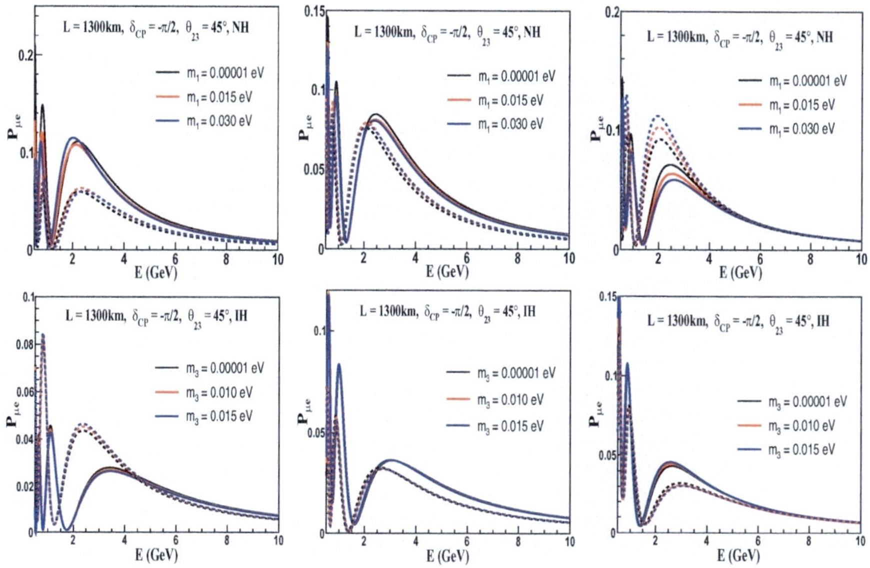

Fig. 4.5 Appearance probabilities for different values of lightest neutrino mass with $\eta_{ee} = \pm 0.1$ (left panel), $\eta_{\mu\mu} = \pm 0.1$ (middle panel), and $\eta_{\tau\tau} = \pm 0.1$ (right panel) for normal hierarchy (top panel) and inverted hierarchy (bottom panel). The solid (dashed) line represents a positive (negative) value of $\eta_{\alpha\beta}$. Also, we consider δ_{CP} = -π/2 and $\theta_{23} = 45°$ to generate the plots.

4.4 Summary and Concluding Remarks

The advanced progress in neutrino physics, coupled with cutting-edge experimental setups, aims to precisely measure neutrino oscillation parameters. Ambitious upcoming flagship neutrino experiments focus on achieving the highest accuracy in measuring these parameters, particularly δ_{CP} and the octant of mixing angle θ_{23} currently the least constrained in neutrino physics. Identifying these subdominant neutrino effects is crucial for understanding their impact on the physics reach of experiments. This study primarily explores the impact of scalar NSI in three upcoming Long-Baseline experiments, with ongoing efforts to investigate NSI effects across different physics sensitivities in various neutrino experiments. Collaborative endeavors involving solar, atmospheric, reactor experiments, etc., are essential to comprehensively grasp NSI impacts. Stringent constraints on scalar NSI effects are equally important for accurately interpreting data from diverse neutrino experiments.

References

1. Y. Fukuda et al., Super-Kamiokande. Phys. Rev. Lett. **81**, 1562–1567 (1998). https://doi.org/10.1103/PhysRevLett.81.1562. [arXiv:hep-ex/9807003]
2. Q.R. Ahmad et al., SNO. Phys. Rev. Lett. **87**, 071301 (2001). https://doi.org/10.1103/PhysRevLett.87.071301. [arXiv:nucl-ex/0106015] [nucl-ex]
3. Super-Kamiokande collaboration, Evidence for an oscillatory signature in atmospheric neutrino oscillation, Phys. Rev. Lett. **93**, 101801 (2004). https://doi.org/10.1103/PhysRevLett.93.101801. [arXiv:hep-ex/0404034] [hep-ex]
4. KamLAND collaboration, Measurement of neutrino oscillation with KamLAND: Evidence of spectral distortion. Phys. Rev. Lett. **94**, 081801 (2005). https://doi.org/10.1103/PhysRevLett.94.081801. [arXiv:hep-ex/0406035] [hep-ex]
5. MINOS collaboration, Measurement of neutrino oscillations with the MINOS detectors in the NuMI beam. Phys. Rev. Lett. **101**, 131802 (2008). https://doi.org/10.1103/PhysRevLett.101.131802. [arXiv:hep-ex/0806.2237] [hep-ex]
6. MINOS collaboration, Measurement of the neutrino mass splitting and flavor mixing by MINOS. Phys. Rev. Lett. **106**, 181801 (2011). https://doi.org/10.1103/PhysRevLett.106.181801. [arXiv:1103.0340] [hep-ex]
7. DUNE collaboration, *Deep Underground Neutrino Experiment (DUNE), Far Detector Technical Design Report, Volume IV: Far Detector Single-phase Technology, 2020 JINST 15 T08010*. https://doi.org/10.1088/1748-0221/15/08/T08010. [arXiv:2002.03010] [hep-ex]
8. Hyper-Kamiokande Proto-collaboration, *Physics Potential of a Long-Baseline Neutrino Oscillation Experiment Using a J-PARC Neutrino Beam and Hyper-Kamiokande*. PTEP 2015 (2015), p. 053C02. https://doi.org/10.1093/ptep/ptv061. [arXiv:1502.05199] [hep-ex]
9. Hyper-Kamiokande collaboration, *Physics Potentials with the Second Hyper-Kamiokande Detector in Korea* (PTEP 2018, 2018), p. 063C01. https://doi.org/10.1093/ptep/pty044. [arXiv:1611.06118] [hep-ex]
10. F. Capozzi, E. Di Valentino, E. Lisi, A. Marrone, A. Melchiorri, A. Palazzo, Global constraints on absolute neutrino masses and their ordering. Phys. Rev. D **95**, 096014 (2017). [Addendum ibid. 101 (2020) 116013]. https://doi.org/10.1103/PhysRevD.95.096014. [arXiv:2003.08511] [hep-ph]
11. S.K. Agarwalla, S. Prakash, S.U. Sankar, Resolving the octant of theta23 with T2K and NOvA. JHEP **07**, 131 (2013). https://doi.org/10.1007/JHEP07(2013)131. [(arXiv:1301.2574] [hep-ph])
12. M. Kobayashi, T. Maskawa, CP violation in the renormalizable theory of weak interaction. Prog. Theor. Phys. **49**, 652 (1973). https://doi.org/10.1143/PTP.49.652
13. B. Pontecorvo, Mesonium and anti-mesonium. Sov. Phys. JETP **6**, 429 (1957)
14. B. Pontecorvo, Inverse beta processes and nonconservation of lepton charge. Zh. Eksp. Teor. Fiz. **34**, 247 (1957)
15. B. Pontecorvo, Neutrino experiments and the problem of conservation of leptonic charge. Zh. Eksp. Teor. Fiz. **53**, 1717 (1967)
16. Particle Data Group collaboration, Rev. Particle Phys. PTEP 2020, 083C01 (2020). https://doi.org/10.1093/ptep/ptaa104
17. L. Wolfenstein, Phys. Rev. D **17**, 2369–2374 (1978). https://doi.org/10.1103/PhysRevD.17.2369
18. J.F. Nieves, P.B. Pal, Generalized fierz identities. Am. J. Phys. **72**, 1100 (2004). https://doi.org/10.1119/1.1757445. [arXiv:hep-ph/0306087]
19. C.C. Nishi, Simple derivation of general Fierz-like identities. Am. J. Phys. **73**, 1160 (2005). https://doi.org/10.1119/1.2074087. [arXiv:hep-ph/0412245] [hep-ph]
20. Z. Maki, M. Nakagawa, S. Sakata, Remarks on the unified model of elementary particles. Prog. Theor. Phys. **28**, 870 (1962). https://doi.org/10.1143/PTP.28.870
21. S.F. Ge, S.J. Parke, Phys. Rev. Lett. **122**(21), 211801 (2019). https://doi.org/10.1103/PhysRevLett.122.211801. [arXiv:1812.08376] [hep-ph]

22. K.S. Babu, G. Chauhan, P.S. Bhupal Dev, Phys. Rev. D **101**(9), 095029 (2020). https://doi.org/10.1103/PhysRevD.101.095029. [arXiv:1912.13488] [hep-ph]
23. A. Medhi, D. Dutta, M.M. Devi, JHEP **06**, 129 (2022). https://doi.org/10.1007/JHEP06(2022)129. [arXiv:2111.12943] [hep-ph]
24. A. Medhi, M.M. Devi, D. Dutta, JHEP **01**, 079 (2023). https://doi.org/10.1007/JHEP01(2023)079. [arXiv:2209.05287] [hep-ph]
25. A. Medhi, A. Sarker, M.M. Devi, [arXiv:2307.05348] [hep-ph]
26. A. Sarker, A. Medhi, D. Bezboruah, M.M. Devi, D. Dutta, [arXiv:2309.12249] [hep-ph]
27. P. Huber, M. Lindner, W. Winter, Simulation of long-baseline neutrino oscillation experiments with GLoBES (General Long Baseline Experiment Simulator). Comput. Phys. Commun. **167**, 195 (2005). https://doi.org/10.1016/j.cpc.2005.01.003. [arXiv:hep-ph/0407333] [hep-ph]
28. J. Kopp, Efficient numerical diagonalization of Hermitian 3×3 matrices. Int. J. Mod. Phys. C **19**, 523 (2008). https://doi.org/10.1142/S0129183108012303. [arXiv:physics/0610206]
29. P. Huber, J. Kopp, M. Lindner, M. Rolinec, W. Winter, New features in the simulation of neutrino oscillation experiments with GLoBES 3.0: general long baseline experiment simulator. Comput. Phys. Commun. **177**, 432 (2007). https://doi.org/10.1016/j.cpc.2007.05.004. [arXiv:hep-ph/0701187]
30. M. Bustamante, NuOscProbExact: a general-purpose code to compute exact two-flavor and three-flavor neutrino oscillation probabilities. arXiv:1904.12391
31. I. Esteban, M.C. Gonzalez-Garcia, M. Maltoni, T. Schwetz, A. Zhou, The fate of hints: updated global analysis of three-flavor neutrino oscillations. JHEP **09**, 178 (2020). https://doi.org/10.1007/JHEP09(2020)178. [arXiv:2007.14792] [hep-ph]

Chapter 5
Effect of Texture Zeros in Modular Left-Right Symmetric Model (LRSM)

Ankita Kakoti, Happy Borgohain, and Mrinal Kumar Das

Abstract In this work, Left-Right Symmetric Model (LRSM) has been realized with the modular group of level 3, that is, $\Gamma(3)$ and weight 2, which is isomorphic to non-abelian discrete symmetry group A_4. Here, we take into consideration the possibilities of texture zeros in the mass matrices of the modular symmetric left-right model. As the type-I and type-II seesaw masses appear naturally in LRSM, the texture zero study has been carried out for both the Dirac and heavy right-handed Majorana mass matrices, respectively. The advantage of using modular symmetry is that we do not require the use of any extra particles (flavons) for obtaining the desired results within the realization of the model. As LRSM provides an environment for the study and analysis of beyond Standard Model phenomenology, in this work, we take into consideration the study of contributions of Neutrinoless Double Beta Decay within the framework of minimal LRSM with A_4 modular symmetry.

5.1 Introduction

The Standard Model (SM) of particle physics is one of the most successful and well-established theories of particle physics to date. However, in spite of its huge achievements, the model does not provide answers to many questions regarding the absolute mass scale of neutrinos, the exact nature of it (Dirac or Majorana), the hierarchical pattern of the mass spectrum (Normal or Inverted) and leptonic CP violation. The absolute mass scale of the neutrinos is not yet known. However, experiments like Planck have given an upper bound on the sum of the light neutrino masses to be $\Sigma|m_{\nu_i}| < 0.23$ eV in 2012 and recently the bound has been constrained to $\Sigma|m_{\nu_i}| < 0.11$ eV. The most successful data pertaining to neutrino oscillation parameters is found in the 3σ global fit data.

A. Kakoti (✉) · M. K. Das
Department of Physics, Tezpur University, Napaam, Tezpur, India
e-mail: kakotiankita97@gmail.com

H. Borgohain
Department of Physics, Silapathar College, Silapathar 787059, India

J. W. F. Valle et al. (eds.), *Proceedings of the International Conference on Future Prospects in Neutrino and Astroparticle Physics*, Springer Proceedings in Physics 361,
https://doi.org/10.1007/978-981-97-7441-8_5

The simplest way to look for neutrino masses is by the seesaw mechanism. The mechanism may be of type-I [1, 2], type-II [3, 4], type-III [5] and Inverse Seesaw [6] and these are extensions of the SM where extra particles are incorporated like right-handed fermions, scalar fermion triplets, gauge singlet neutral fermions, etc. The Standard Model of particle physics also cannot shed any light upon the phenomena like baryon asymmetry of the universe (BAU) [7], Lepton Number Violation (LNV), Lepton Flavor violation (LFV) [8], existence of dark matter [9, 10], etc. These are also some of the reasons why a BSM framework of physics is necessary. A BSM framework which has been successful in explaining the first three of the phenomenologies is the Left-Right Symmetric Model, an extension of the SM corresponding to the addition of $SU(2)_R$ group into the theory. The gauge group of LRSM is $SU(3)_C \otimes SU(2)_R \otimes SU(2)_L \otimes U(1)_{B-L}$. The type-I and type-II seesaw masses appear naturally in the model. The right-handed neutrinos are an essential part of the model, which acquires Majorana mass when $SU(2)_R$ symmetry is broken. LRSM provides a natural framework to understand the spontaneous breaking of parity and origin of small neutrino masses by seesaw mechanism [11].

Also, in our work, we have used A_4 modular symmetry to study neutrino masses and mixings and hence study phenomenology within the model. The advantage of using modular symmetry over discrete flavor symmetries is that the study of the model using symmetries can be done without the introduction of extra particles called 'flavons'. Given below is a very brief overview of modular symmetry as given in [12, 13]. An element q of the modular group acts on a complex variable τ which belongs to the upper-half of the complex plane given as

$$q\tau = \frac{a\tau + b}{c\tau + d} \tag{5.1}$$

where a, b, c, d are integers and $ad - bc = 1$, Imτ >0.

The modular group is isomorphic to the projective special linear group PSL(2, Z) = SL(2, Z)/Z_2 where SL(2, Z) is the special linear group of integer 2×2 matrices having determinant unity and $Z_2 = (I, -I)$ is the center, I being the identity element. The modular group can be represented in terms of two generators S and T which satisfies $S^2 = (ST)^3 = I$. S and T satisfy the following matrix representations:

$$S = \begin{pmatrix} 0 & 1 \\ -1 & 0 \end{pmatrix} \tag{5.2}$$

$$T = \begin{pmatrix} 1 & 1 \\ 0 & 1 \end{pmatrix} \tag{5.3}$$

corresponding to the transformations,

$$S, \tau \rightarrow -\frac{1}{\tau}, T, \tau \rightarrow \tau + 1 \tag{5.4}$$

5.2 Left-Right Symmetric Model (LRSM)

LRSM is an extension of the Standard Model of particle physics, the gauge group being $SU(3)_C \otimes SU(2)_R \otimes SU(2)_L \otimes U(1)_{B-L}$, which has been studied by several groups since 1970s [11, 14]. The usual type-I and type-II seesaw neutrino masses arise naturally in the model. In this model, the electric charge is given by $Q = T_{3L} + T_{3R} + \frac{B-L}{2}$, where T_{3L} and T_{3R} are the generators of $SU(2)_L$ and $SU(2)_R$, respectively. $B - L$ refers to baryon number minus lepton number.

The quarks and leptons that transform in the Left-Right symmetric gauge group are given by

$$\phi = \begin{bmatrix} u \\ d \end{bmatrix}_L, Q_R = \begin{bmatrix} u \\ d \end{bmatrix}_R, \Psi_L = \begin{bmatrix} \nu_l \\ l \end{bmatrix}_L, \Psi_R = \begin{bmatrix} \nu_l \\ l \end{bmatrix}_R. \tag{5.5}$$

The Yukawa Lagrangian of the leptonic sector in LRSM is given by

$$\begin{aligned} \mathcal{L} = & h_{ij}\overline{\Psi}_{L,i}\phi\Psi_{R,j} + \widetilde{h_{ij}}\overline{\Psi}_{L,i}\widetilde{\phi}\Psi_{R,j} + f_{L,ij}{\Psi_{L,i}}^T Ci\sigma_2\Delta_L\Psi_{L,j} \\ & + f_{R,ij}{\Psi_{R,i}}^T Ci\sigma_2\Delta_R\Psi_{R,j} + h.c \end{aligned} \tag{5.6}$$

where $\widetilde{\phi} = \tau_2\phi^*\tau_2$ and $C = i\gamma_2\gamma_0$ is the charge conjugation operator. The Higgs sector in LRSM consists of $\phi(1, 2, 2, 0)$, $\Delta_R(1, 1, 3, 2)$ and $\Delta_L(1, 3, 1, 2)$. Our model is minimal as the Higgs sector does not contain more than three scalar multiplets. The matrix representation for the scalar sector is given by

$$\phi = \begin{pmatrix} \phi_1^0 & \phi_1^+ \\ \phi_2^- & \phi_2^0 \end{pmatrix} \tag{5.7}$$

$$\Delta_{L,R} = \begin{pmatrix} \frac{\delta_{L,R}^+}{\sqrt{2}} & \delta_{L,R}^{++} \\ \delta_{L,R}^0 & -\frac{\delta_{L,R}^+}{\sqrt{2}} \end{pmatrix} \tag{5.8}$$

where the family indices i, j are summed over, the indices $i, j = 1, 2, 3$ represent the three generations of fermions. $C = i\gamma_2\gamma_0$ is the charge conjugation operator, $\widetilde{\phi} = \tau_2\phi^*\tau_2$ and γ_μ are the Dirac matrices. Considering discrete parity symmetry, the Majorana Yukawa couplings $f_L = f_R$ (for left-right symmetry) give rise to Majorana neutrino mass after electroweak symmetry breaking.

The light neutrino mass, generated within a type-I+II seesaw, can be written as

$$M_\nu = {M_\nu}^I + {M_\nu}^{II}, \tag{5.9}$$

$$M_\nu = M_{LL} + M_D {M_{RR}}^{-1} {M_D}^T = \sqrt{2}v_L f_L + \frac{k^2}{\sqrt{2}v_R} h_D {f_R}^{-1} {h_D}^T \tag{5.10}$$

where the first and second terms in equation correspond to type-II seesaw and type-I seesaw mediated by RH neutrino, respectively. In the context of LRSM both type-I and type-II seesaw terms can be written in terms of M_{RR} which arises naturally at a high energy scale as a result of spontaneous symmetry breaking. Thus, the above equation can be written as

$$M_\nu = \gamma\left(\frac{M_W}{v_R}\right) M_{RR} + M_D {M_{RR}}^{-1} {M_D}^T \tag{5.11}$$

where γ is a dimensionless parameter given by

$$\gamma = \frac{\beta_1 k_1 k_2 + \beta_2 k_1^2 + \beta_3 k_2^2}{(2\rho_1 - \rho_3)k^2} \tag{5.12}$$

In our work, we have fine-tuned the dimensionless parameter to $\gamma \approx 10^{-10}$ and $v_R \approx TeV$.

5.2.1 *Modular Symmetric LRSM*

In our work, we will basically be incorporating modular symmetry into LRSM. The advantage of using modular symmetry rather than flavor symmetry is the minimal use of extra particles (flavons). The model contains the usual particle content of LRSM [15]. As we will consider modular symmetry, we will have to assign modular weights to the particles, keeping in mind that matter multiplets corresponding to the model can have negative modular weights, but the modular forms cannot be assigned negative weights. In modular symmetry, the Yukawa couplings are given by the following relations:

$$\begin{aligned} Y_1 = {} & 1 + 12q + 36q^2 + 12q^3 + 84q^4 + 72q^5 + 36q^6 \\ & + 96q^7 + 180q^8 + 12q^9 + 216q^{10} \end{aligned} \tag{5.13}$$

$$\begin{aligned} Y_2 = {} & -6q^{1/3}(1 + 7q + 8q^2 + 18q^3 + 14q^4 + 31q^5 \\ & + 20q^6 + 36q^7 + 31q^8 + 56q^9) \end{aligned} \tag{5.14}$$

$$\begin{aligned} Y_3 = {} & -18q^{2/3}(1 + 2q + 5q^2 + 4q^3 + 8q^4 + 6q^5 \\ & + 14q^6 + 8q^7 + 14q^8 + 10q^9) \end{aligned} \tag{5.15}$$

where q is given by $q = e^{(2\pi i\tau)}$ (Table 5.1).

Table 5.1 Charge assignments for the particle content of the model

Gauge group	l_L	l_R	ϕ	Δ_L	Δ_R
$SU(3)_C$	1	1	1	1	1
$SU(2)_L$	2	1	2	3	1
$SU(2)_R$	1	2	2	1	3
$U(1)_{B-L}$	−1	−1	0	2	2
A_4	3	3	1	1	1
k_I	0	−2	0	−2	2

Table 5.2 Charge assignment and modular weight for the corresponding modular Yukawa forms for the model

	Y (modular forms)
A_4	3
k_I	2

The Yukawa couplings $Y = (Y_1, Y_2, Y_3)$ are expressed as modular forms of level 3 (Table 5.2).

The Yukawa Lagrangian in the lepton sector corresponding to LRSM incorporating modular symmetry is given by

$$\mathcal{L_Y} = \overline{l_L}\phi l_R Y + \overline{l_L}\tilde{\phi} l_R Y + {l_R}^T C \iota \tau_2 \Delta_R l_R Y + {l_L}^T C \iota \tau_2 \Delta_L l_L Y \tag{5.16}$$

The Dirac mass matrix in our model is given by

$$M_D = < v > \begin{pmatrix} 2Y_1 & -Y_3 & -Y_2 \\ -Y_2 & -Y_1 & 2Y_3 \\ -Y_3 & 2Y_2 & -Y_1 \end{pmatrix} \tag{5.17}$$

where $< v >$ is considered to the VEV for the Higgs bidoublet.

And, the right-handed Majorana mass matrix is given as

$$M_R = < v_R > \begin{pmatrix} 2Y_1 & -Y_3 & -Y_2 \\ -Y_3 & 2Y_2 & -Y_1 \\ -Y_2 & -Y_1 & 2Y_3 \end{pmatrix} \tag{5.18}$$

where $< v_R >$ is the VEV for the scalar triplet Δ_R.

5.2.2 Neutrinoless Double Beta Decay in LRSM

Neutrinoless double beta decay ($0\nu\beta\beta$) is a lepton number violating process, which if proven to exist will directly imply the Majorana nature of neutrinos.

$$N(A, Z) \rightarrow N(A, Z+2) + e^- + e^- \quad (5.19)$$

Many groups have however already done a lot of work on NDBD in the model, [14, 16, 17]. In LRSM [18], there are several contributions to NDBD in addition to the standard contribution. In Standard light neutrino contribution, the intermediate particles are the W_L bosons and light neutrinos, the process in which the amplitude depends upon the leptonic mixing matrix elements and light neutrino masses. In this work, we have tried to analyze only the Standard light neutrino contribution.

5.3 Modular Symmetric LRSM with Texture Zeros

Implementing the concept of texture zeros with the context of modular LRSM, we obtain that for 1–0 texture of neutrino mass matrices for both Dirac and Majorana mass matrices, the light neutrino mass matrix where $M_{ee} = 0$ can be taken into consideration. And for 2–0 texture, we obtain the light neutrino mass matrix only for the classes $A1$ and $A2$ as shown below.

5.3.1 1–0 Texture

When $M_{ee} = 0$, the mass matrices obtained are

$$M_D = \begin{pmatrix} 0 & -Y_3 & -Y_2 \\ -Y_2 & 0 & 2Y_3 \\ -Y_3 & 2Y_2 & 0 \end{pmatrix} \quad (5.20)$$

$$M_R = \begin{pmatrix} 0 & -Y_3 & -Y_2 \\ -Y_3 & 2Y_2 & 0 \\ -Y_2 & 0 & 2Y_3 \end{pmatrix} \quad (5.21)$$

5.3.2 2–0 Texture

Class A1 (When $M_{ee} = 0$ and $M_{e\mu} = 0$)

$$M_D = \begin{pmatrix} 0 & 0 & -Y_2 \\ -Y_2 & 0 & 0 \\ 0 & 2Y_2 & 0 \end{pmatrix} \tag{5.22}$$

$$M_R = \begin{pmatrix} 0 & 0 & -Y_2 \\ 0 & 2Y_2 & 0 \\ -Y_2 & 0 & 0 \end{pmatrix} \tag{5.23}$$

Class A2 (When $M_{ee} = 0$ and $M_{e\tau} = 0$)

$$M_D = \begin{pmatrix} 0 & -Y_3 & 0 \\ 0 & 0 & 2Y_3 \\ -Y_3 & 0 & 0 \end{pmatrix} \tag{5.24}$$

$$M_R = \begin{pmatrix} 0 & -Y_3 & 0 \\ -Y_3 & 0 & 0 \\ 0 & 0 & 2Y_3 \end{pmatrix} \tag{5.25}$$

As it is evident from the above matrices that, corresponding to the elements which are zero, we get that either one or two of the modular forms goes to zero. However, it is to be noted here that goes to zero does not imply that the modular form (e.g., say Y_1, Y_2, Y_3) is zero. Because, in mathematical terms texture zeros does not actually imply that the element is completely zero, but it means that the contribution of that particular element in neutrino masses and mixings in taken to be negligible. So, as evident from Eqs. 5.13, 5.14 and 5.15 the modular forms are expressions of q that depend upon τ, so considering texture zeros implies their contribution can be taken to be negligible corresponding to the element that is zero, but it might not be correct to state that the corresponding modular form is completely zero.

5.4 Numerical Analysis and Results

We have considered the above-mentioned textures for the light neutrino mass matrix. We have implemented both 1–0 and 2–0 textures for both Dirac and Majorana mass matrices and then we have obtained the resulting light neutrino mass matrix. After the determination of the mass matrix, we have calculated the values of the modular forms (Y_1, Y_2, Y_3) and their substitution in the matrices leads to the desired result. We have then calculated the sum of the neutrino masses and plotted it against the neutrino parameters α and δ and also with the associated modular forms in each of

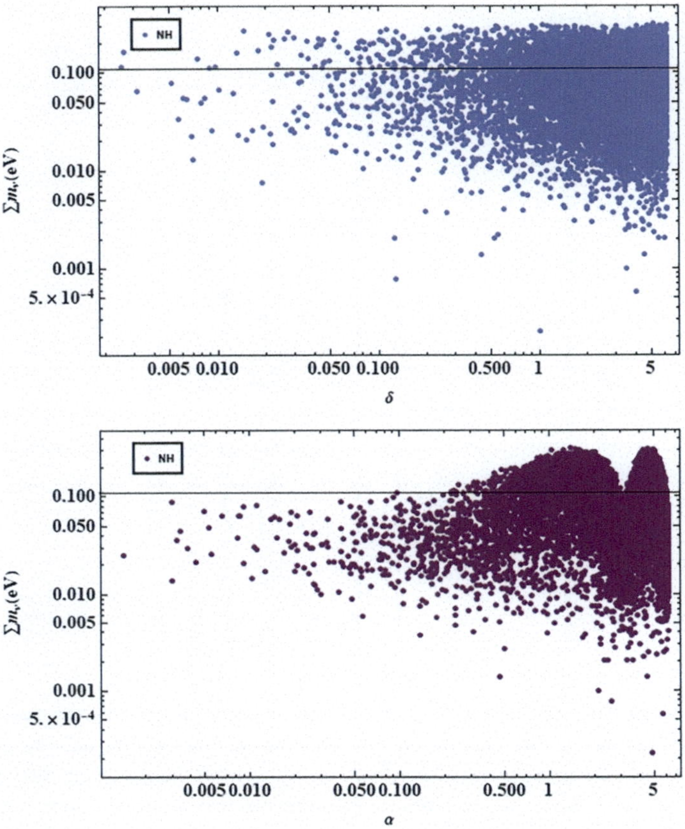

Fig. 5.1 Variation of sum of neutrino masses with δ(above) and α (below) for **Class A1** and normal hierarchy. The black line indicates the Planck bound on sum of the neutrino masses

the texture zero cases. For **Class A1** of 2–0 texture of neutrino mass, the plots are shown in Figs. 5.1 and 5.2

For **Class A2** of 2–0 texture of neutrino mass, the plots are shown in Figs. 5.3 and 5.4.

For 1–0 texture of neutrino mass, the plots are shown in Figs. 5.5 and 5.6.

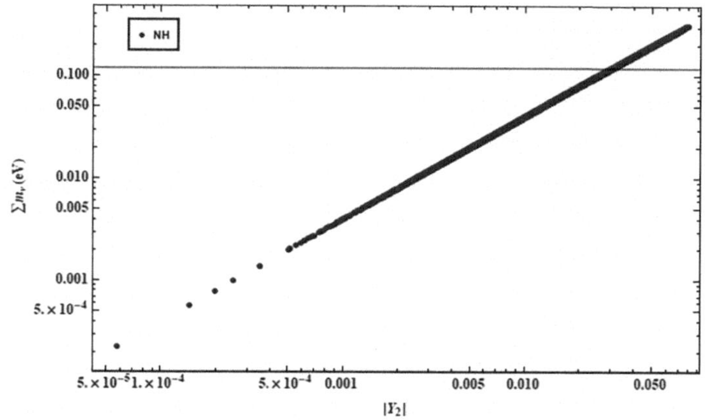

Fig. 5.2 Variation of sum of neutrino masses with modular form Y_2 for **Class A1** and normal hierarchy. The black line indicates the Planck bound on sum of the neutrino masses

5.5 Conclusion

We have determined the textures of the mass matrices taking into account both 1–0 and 2–0 textures for both Dirac and Majorana mass matrices and then we have determined the resulting light neutrino mass matrix. It has been seen that for 2-0 texture the allowed classes are **Class A1** and **Class A2**, and for 1–0 texture, the allowed class is where the element $M_{ee} = 0$. The sum of the neutrino masses has been calculated and its variations have been shown with different parameters as shown in the plots. It has been found that a considerable amount of data lies within the desired experimental bounds.

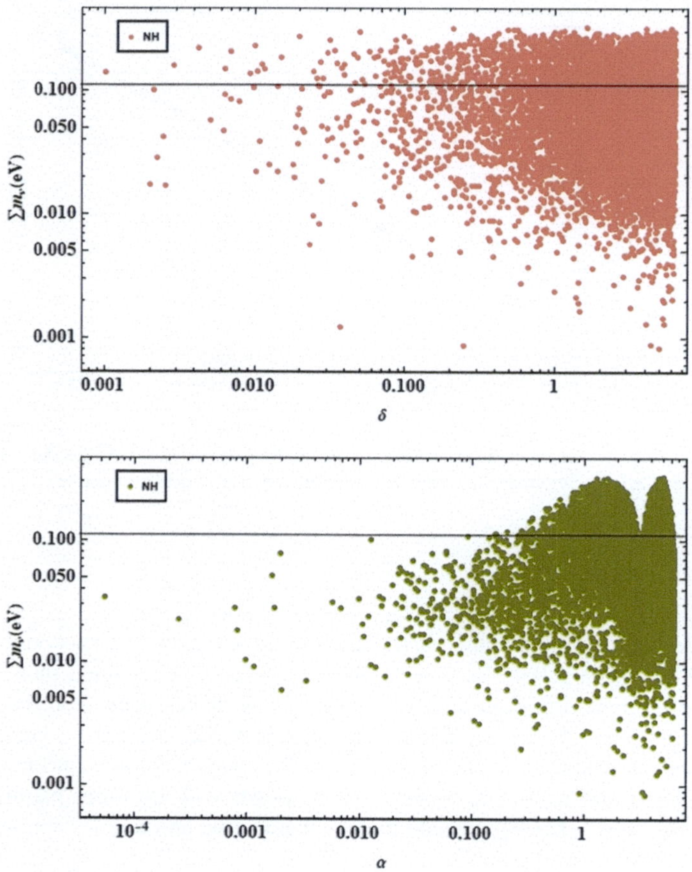

Fig. 5.3 Variation of sum of neutrino masses with δ (above) and α(below) for **Class A2** and normal hierarchy. The black line indicates the Planck bound on sum of the neutrino masses

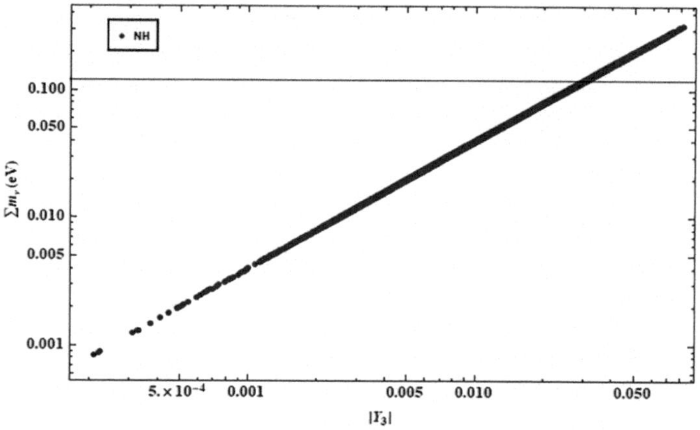

Fig. 5.4 Variation of sum of neutrino masses with modular form Y_3 for **Class A2** and normal hierarchy. The black line indicates the Planck bound on sum of the neutrino masses

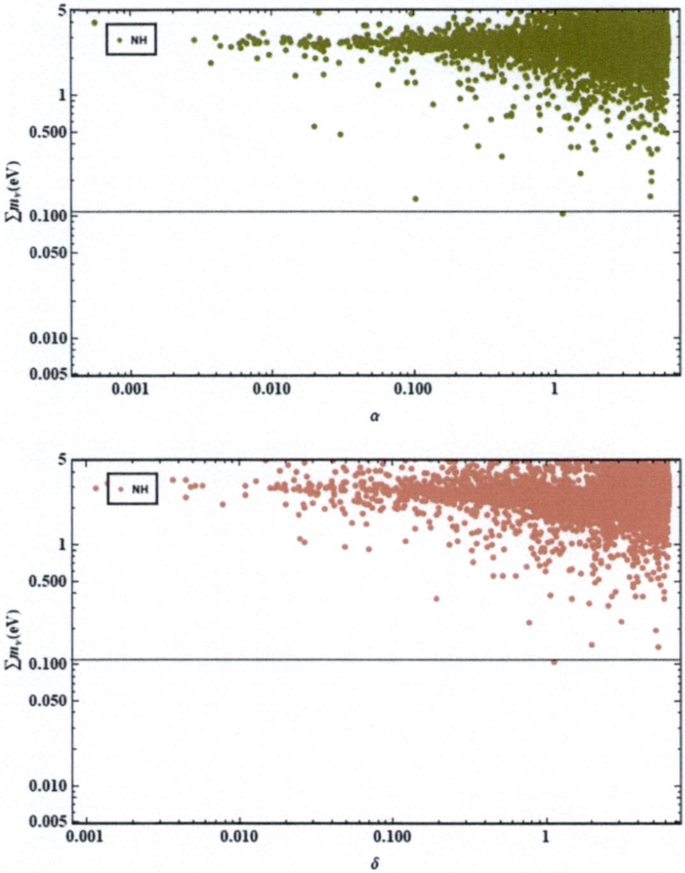

Fig. 5.5 Variation of sum of neutrino masses with α (above) and δ(below) for normal hierarchy. The black line indicates the Planck bound on sum of the neutrino masses

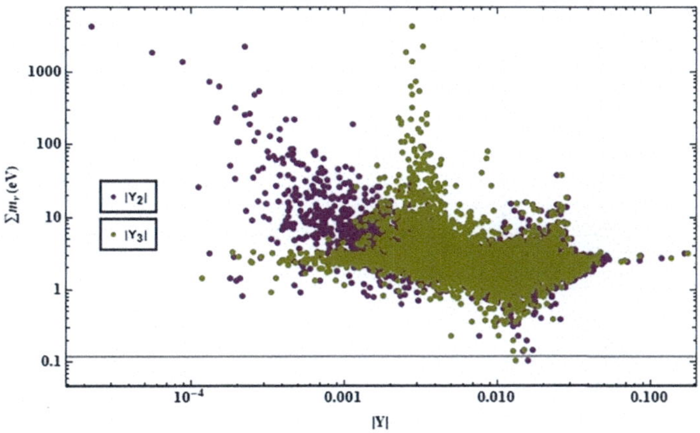

Fig. 5.6 Variation of sum of neutrino masses with modular form Y_2 and $Y_3$for normal hierarchy. The black line indicates the Planck bound on sum of the neutrino masses

Acknowledgements Ankita Kakoti acknowledges the Department of Science and Technology (DST), India (Grant DST/INSPIRE Fellowship/2019/IF190900), for the financial assistantship.

References

1. J. Schechter, J.W.F. Valle, Neutrino masses in SU(2) x U(1) theories. Phys. Rev. D **22**, 2227 (1980)
2. R.N. Mohapatra, G. Senjanovic, Neutrino mass and spontaneous parity nonconservation. Phys. Rev. Lett. **44**, 912 (1980)
3. C. Wetterich, Neutrino masses and the scale of B-L violation. Nucl. Phys. B **187**, 343–375 (1981)
4. S. Antusch, S.F. King, Type II Leptogenesis and the neutrino mass scale. Phys. Lett. B **597**, 199–207 (2004)
5. R. Foot, H. Lew, X.G. He, G.C. Joshi, Seesaw neutrino masses induced by a triplet of leptons. Z. Phys. C **44**, 441 (1989)
6. N. Haba, H. Ishida, Y. Yamaguchi, Naturalness and lepton number/flavor violation in inverse seesaw models. JHEP **11**, 003 (2016)
7. E.W. Kolb, S. Wolfram, The development of baryon asymmetry in the early universe. Phys. Lett. B **91**, 217–221 (1980)
8. G. Altarelli, F. Feruglio, L. Merlo, E. Stamou, Discrete flavour groups, $theta_{13}$ and lepton flavour violation. JHEP **08**, 021 (2012)
9. M. Taoso, G. Bertone, A. Masiero, Dark matter candidates. A ten-point test. JCAP **03**, 022 (2008)
10. M.S. Turner, Dark matter. Theoretical perspectives. Proc. Nat. Acad. Sci. **90**, 4827 (1993)
11. W. Grimus, Introduction to left-right symmetric models, in *4th Hellenic School on Elementary Particle Physics*, vol. 9 (1993), pp. 619–632
12. F. Feruglio, *Are Neutrino Masses Modular Forms?* (2019), pp. 227–266
13. P.P. Novichkov, J.T. Penedo, S.T. Petcov, A.V. Titov, Generalised CP symmetry in modular-invariant models of flavour. JHEP **07**, 165 (2019)

14. J.C. Pati, A. Salam, Lepton number as the fourth color. Phys. Rev. D **10**, 275–289 (1974). [Erratum, Phys. Rev. D **11**, 703–703 (1975)]
15. P. Sahu, S. Patra, P. Pritimita, A_4 *Realization of Left-Right Symmetric Linear Seesaw*, vol. 2 (2020)
16. H. Borgohain, M.K. Das, Neutrinoless double beta decay and lepton flavour violation in broken $\mu - \tau$ symmetric neutrino mass models. Int. J. Theor. Phys. **56**(9), 2911–2934 (2017)
17. B.B. Boruah, M.K. Das, Neutrinoless double-beta decay and lepton flavor violation in discrete flavor symmetric left–right symmetric model. Int. J. Mod. Phys. A **37**(06), 2250026 (2022)
18. G. Senjanovic, R.N. Mohapatra, Exact left-right symmetry and spontaneous violation of parity. Phys. Rev. D **12**, 1502 (1975)

Chapter 6
Inelastic Self-interacting Dark Matter in Non-standard Cosmology

Manoranjan Dutta

Abstract Inconsistencies between the predictions based on the ΛCDM model and astrophysical observations at dwarf and galaxy scales motivate alternatives to WIMPs. A compelling alternative is the self-interacting dark matter (SIDM), which not only addresses the anomalies on small scales but also aligns seamlessly with the large-scale ΛCDM predictions. Nevertheless, the thermal relic of SIDM tends to be under-abundant within the standard cosmological framework, which considers dark matter decoupling from the cosmic soup in a radiation-dominated epoch. To tackle this challenge, our approach involves introducing a non-standard expansion history of the universe by incorporating a non-radiation-like component into the cosmic soup. In this modified scenario, SIDM decouples at an early epoch compared to the standard model, resulting in an enhanced relic density that matches observed values. We consider our SIDM candidate to be a pseudo-Dirac fermion to nullify its diagonal vector interaction, thereby circumventing direct search constraints. Through this, we identify a parameter space that remains consistent with astrophysical observations and relic density criteria.

6.1 Introduction

There exists compelling support for the existence of dark matter (DM), e.g. galaxy rotation curves, gravitational lensing, and cosmic microwave background anisotropies [1–3]. According to the most recent data from the Planck collaboration, DM comprises approximately 26.8% of the current energy density of the universe [4]. Despite this, the Standard Model (SM) of particle physics falls short in providing a good candidate for DM. In extensions beyond the SM, weakly interacting massive particles (WIMPs), a class of cold dark matter with mass and couplings typically at the weak scale, have been proposed as potential DM candidates. However, null

M. Dutta (✉)
Department of Physics, North Lakhimpur University, Khelmati, North Lakhimpur, Lakhimpur, Assam 787031, India
e-mail: dutta.manoranjan@nlc.ac.in

J. W. F. Valle et al. (eds.), *Proceedings of the International Conference on Future Prospects in Neutrino and Astroparticle Physics*, Springer Proceedings in Physics 361,
https://doi.org/10.1007/978-981-97-7441-8_6

results at terrestrial DM search experiments have raised doubt about conventional WIMP candidates [5]. Besides, several small-scale anomalies such as the core-cusp problem, missing satellites, and too-big-to-fail problem [6, 7] cast doubt on cold dark matter hypothesis of the ΛCDM model. To address these anomalies, an alternative called self-interacting dark matter (SIDM) was proposed [8–10]. DM having self-scattering cross-section of approximately $\sigma/m \sim 1\,\text{cm}^2/\text{g} \approx 2 \times 10^{-24}\,\text{cm}^2/\text{GeV}$ [11–16] has been shown to be suitable for the purpose. DM self-interaction is also expected to be more pronounced at lower velocities, influencing small-scale structures and gradually decreasing towards large scale in a way to align with CDM predictions on larger scales, particularly at the cluster scale [11–28]. Such velocity-dependent self-interaction is realized in SIDM models with a light mediator. Spontaneously broken $U(1)$ models are suitable for ensuring DM stability through charge conservation [29–33]. However, due to a substantial $\mathcal{O}(1)$ coupling between SIDM and the light mediator, efficient annihilation into the latter leads to an under-abundant SIDM relic.

An important consideration regarding the under-abundant SIDM thermal relic is that existing literature usually assumes DM decoupling to occur during a radiation-dominated epoch, although there is no strict justification for this assumption. The introduction of a non-radiation-like component alters the Hubble parameter, significantly impacting the freeze-out dynamics [34–73]. This study investigates the possibility of DM freeze-out occurring during a matter-dominated phase of the universe to check if this can produce the correct thermal relic abundance for SIDM.

SIDM with diagonal self-interaction in non-standard cosmology was explored in [74], where it was found that the required reheating temperature to obtain the correct relic density is much above the mass scale of SIDM, which can potentially bring frozen out DM back into thermal equilibrium prior to BBN. The only way to avoid this is to make the portal between the dark and visible sector inactive at the epoch of reheating. This means the light BSM mediator has to decay before reheating as the portal between the dark and the visible vector is via the kinetic mixing between the light BSM mediator and Z boson. The kinetic mixing between the two sectors also makes SIDM detectable at the direct search experiments and these experiments give a stringent upper bound on the mixing. The same tiny mixing also makes the light mediator very long live. To alleviate this problem, here we consider the SIDM to be of Majorana nature so that its diagonal vector interaction vanishes. Based on inelastic DM scenario [75, 76], its direct detection (and hence the constraint on the kinetic mixing) can be evaded if the mass difference between the two pseudo-Dirac states is kept above MeV scale. Hence, the light mediator can decay much earlier to the epoch of reheating.

The structure of this paper is outlined as follows: Sect. 6.2 presents the introduction of the $U(1)_D$ extended model essential for realizing self-interacting dark matter (SIDM) phenomenology. Section 6.3 explores the details of SIDM phenomenology. In Sect. 6.4, we examine the freeze-out abundance of SIDM within a non-standard cosmology scenario and conduct a comparative analysis with the standard radiation-dominated case. Finally, we summarize and conclude in Sect. 6.5.

6.2 Self-interaction with Inelastic Dark Matter

We consider a simple $U(1)$ extension of the SM. In this extension, a vector-like Dirac fermion Ψ, which is an SM singlet, carries a charge of 1, while an SM singlet scalar Φ bears a charge of -2 under the $U(1)_D$ gauge symmetry. Upon acquiring a vacuum expectation value (VEV), the scalar Φ not only breaks the $U(1)_D$ symmetry spontaneously but also causes the Dirac fermion to split into two pseudo-Dirac states. The Lagrangian for this model is expressed as follows:

$$\begin{aligned}\mathcal{L} &= \mathcal{L}_{\text{SM}} + i\overline{\Psi}\gamma^{\mu}D_{\mu}\Psi - M\left(\overline{\Psi}_L\Psi_R + \overline{\Psi}_R\Psi_L\right) \\ &- \left(y_L\Phi\overline{(\Psi_L)^c}\Psi_L + y_R\Phi\overline{(\Psi_R)^c}\Psi_R + h.c.\right) + \mathcal{L}_{\Phi} + \frac{\epsilon}{2}B^{\alpha\beta}Y_{\alpha\beta}\end{aligned} \tag{6.1}$$

where $D_\mu = \partial_\mu - ig_D Z'_\mu$ and $B^{\alpha\beta}$, $Y_{\alpha\beta}$ are the field strength tensors of $U(1)_D$ and $U(1)_Y$, ϵ being the kinetic mixing between them. The Lagrangian involving singlet scalar can be written as

$$\mathcal{L}_{\Phi} = (D_\mu\Phi)^\dagger(D^\mu\Phi) + m_\Phi^2\Phi^\dagger\Phi - \lambda_\phi(\Phi^\dagger\Phi)^2 - \lambda_{\Phi H}(\Phi^\dagger\Phi)(H^\dagger H) \tag{6.2}$$

where H is the SM Higgs doublet. The singlet scalar Φ acquires a non-zero VEV $\langle\Phi\rangle = u$ giving rise to $U(1)_D$ gauge boson mass $M_{Z'} = 2g_D u$. As Φ acquires a VEV, the $U(1)D$ symmetry undergoes spontaneous breaking, resulting in a residual Z_2 symmetry. Under this Z_2 symmetry, $\Psi_{L,R}$ are odd while the rest are even. The combination of Ψ_L and Ψ_R forms a stable DM candidate in the low-energy effective theory. The VEV of the singlet scalar also generates Majorana masses for the fields Ψ_L and Ψ_R: $m_L = y_L u/\sqrt{2}$ and $m_R = y_R u/\sqrt{2}$. Assuming $m_L, m_R \ll M$, the Dirac fermion $\Psi = \Psi_L + \Psi_R$ splits into two pseudo-Dirac states, denoted as ψ_1 and ψ_2, with masses $M_1 = M - m_+$ and $M_2 = M + m_+$, where $m_\pm = (m_L \pm m_R)/2$. The Lagrangian for the dark matter, post-symmetry breaking, is then given by

$$\begin{aligned}\mathcal{L}_{DM} \supset &\frac{1}{2}\overline{\psi_1}\gamma^\mu\partial_\mu\psi_1 + \frac{1}{2}\overline{\psi_2}\gamma^\mu\partial_\mu\psi_2 - \frac{1}{2}M_1\overline{\psi_1}\psi_1 - \frac{1}{2}M_2\overline{\psi_2}\psi_2 + \frac{\epsilon}{2}B^{\alpha\beta}Y_{\alpha\beta} \\ &+ ig_D Z'_\mu\overline{\psi_1}\gamma^\mu\psi_2 + \frac{1}{2}g_D Z'_\mu\frac{m_-}{M}(\overline{\psi_2}\gamma^\mu\gamma^5\psi_2 - \overline{\psi_1}\gamma^\mu\gamma^5\psi_1) \\ &+ \frac{1}{2}(y_L\cos^2\theta - y_R\sin^2\theta)\overline{\psi_1}\psi_1\phi + \frac{1}{2}(y_R\cos^2\theta - y_L\sin^2\theta)\overline{\psi_2}\psi_2\phi\end{aligned} \tag{6.3}$$

where $\sin\theta \approx m_-/M$. The mass splitting between the two mass eigenstates is given by $\Delta m = M_2 - M_1 = 2m_+ = (y_L + y_R)\frac{u}{\sqrt{2}}$. We keep this mass difference greater than a MeV so that the up-scattering process is kinematically forbidden and the direct detection constraint is simply evaded. The SIDM candidates can interact among themselves either via scalar (Φ) or vector mediators (Z'). We consider only the vector-mediated interaction to be dominant ($y_L, y_R \ll g_D$). The elastic scattering term via Z' is also suppressed by a factor of m_-/M. Thus, the relevant Lagrangian for self-interactions via Z' is

$$\mathcal{L} \supset i g_D Z'_\mu \overline{\psi_1} \gamma^\mu \psi_2 \tag{6.4}$$

For sufficient self-interaction with desired velocity dependence, the model needs to fulfil two conditions: Z' should be sufficiently lighter than the SIDM and the coupling g_D must be sizeable.

6.3 Dark Matter Self-interaction

From the Lagrangian term given in Eq. 6.4, the Feynman diagram depicting DM self-scattering is shown in Fig. 6.1. Considering non-relativistic dark matter particle scatterings, the effective potential for two Majorana fermions with a light mediator is given by [77–80]

$$V(r) = \begin{pmatrix} 0 & -\alpha \frac{e^{M_{Z'} r}}{r} \\ -\alpha \frac{e^{M_{Z'} r}}{r} & 2\Delta m \end{pmatrix}. \tag{6.5}$$

The corresponding two-body Schroedinger equation is given by

$$\frac{1}{M_\psi} \nabla^2 \Psi(\mathbf{r}) = \left(V(r) - M_\psi v^2\right) \Psi(\mathbf{r}) \tag{6.6}$$

where M_ψ is the SIDM mass ignoring the tiny mass splitting Δm and v is the individual velocity of SIDM in the centre of mass frame, $\Psi(\mathbf{r})$ being the DM wave function. Defining dimensionless parameters $\epsilon_v = \frac{v}{\alpha}$, $\epsilon_\delta = \sqrt{\frac{2\Delta m}{M_\psi \alpha^2}}$, $\epsilon_{Z'} = \frac{M_{Z'}}{M_\psi \alpha}$, and $r\Psi(\mathbf{r}) = \psi(r)$, the s-wave Schroedinger equation is given by

$$\psi''(r) = \begin{pmatrix} -\epsilon_v^2 & -\frac{e^{\epsilon_{Z'} r}}{r} \\ -\frac{e^{\epsilon_{Z'} r}}{r} & \epsilon_\delta^2 - \epsilon_v^2 \end{pmatrix} \psi(r) \tag{6.7}$$

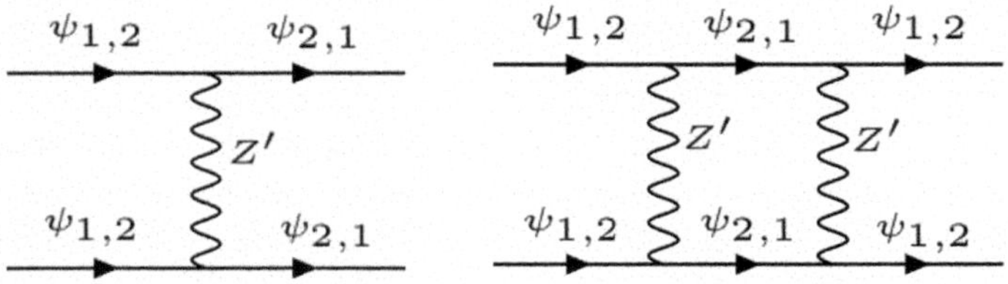

Fig. 6.1 Feynman diagrams for DM self-interaction

As DM velocity is much lower in dwarf galaxies ($v \sim 10^{-4}$) than in the local DM wind, naively if the above mass splitting Δm is kept large enough to forbid the up scattering in direct detection, the tree level up-scattering processes $\psi_1\psi_1 \to \psi_2\psi_2$ as shown in the left panel of Fig. 6.1 will also be kinematically forbidden. However, it is still possible to achieve elastic scattering at one loop level as shown in the right panel of Fig. 6.1. In principle, there can be multiple mediator exchanges and all sorts of scatterings are possible. Such loop diagrams are not inherently suppressed beyond the perturbative regime, necessitating the resumption of ladder diagrams with arbitrary numbers of vector boson exchanges (for a more comprehensive analysis, refer to [77]). The relevant cross-sections are provided in 6.5. Utilizing these findings, we constrain the self-interaction favoured parameter space in terms of dark matter and mediator masses, considering the required self-interactions at different scales while maintaining a mass splitting of 1 MeV between the two pseudo-Dirac states.

In Fig. 6.2, we present the allowed parameter space with self-interaction cross-section (σ/m) in the range $0.1 - 1\ \text{cm}^2/\text{g}$ for clusters ($v \sim 1000$ km/s) and galaxies ($v \sim 200$ km/s) and $0.1 - 10\ \text{cm}^2/\text{g}$ for dwarf galaxies ($v \sim 10$ km/s) keeping the coupling constant $g_D = 0.1$ and $\Delta m = 1$ MeV. We note that across all scales, there is no sufficient self-interaction (σ/m) from up scattering in the entire parameter space. This is due to the fact that the lighter dark matter, given its low velocities, lacks the necessary kinetic energies to efficiently scatter into heavier dark matter, and self-interaction is primarily contributed by elastic scattering. We also plot in the left panel of Fig. 6.2 σ/m as a function of v, which fits the data from dwarfs (red), LSB galaxies (blue), and clusters (green) [18, 81]. Different combinations of M_ψ and $M_{Z'}$ are depicted in colour codes keeping $g_D = 0.1$. Thus, the model gives rise to sufficient DM self-interaction with desired velocity dependence.

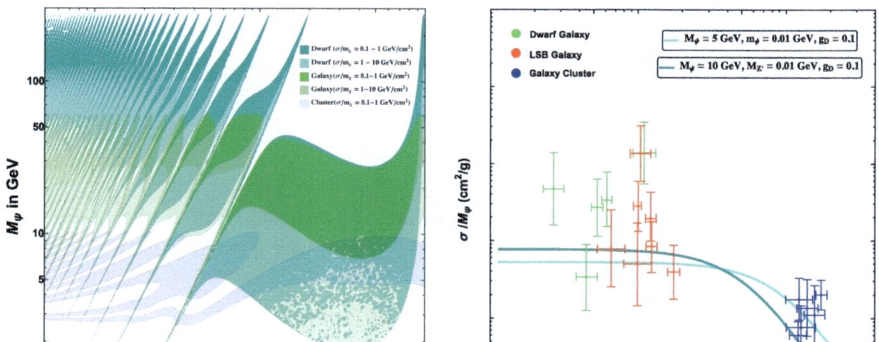

Fig. 6.2 [Left]: Self-interaction cross-section (σ/M_ψ) in the required range for different astrophysical objects with $g_D = 0.1$ and $\Delta m = 1 MeV$. [Right]: The self-interaction cross-section per unit mass of DM as a function of average collision velocity fitted to available data from astrophysical observations.

6.4 Relic Density of SIDM

The evolution of the DM number density n_ψ is governed by the Boltzmann equation, which in the standard cosmology is given by

$$\dot{n}_\psi + 3Hn_\psi = -\langle\sigma v\rangle[n_\psi^2 - (n_\psi^{\text{eq}})^2] \tag{6.8}$$

Defining dimensionless parameters $x = M_\psi/T$ and $Y_\psi = n_\psi/s(T))$ where $s(T) == \frac{2\pi^2}{45} g_{*S}T^3$ is the energy density of the universe, the Boltzmann equations can also be written in term of the co-moving DM number density Y_ψ as

$$\frac{dY_\psi}{dx} = -\frac{s(M_\psi)}{x^2 H(M_\psi)}\langle\sigma v\rangle\left(Y_\psi^2 - \left(Y_\psi^{eq}\right)^2\right) \tag{6.9}$$

The number density of SIDM depletes mainly by two number changing processes depicted by Feynman graphs in Fig. 6.3. The left panel diagram is the dominant one as it is unsuppressed, and the thermally averaged cross-section for this annihilation is given by

$$\langle\sigma v\rangle = \frac{\pi\alpha^2}{M_\psi^2}\sqrt{1 - \frac{M_{Z'}^2}{M_\psi^2}} \tag{6.10}$$

where $\alpha_D = g_D^2/(4\pi)$. For $g_D = 0.1$ and DM (mediator) mass in the GeV (MeV) scale, the cross-section obtained from Eq. 6.10 is much larger than the one governing typical WIMP annihilation, resulting in an under-abundant relic density.

As the standard freeze-out results in an under-abundance of thermal relic, we invoke a non-standard expansion history by adding a heavy scalar η with energy density ρ_η besides the components considered in the standard cosmology.

The Hubble parameter (H) in this modified scenario is given by

$$H^2 = \frac{8\pi}{3M_{\text{pl}}^2}\left(\rho_R + \rho_\psi + \rho_\eta\right) \tag{6.11}$$

where ρ_R and ρ_ψ are the energy densities of the SM radiation bath and that of DM, respectively. We define a critical temperature $T_\star$, such that η evolves as

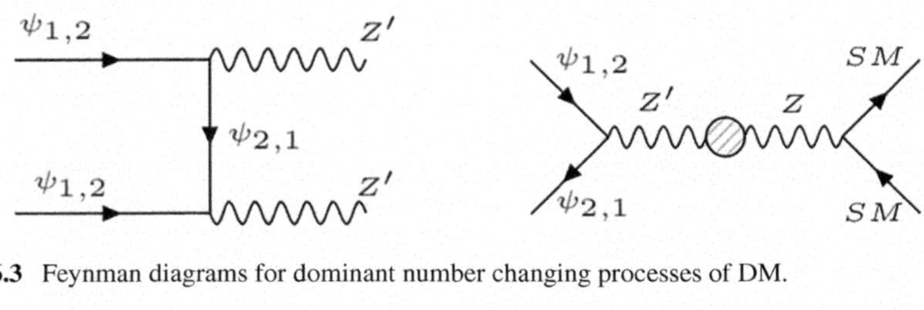

Fig. 6.3 Feynman diagrams for dominant number changing processes of DM.

$$\eta \propto \begin{cases} a^{-4} \text{ for } T > T_\star \\ a^{-3} \text{ for } T < T_\star \end{cases} \tag{6.12}$$

For $T_\star > T \gg M_Z, M_\psi$, Eq. 6.11 can be rewritten as

$$\frac{H^2}{H_\star^2} = \frac{g_* r}{g_* + g_\psi}\left(\frac{a_\star}{a}\right)^4 + (1-r)\left(\frac{a_\star}{a}\right)^3 + \frac{g_\psi r}{g_* + g_\psi}\left(\frac{a_\star}{a}\right)^4 \tag{6.13}$$

where g_*, g_ψ, and g_η are the effective number of relativistic degrees of freedom (DoF) of the SM, internal DoF of DM, and internal DoF of η, respectively. The quantity r represents the fraction of the energy in radiation at temperature $T_\star$ and is defined as

$$r \equiv \left.\frac{\rho_R + \rho_\psi}{\rho_R + \rho_\psi + \rho_\eta}\right|_{T=T_\star}. \tag{6.14}$$

In terms of $T_\star$, $H_\star$ can be re-expressed as

$$H_\star^2 = \frac{8\pi^3}{90 M_{\rm pl}^2}(g_*(T_\star) + g_\psi + g_\eta)T_\star^4 \tag{6.15}$$

Therefore, initially the universe was radiation dominated ($H \propto T^2$) for $r \sim 1$, but becomes matter dominated ($H \propto T^{\frac{3}{2}}$) at $T_\star$ for $r \ll 1$.

Assuming that until DM decouples, the entropy is conserved in the SM bath and $g_\star$ remains constant, the scale factor relates to temperature as

$$\left(\frac{a_\star}{a}\right) \simeq \left(\frac{T}{T_\star}\right) \tag{6.16}$$

Neglecting the small DM contribution and making use of dimensionless variables $x \equiv M_\psi/T$ and $x_\star \equiv x(T_\star)$, the Friedmann equation can be written as

$$H \equiv H_\star \left(\frac{x_\star}{x}\right)^{\frac{3}{2}} \left(r\frac{x_\star}{x} + 1 - r\right)^{\frac{1}{2}} \tag{6.17}$$

We see that for $r = 1$, the modified Hubble parameter reduces to the Standard Hubble parameter. In the light of this modified Hubble expansion rate, the cosmological Boltzmann equation also modifies as

$$\frac{dY}{dx} = -\frac{s(M_\psi)\langle\sigma v\rangle}{x^2 H(M_\psi)\left(\frac{x}{x_\star}\right)^{\frac{1}{2}}\left(1 - r + r\frac{x_\star}{x}\right)^{\frac{1}{2}}}\left(Y_\psi^2 - (Y_\psi^{\rm eq})^2\right) \tag{6.18}$$

As expected, in the limit $r = 1$, the modified Boltzmann equation reduces to the one obtained in the standard cosmology. The relic abundance for a DM of mass 10 GeV using the modified Boltzmann equation is shown in Fig. 6.4, with $T_\star = 10^6$ GeV

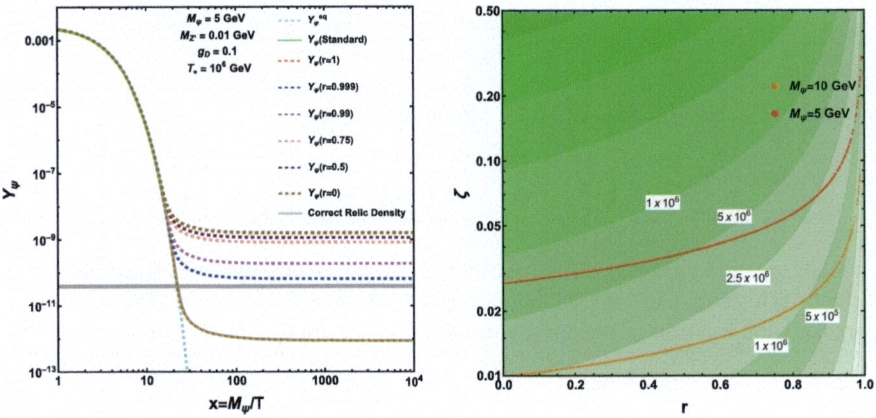

Fig. 6.4 [Left]: Evolution of SIDM relic density in non-standard cosmology for different values of the parameter r. [Right]: Contours of correct relic density in ζ vs r plane for different DM masses, along with contours of $T_{\rm RH}/\text{GeV}$.

and $M_{Z'} = 0.01$ GeV. We see that the relic increases as r decreases from 1 gradually, with the case $r = 1$ matching with the result obtained from standard Boltzmann equation. The case $r = 0$ depicts maximum matter domination and yields maximum relic density.

Now, we enforce the requirement that the universe transitions back to radiation domination before Big Bang Nucleosynthesis (BBN) to preserve the predictions of BBN. This is achieved through the decay of the additional component η into Standard Model (SM) particles prior to the initiation of BBN. For simplicity, we assume η to be a gauge singlet scalar which can decay into SM states via its mixing with SM Higgs ($\sim \lambda_{\eta H}\eta^2 H^2$). Other possible decays may occur depending on its non-minimal charge assignments under SM gauge group as well as any extensions of SM, which are model dependent. Throughout this work, we will assume that η has no coupling with DM. This can be achieved by assigning a charge -1 to it under the $U(1)_D$ symmetry. Hence η decays solely to SM sector, which introduces additional entropy into the radiation bath. Consequently, after the dark matter (DM) freezes out, its relic abundance undergoes dilution by a factor ζ, signifying the change in the entropy of the radiation bath. Assuming the sudden decay approximation, the magnitude of the entropy injection can be quantified as

$$\zeta \equiv \frac{s_{\rm before}}{s_{\rm after}} \simeq \left(\frac{T_{\Gamma_\eta}}{T_{\rm RH}}\right)^3 \tag{6.19}$$

where $T_{\rm RH}$ is the Reheating temperature of the universe and Γ_η is the decay width of η. In terms of the Planck mass and the decay width of η, we can define

$$T_{\rm RH} \equiv \sqrt{\Gamma_\eta M_{Pl}} \gtrsim 10 \text{ MeV} \tag{6.20}$$

Here, we neglect any changes in degrees of freedom. T_{Γ_η} is obtained by evolving $T_\star$ using $(a_\star/a_{\Gamma_\eta})$, where a_{Γ_η} is defined by the condition $H(a_{\Gamma_\eta}) \simeq \Gamma_\eta$. Assuming that η decays after it dominates the energy density, using Eq. 6.17,

$$\frac{a_\star}{a_\Gamma} \approx \frac{1}{(1-r)^{1/3}} \left(\frac{\Gamma_\eta}{H_\star}\right)^{2/3} \tag{6.21}$$

Then using Eqs. 6.16 and 6.20, we obtain

$$T_{\Gamma_\phi} \simeq \left(\frac{45}{4\pi^3 g_*(1-r)} \frac{T_{\rm RH}^4}{T_\star}\right)^{1/3} \tag{6.22}$$

Substituting ζ as a function of $T_{\rm RH}$, as in eq. (6.19), we get

$$\zeta \simeq \frac{45}{4\pi^3}(1-r)\frac{g_\star T_\star}{T_{\rm RH}} \tag{6.23}$$

The final diluted relic density due to this change in entropy is given by

$$Y_{\rm relic} = \frac{n_\psi}{s_{\rm after}} = \zeta \frac{n_\psi}{s_{\rm before}} = \zeta Y_{\rm FO}. \tag{6.24}$$

Hence, the final DM relic abundance is

$$\Omega_\psi^{\rm relic} = \zeta \Omega_\psi^{\rm FO} = \zeta \times \frac{s_0 M_\psi Y_{\rm FO}}{\rho_c} \tag{6.25}$$

where ρ_c and s_0 are the critical density and the present entropy density of the universe, respectively.

In the right panel of Fig. 6.4, we show the contours of correct relic density as a function of dilution factor ζ and the parameter r for different SIDM masses indicated by different colour codes. In the same plane, the contours of $T_{\rm RH}/\rm GeV$ using Eq. 6.23 are also depicted in green colour with the value indicated inside white boxes.

Hence, it is clear that the correct relic abundance for SIDM, suitable for addressing small-scale anomalies, can be achieved through its freeze-out in this modified cosmological scenario. Note that although the reheating temperature is quite high, it cannot bring SIDM back to thermal equilibrium because the light mediator Z' decays well before the decay of η and the dark sector is no longer connected to the visible sector when η decays and the universe reheats. This is the advantage of having a pseudo-Dirac state as SIDM candidate, as the direct search constraints can be evaded in this scenario and hence the kinetic mixing between $U(1)_D$ and $U(1)_Y$ can be sufficiently large making Z' decay much earlier, in contrast to the case for pure Dirac SIDM [31, 74].

Throughout this work, the DM does not couple to η due to $U(1)_D$ charge assignment. Therefore, DM relic is set during matter domination and there is no subsequent

DM production. In scenarios where η couples to DM via a Yukawa term like $\overline{\psi}\psi\eta$, one has to take into account late time DM production by η decay. This is possible if η carries a $U(1)_D$ charge zero. If the decay branching ratio to DM is sufficiently large, then DM production due to η decay cannot be neglected in a epoch of significant η decays and the decay contribution has to be added to the Boltzmann equation. DM-η coupling may arise via loop-induced processes even if DM-η coupling is forbidden at tree level. We shall explore such scenarios in future works.

6.5 Summary and Conclusion

In this study, we explore the feasibility of attaining the correct relic abundance for SIDM within a non-standard cosmological framework. Our examination centres on a generic $U(1)_D$ extension, utilizing a pseudo-Dirac fermion as the SIDM candidate and the corresponding $U(1)_D$ gauge boson Z' as the mediator facilitating DM self-scattering. To achieve the appropriate SIDM relic density, we propose a scenario where the universe's expansion history is altered by the presence of an additional component, denoted as η, leading to a significant enhancement of the dark matter relic. We also consider the change in the universe's entropy density resulting from the decay of η into standard model (SM) particles before the initiation of Big Bang Nucleosynthesis (BBN). The pseudo-Dirac nature of SIDM helps it evade the direct search constraints on kinetic mixing that would arise if SIDM were purely Dirac. A substantial kinetic mixing ensures the decay of Z' not only before BBN but also prior to the decay of η thus avoiding any conflict with the reheating of the universe. Regarding indirect detection, even with the maximum Sommerfeld enhancement, the present-day annihilation cross-section of SIDM to SM final states remains well below current limits from indirect search experiments [32, 74]. It would be concerning if SIDM were entirely elusive. Therefore, advancements in techniques and sensitivities across all dark matter detection strategies are in order. As several light DM search experiments with better sensitivities are expected to become operational in the coming decade, and given the natural alignment of the SIDM scenario within the light DM category, we foresee a testing period for the SIDM paradigm.

A. Relevant Cross-Section and Decay Widths

The scattering cross-sections can be derived as [77]

$$\sigma_{\psi_1\psi_1\to\psi_1\psi_1} = \frac{\pi}{\epsilon_v^2}\left|1 + \left(\frac{V_0}{4\mu^2}\right)^{-\frac{2i\epsilon_v}{\mu}}\right.$$

$$\left(\frac{\Gamma_v}{\Gamma_v^*}\right)\left[\frac{\cosh(\frac{\pi(\epsilon_\Delta+\epsilon_v)}{2\mu})\sinh(\frac{\pi(\epsilon_v-\epsilon_\Delta)}{2\mu}+i\varphi)}{\cosh(\frac{\pi(\epsilon_\Delta-\epsilon_v))}{2\mu}\sinh(\frac{\pi(\epsilon_v+\epsilon_\Delta)}{2\mu}-i\varphi)}\right]\Bigg|^2 \tag{6.26}$$

$$\sigma_{\psi_2\psi_2\to\psi_2\psi_2} = \frac{\pi}{\epsilon_\Delta^2}\Bigg|1+\left(\frac{V_0}{4\mu^2}\right)^{-\frac{2i\epsilon_\Delta}{\mu}} \left(\frac{\Gamma_\Delta}{\Gamma_\Delta^*}\right)\left[\frac{\cosh(\frac{\pi(\epsilon_\Delta+\epsilon_v)}{2\mu})\sinh(\frac{\pi(\epsilon_v-\epsilon_\Delta)}{2\mu}+i\varphi)}{\cosh(\frac{\pi(\epsilon_\Delta-\epsilon_v))}{2\mu}\sinh(\frac{\pi(\epsilon_v+\epsilon_\Delta)}{2\mu}-i\varphi)}\right]\Bigg|^2 \tag{6.27}$$

$$\sigma_{\psi_1\psi_1\to\psi_2\psi_2} = \frac{2\pi\cos^2\varphi\sinh\left(\frac{\pi\epsilon_v)}{\mu}\right)\sinh\left(\frac{\pi\epsilon_\Delta}{\mu}\right)}{\epsilon_v^2\cosh^2\left(\frac{\pi(\epsilon_\Delta-\epsilon_v)}{2\mu}\right)\Big(\cosh\left(\frac{\pi(\epsilon_v+\epsilon_\Delta)}{\mu}\right)-\cosh(2\varphi)\Big)} \tag{6.28}$$

$$\sigma_{\psi_2\psi_2\to\psi_1\psi_1} = \frac{2\pi\cos^2\varphi\sinh\left(\frac{\pi\epsilon_v)}{\mu}\right)\sinh\left(\frac{\pi\epsilon_\Delta}{\mu}\right)}{\epsilon_\Delta^2\cosh^2\left(\frac{\pi(\epsilon_\Delta-\epsilon_v)}{2\mu}\right)\Big(\cosh\left(\frac{\pi(\epsilon_v+\epsilon_\Delta)}{\mu}\right)-\cosh(2\varphi)\Big)} \tag{6.29}$$

where we have defined $\epsilon_\Delta = \sqrt{\epsilon_v^2-\epsilon_\delta^2}$, μ and V_0 are defining parameters for the exponential potential $V_0 e^{-\mu r}$, given by

$$\mu = \epsilon_Z\left(\frac{1}{2}+\frac{1}{2}\sqrt{1+\frac{4}{\epsilon_Z r_M}}\right), \qquad V_0 = \frac{e^{\epsilon_Z r_M\left(-\frac{1}{2}+\frac{1}{2}\sqrt{1+\frac{4}{\epsilon_Z r_M}}\right)}}{r_M}. \tag{6.30}$$

Here r_M is chosen from the relation $e^{-\epsilon_\phi r_M}/r_M = \max(\epsilon_\delta^2/2, \epsilon_\phi^2)$. The terms Γ_v, Γ_Δ are given by

$$\Gamma_v = \Gamma\left(1+i\frac{\epsilon_v}{\mu}\right)\Gamma\left(i\frac{\epsilon_v-\epsilon_\Delta}{2\mu}+\frac{1}{2}\right)\Gamma\left(i\frac{\epsilon_v+\epsilon_\Delta}{2\mu}+\frac{1}{2}\right) \tag{6.31}$$

$$\Gamma_\Delta = \Gamma\left(1+i\frac{\epsilon_\Delta}{\mu}\right)\Gamma\left(i\frac{\epsilon_\Delta-\epsilon_v}{2\mu}+\frac{1}{2}\right)\Gamma\left(i\frac{\epsilon_v+\epsilon_\Delta}{2\mu}+\frac{1}{2}\right) \tag{6.32}$$

with Γ denoting the gamma function.

References

1. F. Zwicky, Die Rotverschiebung von extragalaktischen Nebeln. Helv. Phys. Acta **6**, 110–127 (1933). [Gen. Rel. Grav. 41, 207 (2009)]
2. V.C. Rubin, W.K. Ford Jr., Rotation of the andromeda nebula from a spectroscopic survey of emission regions. Astrophys. J. **159**, 379–403 (1970)
3. D. Clowe, M. Bradac, A.H. Gonzalez, M. Markevitch, S.W. Randall, C. Jones, D. Zaritsky, A direct empirical proof of the existence of dark matter. Astrophys. J. **648**, L109–L113 (2006). ([astro-ph/0608407])
4. Planck Collaboration, N. Aghanim et al., *Planck 2018 Results. VI. Cosmological Parameters*. (arXiv:1807.06209)
5. E. Aprile et al., Dark Matter Search Results from a One Tonne$\times$ Year Exposure of XENON1T. (arXiv:1805.12562)
6. S. Tulin, H.-B. Yu, Dark matter self-interactions and small scale structure. Phys. Rept. **730**, 1–57 (2018). (arXiv:1705.02358)
7. J.S. Bullock, M. Boylan-Kolchin, Small-scale challenges to the LambdaCDM paradigm. Ann. Rev. Astron. Astrophys. **55**, 343–387 (2017). (arXiv:1707.04256)
8. D.N. Spergel, P.J. Steinhardt, Observational evidence for self-interacting cold dark matter. Phys. Rev. Lett. **84**, 3760–3763 (2000). ([astro-ph/9909386])
9. E.D. Carlson, M.E. Machacek, L.J. Hall, Self-interacting dark matter. Astrophys. J. **398**, 43–52 (1992)
10. A.A. de Laix, R.J. Scherrer, R.K. Schaefer, Constraints of selfinteracting dark matter. Astrophys. J. **452**, 495 (1995). ([astro-ph/9502087])
11. M.R. Buckley, P.J. Fox, Dark matter self-interactions and light force carriers. Phys. Rev. D **81**, 083522 (2010). (arXiv:0911.3898)
12. J.L. Feng, M. Kaplinghat, H.-B. Yu, Halo shape and relic density exclusions of Sommerfeld-enhanced dark matter explanations of cosmic ray excesses. Phys. Rev. Lett. **104**, 151301 (2010). (arXiv:0911.0422)
13. J.L. Feng, M. Kaplinghat, H. Tu, H.-B. Yu, Hidden charged dark matter. JCAP **07**, 004 (2009). ([arXiv:0905.3039])
14. A. Loeb, N. Weiner, Cores in dwarf galaxies from dark matter with a Yukawa potential. Phys. Rev. Lett. **106**, 171302 (2011). (arXiv:1011.6374)
15. J. Zavala, M. Vogelsberger, M.G. Walker, Constraining self-interacting dark matter with the milky way's dwarf spheroidals. Mon. Not. Roy. Astron. Soc. **431**, L20–L24 (2013). ([arXiv:1211.6426])
16. M. Vogelsberger, J. Zavala, A. Loeb, Subhaloes in self-interacting galactic dark matter haloes. Mon. Not. Roy. Astron. Soc. **423**, 3740 (2012). ([arXiv:1201.5892])
17. T. Bringmann, F. Kahlhoefer, K. Schmidt-Hoberg, P. Walia, Strong constraints on self-interacting dark matter with light mediators. Phys. Rev. Lett. **118**(14), 141802 (2017). (arXiv:1612.00845)
18. M. Kaplinghat, S. Tulin, H.-B. Yu, Dark matter halos as particle colliders: unified solution to small-scale structure puzzles from dwarfs to clusters. Phys. Rev. Lett. **116**(4), 041302 (2016). (arXiv:1508.03339)
19. L.G. van den Aarssen, T. Bringmann, C. Pfrommer, Is dark matter with long-range interactions a solution to all small-scale problems of Λ CDM cosmology? Phys. Rev. Lett. **109**, 231301 (2012). (arXiv:1205.5809)
20. S. Tulin, H.-B. Yu, K.M. Zurek, Beyond collisionless dark matter: particle physics dynamics for dark matter halo structure. Phys. Rev. D **87**(11), 115007 (2013). (arXiv:1302.3898)
21. M. Kaplinghat, S. Tulin, H.-B. Yu, Direct detection portals for self-interacting dark matter. Phys. Rev. D **89**(3), 035009 (2014). (arXiv:1310.7945)
22. E. Del Nobile, M. Kaplinghat, H.-B. Yu, Direct detection signatures of self-interacting dark matter with a light mediator. JCAP **10**, 055 (2015). ([arXiv:1507.04007])
23. C. Kouvaris, I.M. Shoemaker, K. Tuominen, Self-interacting dark matter through the higgs portal. Phys. Rev. D **91**(4), 043519 (2015). (arXiv:1411.3730)

24. N. Bernal, X. Chu, C. Garcia-Cely, T. Hambye, B. Zaldivar, Production regimes for self-interacting dark matter. JCAP **03**, 018 (2016). ([arXiv:1510.08063])
25. K. Kainulainen, K. Tuominen, V. Vaskonen, Self-interacting dark matter and cosmology of a light scalar mediator. Phys. Rev. D **93**(1), 015016 (2016). (arXiv:1507.04931). Erratum: Phys. Rev. D 95, 079901 (2017)
26. T. Hambye, L. Vanderheyden, Minimal self-interacting dark matter models with light mediator. JCAP **05**, 001 (2020). ([arXiv:1912.11708])
27. M. Cirelli, P. Panci, K. Petraki, F. Sala, M. Taoso, Dark matter's secret liaisons: phenomenology of a dark U(1) sector with bound states. JCAP **05**, 036 (2017). ([arXiv:1612.07295])
28. F. Kahlhoefer, K. Schmidt-Hoberg, S. Wild, Dark matter self-interactions from a general spin-0 mediator. JCAP **08**, 003 (2017). ([arXiv:1704.02149])
29. M. Dutta, S. Mahapatra, D. Borah, N. Sahu, Self-interacting inelastic dark matter in the light of XENON1T excess. Phys. Rev. D **103**(9), 095018 (2021). (arXiv:2101.06472)
30. D. Borah, M. Dutta, S. Mahapatra, N. Sahu, Boosted self-interacting dark matter and XENON1T excess. Nucl. Phys. B **979**, 115787 (2022). (arXiv:2107.13176)
31. D. Borah, M. Dutta, S. Mahapatra, N. Sahu, Self-interacting dark matter via right handed neutrino portal. Phys. Rev. D **105**(1), 015004 (2022). (arXiv:2110.00021)
32. D. Borah, M. Dutta, S. Mahapatra, N. Sahu, Singlet-doublet self-interacting dark matter and radiative neutrino mass. Phys. Rev. D **105**(7), 075019 (2022). (arXiv:2112.06847)
33. M. Dutta, N. Narendra, N. Sahu, S. Shil, Asymmetric self-interacting dark matter via Dirac leptogenesis. Phys. Rev. D **106**(9), 095017 (2022). (arXiv:2202.04704)
34. J.A. Evans, A. Ghalsasi, S. Gori, M. Tammaro, J. Zupan, Light dark matter from entropy dilution. JHEP **02**, 151 (2020). ([arXiv:1910.06319])
35. S. Hamdan, J. Unwin, Dark matter freeze-out during matter domination. Mod. Phys. Lett. A **33**(29), 1850181 (2018). (arXiv:1710.03758)
36. R. Allahverdi, J.K. Osiński, Freeze-in production of dark matter prior to early matter domination. Phys. Rev. D **101**(6), 063503 (2020). (arXiv:1909.01457)
37. R. Allahverdi, J.K. Osiński, Nonthermal dark matter from modified early matter domination. Phys. Rev. D **99**(8), 083517 (2019). (arXiv:1812.10522)
38. M. Drees, F. Hajkarim, Dark matter production in an early matter dominated era. JCAP **02**, 057 (2018). ([arXiv:1711.05007])
39. H. Davoudiasl, D. Hooper, S.D. McDermott, Inflatable dark matter. Phys. Rev. Lett. **116**(3), 031303 (2016). (arXiv:1507.08660)
40. M. Sten Delos, T. Linden, A.L. Erickcek, Breaking a dark degeneracy: the gamma-ray signature of early matter domination. (arXiv:1910.08553)
41. L. Randall, J. Scholtz, J. Unwin, Flooded dark matter and S level rise. JHEP **03**, 011 (2016). ([arXiv:1509.08477])
42. R.T. Co, F. D'Eramo, L.J. Hall, D. Pappadopulo, Freeze-in dark matter with displaced signatures at colliders. JCAP **12**, 024 (2015). ([arXiv:1506.07532])
43. J.A. Dror, E. Kuflik, W.H. Ng, Codecaying dark matter. Phys. Rev. Lett. **117**(21), 211801 (2016). (arXiv:1607.03110)
44. F. D'Eramo, N. Fernandez, S. Profumo, When the universe expands too fast: relentless dark matter. JCAP **1705**(05), 012 (2017). (arXiv:1703.04793)
45. F. D'Eramo, N. Fernandez, S. Profumo, Dark matter freeze-in production in fast-expanding universes. JCAP **1802**(02), 046 (2018). (arXiv:1712.07453)
46. B. Barman, P. Ghosh, F.S. Queiroz, A.K. Saha, Scalar multiplet dark matter in a fast expanding universe: resurrection of the desert region. Phys. Rev. D **104**(1), 015040 (2021). (arXiv:2101.10175)
47. M.A.G. Garcia, M.A. Amin, Prethermalization production of dark matter. Phys. Rev. D **98**(10), 103504 (2018). (arXiv:1806.01865)
48. N. Bernal, C. Cosme, T. Tenkanen, V. Vaskonen, Scalar singlet dark matter in non-standard cosmologies. (arXiv:1806.11122)
49. M. Drees, F. Hajkarim, Neutralino dark matter in scenarios with early matter domination. JHEP **12**, 042 (2018). ([arXiv:1808.05706])

50. A. Betancur, O. Zapata, Phenomenology of doublet-triplet fermionic dark matter in non-standard cosmology and multicomponent dark sectors. Phys. Rev. D **98**(9), 095003 (2018). (arXiv:1809.04990)
51. C. Maldonado, J. Unwin, Establishing the dark matter relic density in an era of particle decays. JCAP **1906**(06), 037 (2019). (arXiv:1902.10746)
52. A. Poulin, Dark matter freeze-out in modified cosmological scenarios. Phys. Rev. **D100**(4), 043022 (2019). (arXiv:1905.03126)
53. P. Arias, N. Bernal, A. Herrera, C. Maldonado, Reconstructing non-standard cosmologies with dark matter. JCAP **1910**(10), 047 (2019). (arXiv:1906.04183)
54. J. Bramante, J. Unwin, Super heavy thermal dark matter and primordial asymmetries. JHEP **02**, 119 (2017). ([arXiv:1701.05859])
55. N. Bernal, C. Cosme, T. Tenkanen, Phenomenology of self-interacting dark matter in a matter-dominated universe. Eur. Phys. J. **C79**(2), 99 (2019). (arXiv:1803.08064)
56. E. Hardy, Higgs portal dark matter in non-standard cosmological histories. JHEP **06**, 043 (2018). ([arXiv:1804.06783])
57. K. Kaneta, Y. Mambrini, K.A. Olive, Radiative production of nonthermal dark matter. Phys. Rev. D **99**(6), 063508 (2019). (arXiv:1901.04449)
58. N. Bernal, F. Elahi, C. Maldonado, J. Unwin, Ultraviolet freeze-in and non-standard cosmologies. (arXiv:1909.07992)
59. A. Berlin, D. Hooper, G. Krnjaic, PeV-scale dark matter as a thermal relic of a decoupled sector. Phys. Lett. B **760**, 106–111 (2016). (arXiv:1602.08490)
60. M. Kamionkowski, M.S. Turner, Thermal relics: do we know their abundances? Phys. Rev. D **42**, 3310–3320 (1990)
61. P. Chanda, S. Hamdan, J. Unwin, Reviving Z and Higgs mediated dark matter models in matter dominated freeze-out. JCAP **01**, 034 (2020). ([arXiv:1911.02616])
62. P. Konar, A. Mukherjee, A.K. Saha, S. Show, A dark clue to seesaw and leptogenesis in a pseudo-Dirac singlet doublet scenario with (non)standard cosmology. JHEP **03**, 044 (2021). ([arXiv:2007.15608])
63. P. Ghosh, P. Konar, A.K. Saha, S. Show, Self-interacting freeze-in dark matter in a singlet doublet scenario. JCAP **10**, 017 (2022). ([arXiv:2112.09057])
64. B. Barman, A. Dutta Banik, A. Paul, Implications of NANOGrav results and UV freeze-in in a fast-expanding universe. (arXiv:2012.11969)
65. S. Biondini, Interplay between improved interaction rates and modified cosmological histories for dark matter. **9** (2023). (arXiv:2309.00323)
66. J. McDonald, WIMP densities in decaying particle dominated cosmology. Phys. Rev. D **43**, 1063–1068 (1991)
67. D.J.H. Chung, E.W. Kolb, A. Riotto, Production of massive particles during reheating. Phys. Rev. D **60**, 063504 (1999). ([hep-ph/9809453])
68. G.F. Giudice, E.W. Kolb, A. Riotto, Largest temperature of the radiation era and its cosmological implications. Phys. Rev. D **64**, 023508 (2001). ([hep-ph/0005123])
69. G.B. Gelmini, P. Gondolo, Neutralino with the right cold dark matter abundance in (almost) any supersymmetric model. Phys. Rev. D **74**, 023510 (2006). ([hep-ph/0602230])
70. B. Spokoiny, Deflationary universe scenario. Phys. Lett. B **315**, 40–45 (1993). ([gr-qc/9306008])
71. P.G. Ferreira, M. Joyce, Cosmology with a primordial scaling field. Phys. Rev. D **58**, 023503 (1998). ([astro-ph/9711102])
72. P. Salati, Quintessence and the relic density of neutralinos. Phys. Lett. B **571**, 121–131 (2003). ([astro-ph/0207396])
73. C. Pallis, Quintessential Kination and cold dark matter abundance. JCAP **10**, 015 (2005). ([hep-ph/0503080])
74. M. Dutta, Self-interacting dark matter in non-standard cosmology. ([hep-ph/2310.03909])
75. D. Tucker-Smith, David, N.D. Weiner, Inelastic dark matter. Phys. Rev. D **64**, 043502 (2001). ([arXiv:0101138])

76. Y. Cui, D.E. Morrissey, D. Poland, David, L. Randall, Candidates for inelastic dark matter. JHEP **05**, 076 (2009). (arXiv:0901.0557)
77. K. Schutz, T.R. Slatyer, Self-scattering for dark matter with an excited state. JCAP **01**, 021 (2015). ([arXiv:1409.2867])
78. M. Blennow, S. Clementz, J. Herrero-Garcia, Self-interacting inelastic dark matter: a viable solution to the small scale structure problems. JCAP **03**, 048 (2017). ([arXiv:1612.06681])
79. Y. Zhang, Self-interacting dark matter without direct detection constraints. Phys. Dark Univ. **15**, 82–89 (2017). (arXiv:1611.03492)
80. N. Arkani-Hamed, D.P. Finkbeiner, T.R. Slatyer, N. Weiner, A theory of dark matter. Phys. Rev. D **79**, 015014 (2009). (arXiv:0810.0713)
81. A. Kamada, H.J. Kim, T. Kuwahara, Maximally self-interacting dark matter: models and predictions. JHEP **20**, 202 (2020). ([arXiv:2007.15522])

Chapter 7
A Real Scalar Singlet Pseudo-FIMP in Presence of a Real Scalar WIMP

Dipankar Pradhan, Subhaditya Bhattacharya, and Jayita Lahiri

Abstract In the multi-component dark matter (DM) framework, DM-DM conversion has an important role in DM phenomenology. Depending on its conversion rate and its visible connection, we might be able to reach different kinds of multipartite thermal DM scenarios. Here, we show the dynamics of DM (DM_1), which has a feeble visible connection with the visible sector, in the presence of any thermal DM (DM_2). We propose to classify this new kind of DM as pseudo-FIMP (pFIMP) which has unique characteristics impacting the relic density, direct, and collider search implications. We studied it in a model-independent manner by solving coupled Boltzmann Equations (cBEQ) and also for a toy model with two-component real scalar DM.

7.1 Introduction

The presence of DM is well-motivated, to understand our observed universe, from different kinds of astrophysical and cosmological observations like Bullet Cluster, Rotation curve of cluster of galaxies, CMB anisotropic spectrum, etc. Finally, the WMAP collaboration suggested that DM is present around 26.8% [1] of the energy budget of the universe. The most unknown properties of DM are constituent type (single or multi-component) with mass, coupling with visible sector, etc.; while some are known: electro-magnetic charge neutral, massive, and stable (or lifetime should be larger than universe). The multi-component DM is also well-motivated, although its presence is hard to prove by direct detection (DD), indirect detection

D. Pradhan (✉) · S. Bhattacharya
Department of Physics, Indian Institute of Technology Guwahati, Assam 781039, India
e-mail: d.pradhan@iitg.ac.in

S. Bhattacharya
e-mail: subhab@iitg.ac.in

J. Lahiri
Institut für Theoretische Physik, Universität Hamburg, 22761 Hamburg, Germany
e-mail: jayita.lahiri@desy.de

J. W. F. Valle et al. (eds.), *Proceedings of the International Conference on Future Prospects in Neutrino and Astroparticle Physics*, Springer Proceedings in Physics 361,
https://doi.org/10.1007/978-981-97-7441-8_7

(ID), and collider. There are few theoretical studies [2–5] that try to explain how the multi-component signal would differ from the signal in DD and collider.

In a two-component DM (simplest) setup, there might be an interaction between the two DM components, and if it is present then depending on its strength, it results in two-WIMP, two-FIMP, two-SIMP, WIMP-FIMP, WIMP-SIMP, and FIMP-SIMP scenarios. We have proposed a two-component DM setup where one of the DMs is weakly coupled with the thermal bath while the other one is feebly coupled with the visible sector and never reaches equilibrium with the thermal bath in the absence of DM-DM conversion. If the conversion rate is sufficiently large to maintain the aforesaid equilibrium, since this kind of DM, although has a feeble visible connection, has weak/strong coupling with another WIMP-DM component, the DM component is specified as pseudo-FIMP (pFIMP) in our work. The apparent undetectability of such DM by DD, ID, and collider experiments makes it an intriguing case study. Interestingly, the thermal partner DM may have a detection possibility presence of its visible connection and pFIMP, although has no direct visible connection, but via the thermal DM loop, it can communicate with the visible sector. So, in this new setup, pFIMP with any kind of thermal DM might have a detection possibility, and its phenomenology would be very different and richer than other two-component DM setups. Here we explore this pFIMP detection possibility in the presence of a WIMP-type DM in model-independent and also for a toy model, two-component real scalar DM. For the detailed analysis, please see the following [6, 7].

7.2 Model Independent Study of pFIMP Dynamics

To study the pFIMP dynamics more accurately, we start with a two-component DM setup, where one of the DM is WIMP (DM_1) and another DM (DM_2) is feebly connected with SM. The DM number density evolution in WIMP-FIMP [8] framework is governed by a set of coupled Boltzmann equations (cBEQ) shown in Eq. (7.1), in terms of DM yield ($Y = \frac{n}{s}$) as a function of $x = \mu_{12}/T$, where $\mu_{12} = m_1 m_2/(m_1 + m_2)$ denote the reduced mass of the two DMs, and T denotes the thermal bath temperature.

$$
\begin{aligned}
\frac{dY_1}{dx} &= -\frac{s}{x\,H(x)}\left[\langle\sigma v\rangle_{11\to \mathrm{SM}}\left(Y_1^2 - Y_1^{\mathrm{eq}^2}\right) + \langle\sigma v\rangle_{11\to 22}\left(Y_1^2 - \frac{Y_1^{\mathrm{eq}^2}}{Y_2^{\mathrm{eq}^2}}Y_2^2\right)\right],\\
\frac{dY_2}{dx} &= \frac{2\,s}{x\,H(x)}\left[\frac{1}{s}\langle\Gamma_{\mathrm{SM}\to 22}\rangle\left(Y_{\mathrm{SM}}^{\mathrm{eq}} - \frac{Y_2^2}{Y_2^{\mathrm{eq}^2}}Y_{\mathrm{SM}}^{\mathrm{eq}}\right)\right.\\
&\left.+ \langle\sigma v\rangle_{\mathrm{SM}\to 22}\left(Y_{\mathrm{SM}}^{\mathrm{eq}^2} - \frac{Y_2^2}{Y_2^{\mathrm{eq}^2}}Y_{\mathrm{SM}}^{\mathrm{eq}^2}\right) + \langle\sigma v\rangle_{11\to 22}\left(Y_1^2 - \frac{Y_1^{\mathrm{eq}^2}}{Y_2^{\mathrm{eq}^2}}Y_2^2\right)\right].
\end{aligned}
\tag{7.1}
$$

In the above equation, $s = \frac{2\pi^2}{45} g_*^s (\mu_{12}/x)^3$, $H(x) = 1.67\sqrt{g_*^\rho}\mu_{12}^2 x^{-2} \mathrm{M}_{pl}^{-1}$ with Planck mass $\mathrm{M_{pl}} = 1.22 \times 10^{19}$ GeV, and subscripts 1, 2 in Eq. 7.1 are corresponding to WIMP and FIMP (pFIMP) components, respectively. We numerically solve the cBEQ, for the semi-analytical solution see [6], to understand the dynamics pertinently where we have taken thermal average cross-section as a numerical number. The solutions of cBEQ (Eq. 7.1) are shown in Fig. 7.1. We divided the scenario into two modes: (i) $\mathrm{m_1} > \mathrm{m_2}$ (Fig. 7.1a, c) and (ii) $\mathrm{m_1} < \mathrm{m_2}$ (Fig. 7.1b, d).

In Fig. 7.1a, we have shown the variation of DM yield with μ_{12}/T for different values of DM-DM conversion cross-section ($\sigma^{\mathrm{T}}_{11\to 22}$) represented by different color points, mentioned in figure inset. In Fig. 7.1c, we have shown the variation DM relic with $\sigma^{\mathrm{T}}_{11\to 22}$ for different values of $\sigma^{\mathrm{T}}_{11\to \mathrm{SM\ SM}}$ represented by different color points, mentioned in figure inset. When the conversion is very small compared to $\sigma^{\mathrm{T}}_{\mathrm{SM\ SM}\to 22}$ then there is no contribution in DM yields from conversion and both dark matter show their independent WIMP and pure FIMP behavior, see the violet dot-dashed line in

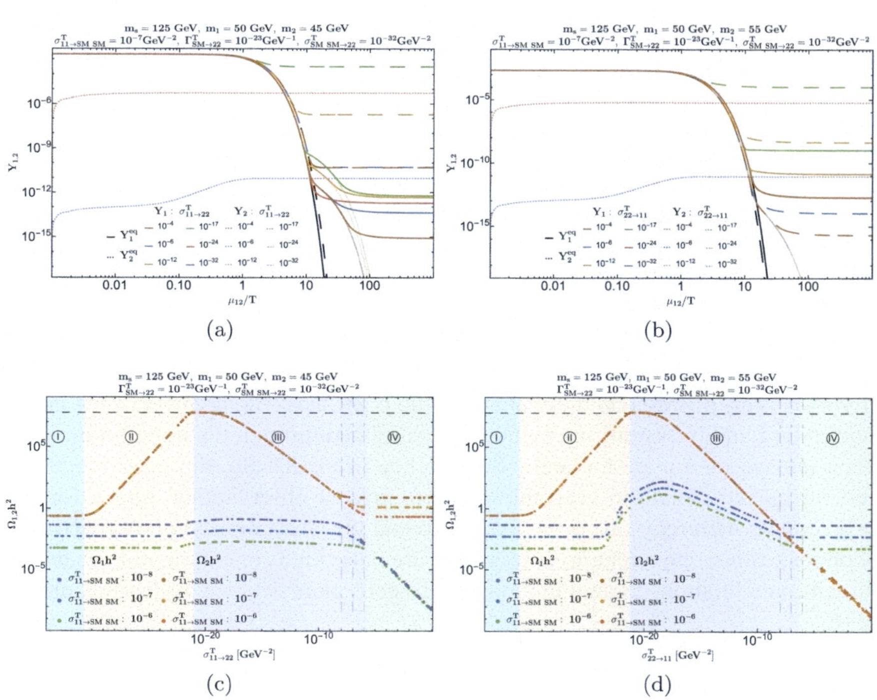

Fig. 7.1 Figure 7.1a, b and c, d represents the DM yields ($\mathrm{Y}_{1,2}(x)$) and relic ($\Omega_{1,2}\mathrm{h}^2(x)$), for mass hierarchy 1 and 2, respectively. In Fig. 7.1a and b, solid (dotted) [dashed] curves denote WIMP (FIMP) [pFIMP] with equilibrium distributions marked in black. Figure 7.1c and d shows the variation of $\Omega_{1,2}\mathrm{h}^2$ as a function of $\sigma^{T}_{11\to 22}$ and $\sigma^{T}_{22\to 11}$ for hierarchy 1 and 2, respectively; red/yellow/orange dotted lines represent FIMP (pFIMP), while green/blue/violet ones represent WIMP

Fig. 7.1a and light blue region Ⓘ in 7.1c. With the enhancement of $\sigma^{T}_{11\to SM\ SM}$ cross-section, only the WIMP relic decreases but the FIMP relic remains unaffected. This regime corresponds to a pure WIMP-FIMP scenario where the conversion effect is still negligible.

If we enhanced the conversion cross-section slightly larger than $\sigma^{T}_{SM\ SM\to 22}$ then DM_2 yield increases gradually, see red dot-dashed (thick) lines correspond to FIMP (WIMP) in 7.1a and light yellow region Ⓘ in 7.1c. However, in Fig. 7.1a and c, DM_1 yield is unaffected due to its large $\sigma^{T}_{11\to SM\ SM}$ cross-section but decreases with enhancement of $\sigma^{T}_{11\to SM\ SM}$. This overabundance region also specifies a WIMP-FIMP regime in the presence of $DM_1 - DM_2$ conversion.

Further sufficient enhancement of conversion cross-section helps DM_2 to reach thermal equilibrium and follow its equilibrium distribution before it freezes out. Now, we classify this DM_2 as pFIMP, see green and yellow dashed (thick) lines for pFIMP (WIMP) in Fig. 7.1a and light violet region Ⓘ in Fig. 7.1c. When the DM_2 enters the thermal bath, there is a slight enhancement in DM_1 yield due to the opening up of the conversion channel where $DM_2 \to DM_1$ conversion rate is sufficiently larger because of large DM_2 number density (due to its large FO temperature) compared to DM_1. In this regime, both yields decrease with the enhancement of the conversion cross-section, concurrently only the DM_1 yield also decreases with the enhancement of $\sigma^{T}_{11\to SM\ SM}$.

Finally, if the conversion cross-section is sufficiently larger than $\sigma^{T}_{11\to SM\ SM}$ which reduces the conversion rate because of the smallness of DM_1 yield and stops the $DM_2 \to DM_1$ conversion but $DM_1 \to DM_2$ still active. So, the DM_2 number density remains unaffected but the DM_1 yield gradually decreases and follows the modified equilibrium distribution before it freezes out, see blue and brown dashed line (thick) corresponds to pFIMP (WIMP) in Fig. 7.1a and region Ⓘ in Fig. 7.1c.

Similarly, in Fig. 7.1b and d, we have shown a similar analysis but for an opposite hierarchy where the mass of pFIMP is larger than WIMP. Another important characteristic of this pFIMP and thermal DM setup is the mass hierarchy between them. The forward and backward interaction rate is related through the equilibrium ratio which introduces a mass-hierarchy-dependency in cBEQ. So, the mass hierarchy needs to be very small, to keep the model in present observational limits, for the positive mass hierarchy ($m_1 > m_2$) scenario and the situation is slightly relaxed in the opposite hierarchy scenario unless there are other kinds of annihilation channels like semi-annihilation, co-annihilation, etc. To get a more detailed discussion, please see [6].

7.3 Two-Component Scalar WIMP-pFIMP Scenario

We would now like to validate our claims of pFIMP characteristics using a specific model example, two-component real scalar DM. This model consists of two real scalar singlets, where ϕ_1 behaves as WIMP and ϕ_2 as FIMP/(pFIMP), both of which are rendered stable under $\mathbb{Z}_2 \otimes \mathbb{Z}'_2$ symmetry. The Lagrangian density is given by

$$\mathcal{L} = \mathcal{L}_{\rm SM} + \mu_{\rm H}^2 {\rm H}^\dagger {\rm H} - \lambda_{\rm H} {\rm H}^\dagger {\rm H} + \frac{1}{2}\partial_\mu \phi_1 \partial^\mu \phi_1 - \frac{1}{2}\mu_1^2 \phi_1^2 - \frac{1}{4!}\lambda_1 \phi_1^4 - \frac{1}{2}\lambda_{1{\rm H}} \phi_1^2 {\rm H}^\dagger {\rm H}$$
$$+ \frac{1}{2}\partial_\mu \phi_2 \partial^\mu \phi_2 - \frac{1}{2}\mu_2^2 \phi_2^2 - \frac{1}{4!}\lambda_2 \phi_2^4 - \frac{1}{2}\lambda_{2{\rm H}} \phi_2^2 {\rm H}^\dagger {\rm H} - \frac{1}{4}\lambda_{12}\phi_1^2\phi_2^2. \quad (7.2)$$

The necessary theoretical constraints on parameters of the model [9], which are relevant to our analysis, Perturbativity : $\lambda_i < 4\pi,\ \lambda_{iH} < 4\pi$; Stability : $\lambda_i \geq 0,\ \lambda_{iH} + 2\sqrt{\lambda_H \lambda_i} \geq 0$; Tree level unitarity : $|\lambda_i| < 8\pi,\ |\lambda_{iH}| < 8\pi$, where $\{i = 1, 2\}$. To find out the DM yields, we solve the governing cBEQ corresponding to ϕ_1 and ϕ_2. Here we use temperature-dependent $\langle \sigma v \rangle$ in terms of the model parameters. Now the model free parameters are $\{{\rm m}_1,\ {\rm m}_2,\ \lambda_{1{\rm H}},\ \lambda_{2{\rm H}},\ \lambda_{12}\}$.

The solution of cBEQ for a two-scalar model is shown in Fig. 7.2 where Fig. 7.2a and b shows the variation of DM yields with μ_{12}/T for $m_{\phi_1} > m_{\phi_2}$ and $m_{\phi_1} < m_{\phi_2}$, respectively. The variation of conversion coupling (λ_{12}) is represented by different color lines; thick lines correspond to WIMPs and dot-dashed (dashed) correspond to lines FIMP (pFIMP), respectively. For small $\lambda_{12} = 10^{-12},\ (10^{-8})$, DM yield Y_{ϕ_1} does not change, see overlapped thick violet (yellow) lines, but Y_{ϕ_2} enhanced due to the production of ϕ_2 by conversion, see dot-dashed violet (yellow) lines. With further enhancement of $\lambda_{12} = 10^{-5},\ (10^{-2})$, Y_{ϕ_2} reach equilibrium before freeze-out (see dashed blue (red) lines), while Y_{ϕ_1} initially enhanced, near to FIMP to pFIMP transition phase, due to large backwards conversion rate but again decreases with the enhancement of λ_{12}. Finally, for $\lambda_{12} > 1$, Y_{ϕ_2} (Y_{ϕ_1}) yield doesn't change, as the conversion rate also reduces significantly with yield, and Y_{ϕ_1} (Y_{ϕ_2}) gradually decreases with λ_{12} for $m_{\phi_1} > m_{\phi_2}$ ($m_{\phi_2} > m_{\phi_1}$), respectively; see olive and warm black thick (dashed) color lines for WIMP(pFIMP).

All observations, DD, ID, or collider experiments are extremely sensitive to WIMP-DM because of their weak-order interaction with the visible sector [10]. But, the DM which is feebly connected with the visible sector, their detection is non-trivial unless the detector sensitivity is increased multi-fold. However, for a pFIMP, the situation is more hopeful. Future DD, ID, and collider experiments can be sensitive to

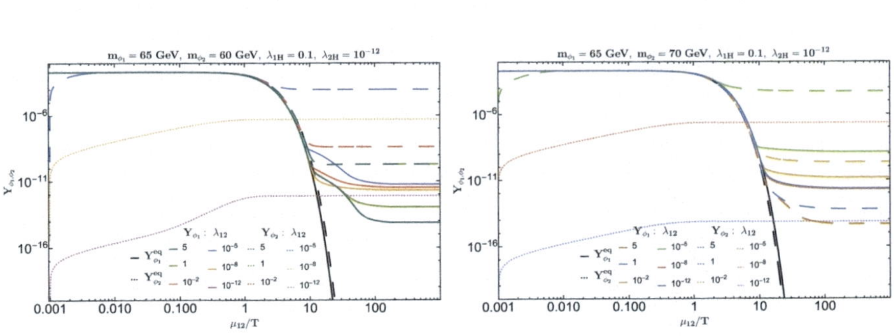

(a) (b)

Fig. 7.2 Figure 7.2a and b represents the DM yields ($\mathrm{Y}_{\phi_1,\phi_2}(x)$) for mass hierarchy 1 and 2, respectively. The solid (dotted) [dashed] curves denote WIMP (FIMP) [pFIMP] with equilibrium distributions marked in black

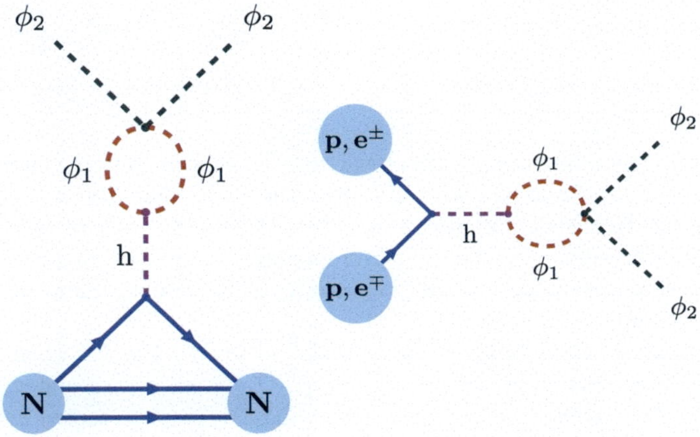

Fig. 7.3 The Feynman diagrams correspond to the direct detection (left) and collider search (right) of pFIMP (ϕ_2) via one-loop WIMP (ϕ_1) interaction

pFIMP owing to its interaction with the visible sector via WIMP loop [7]. In Fig. 7.3, we show the possible Feynman diagrams that are related to DD and collider search. The effective spin-independent pFIMP-nucleon direct detection cross-section in zero transfer momentum ($q_h^2 = t \to 0$) limit turns out to be [6],

$$\sigma_{\phi_2}^{\text{eff}} = \frac{\Omega_{\phi_2} h^2}{\Omega_{\phi_1} h^2 + \Omega_{\phi_2} h^2} \frac{\mu_n^2 m_n^2}{4\pi v^2 m_{\phi_2}^2} \frac{f_n^2}{m_h^4} |\Gamma_{h\phi_2\phi_2}^{\text{total}}|_{t\to 0}^2. \tag{7.3}$$

In the above, $\mu_n = \frac{m_n m_{\phi_2}}{m_n + m_{\phi_2}}$ where m_n is the nucleon mass, $f_n = \frac{2}{9} + \frac{7}{9} \sum_{u,d,s} f_{T_q}^n$ with $f_{T_u}^{p(n)} = 0.018(0.013)$, $f_{T_d}^{p(n)} = 0.027(0.040)$, and $f_{T_s}^{p(n)} = 0.037(0.037)$ and

$$\Gamma_{h\to\phi_2\phi_2}^{\text{total}}\left(q_h^2\right) \approx -\frac{i}{32\pi^2}\left(\lambda_{\phi_1\phi_1\phi_2\phi_2}\lambda_{h\phi_1\phi_1}\right)\int_0^1 dx \ln\left[\frac{m_{\phi_1}^2 - x(1-x)4m_{\phi_2}^2}{m_{\phi_1}^2 - x(1-x)q_h^2}\right]. \tag{7.4}$$

Finally, we conclude the model discussion by providing a benchmark shown in Table 7.1. For more benchmarks see [6], where one can see that only near-resonance regions are available for DM observations.

Table 7.1 Some benchmarks for the two-component scalar singlet model (Eq. 7.2) in WIMP-pFIMP limit, where $\lambda_{2H} = 10^{-12}$. The last two columns depict the WIMP and pFIMP production cross-section at LHC via Gluon fusion yielding mono-jet+$\not{E}_T$ in the final state at $\sqrt{s} = 14$ TeV [6]

m_{ϕ_1} [GeV]	m_{ϕ_2} [GeV]	λ_{1H}	λ_{12}	$\Omega_{\phi_1} h^2$	$\Omega_{\phi_2} h^2$	$\sigma^{\rm eff}_{\phi_1}$ [cm^2]	$\sigma^{\rm eff}_{\phi_2}$ [cm^2]	Gluon fusion (fb)	
								WIMP	pFIMP
63.33	60.50	0.10	4	0.0002	0.1203	3.3×10^{-47}	5.0×10^{-48}	8.4	13.9
104.1	100.0	0.22	6	0.0002	0.1210	7.89×10^{-47}	2.15×10^{-47}	0.69	1.18×10^{-3}
405.0	402.9	0.30	8	0.0101	0.1091	4.06×10^{-46}	7.49×10^{-48}	9×10^{-4}	2.34×10^{-6}

7.4 Summary

The kind of two-component DM setup depends on the interaction rate of DM with the visible sector and between the two DM candidates. Here, we have proposed a kind of two-component DM setup where one of the DMs has a visible connection, which helps it remain in equilibrium with the thermal bath, and another DM is feebly connected with the visible sector but strongly connected with its DM partner so that it can reach in chemical equilibrium, we call this second DM component as pFIMP. Firstly, we show the dynamics of this pFIMP, in the presence of a WIMP first in a model-independent manner and then for a model involving two-scalar DM. By the variation of the DM-DM interaction rate, we show the transition of a FIMP into a pFIMP. This pFIMP possibility is achievable in the presence of any kind of thermal DM. The most important characteristic of this setup is that the mass hierarchy between DMs needs to be very small to satisfy observational constraints on DM when thermal DM mass is greater than pFIMP, the situation gets slightly relaxed in opposite mass hierarchies.

The detection prospect of pFIMP is also studied in this particular model. The detection of both DMs is possible simultaneously, given the interaction between the thermal DM and pFIMP is substantial. The simplest way for the pFIMP to connect with the visible sector is via the partner thermal DM loop. Here, we show that for a two-component scalar model, simultaneous detection is possible near Higgs resonance only, given all DM constraints are satisfied. However, we emphasize that one can think of many other model possibilities in WIMP-pFIMP or [7] or SIMP-pFIMP [6] scenarios, where a richer parameter space can open up.

References

1. N. Aghanim, et al., [Planck], Astron. Astrophys. **641** (2020); A6 [erratum: Astron. Astrophys. **652** (2021) C4]. https://doi.org/10.1051/0004-6361/201833910 [arXiv:1807.06209 [astro-ph.CO]]
2. S. Bhattacharya, P. Ghosh, J. Lahiri, B. Mukhopadhyaya, Phys. Rev. D **108**(11), L111703 (2023). https://doi.org/10.1103/PhysRevD.108.L111703 [arXiv:2211.10749 [hep-ph]]
3. S. Bhattacharya, P. Ghosh, J. Lahiri, B. Mukhopadhyaya, JHEP **12**(049) (2022). https://doi.org/10.1007/JHEP12(2022)049 [arXiv:2202.12097 [hep-ph]]
4. J. Herrero-Garcia, A. Scaffidi, M. White, A.G. Williams, JCAP **11**(021) (2017). https://doi.org/10.1088/1475-7516/2017/11/021 [arXiv:1709.01945 [hep-ph]]
5. J. Herrero-Garcia, A. Scaffidi, M. White, A.G. Williams, JCAP **01**(008) (2019). https://doi.org/10.1088/1475-7516/2019/01/008 [arXiv:1809.06881 [hep-ph]]
6. S. Bhattacharya, D. Pradhan, J. Lahiri, Phys. Rev. D **108**(11), L111702 (2023). https://doi.org/10.1103/PhysRevD.108.L111702 [arXiv:2212.07622 [hep-ph]]
7. S. Bhattacharya, J. Lahiri, D. Pradhan, [arXiv:2212.14846 [hep-ph]]
8. S. Bhattacharya, S. Chakraborti, D. Pradhan, JHEP **07**(091) (2022). https://doi.org/10.1007/JHEP07(2022)091 [arXiv:2110.06985 [hep-ph]]
9. S. Bhattacharya, P. Poulose, P. Ghosh, JCAP **04**(043) (2017).https://doi.org/10.1088/1475-7516/2017/04/043 [arXiv:1607.08461 [hep-ph]]
10. G. Arcadi, M. Dutra, P. Ghosh, M. Lindner, Y. Mambrini, M. Pierre, S. Profumo, F.S. Queiroz, Eur. Phys. J. C **78**(3), 203 (2018). https://doi.org/10.1140/epjc/s10052-018-5662-y [arXiv:1703.07364 [hep-ph]]

Chapter 8
Two Minor Zeros in Neutrino Mass Matrix Using Trimaximal Neutrino Mixing

Iffat Ara Mazumder and Rupak Dutta

Abstract We explore the phenomenological implications of two minor zeros in neutrino mass matrix using trimaximal mixing matrix. In this context, we analyze 15 possible classes of two minor zeros in neutrino mass matrix and found that only two classes, namely class A_1 and class A_2 are allowed. We predict the absolute values of total neutrino masses, effective Majorana masses, the effective electron anti-neutrino mass, and Majorana CP-violating phases for these two classes. Furthermore, we analyze the correlations of neutrino oscillation parameters. We discuss a flavor model within the seesaw model along with Z_8 symmetry group for generating these classes.

8.1 Introduction

Harrison, Perkins, and Scott first proposed the tribimaximal (TB) neutrino flavor mixing [1, 2]. This mixing pattern shows impressive agreement with solar mixing angle (θ_{12}) and atmospheric mixing angles (θ_{23}). According to this mixing pattern, the values of θ_{12} and θ_{23} are 35.26° and 45°, respectively, and the reactor mixing angle (θ_{13}) is precisely zero. The zero value of θ_{13} is an indication of no CP violation. The various experiments reported θ_{13} to be non-zero. The T2K [3] experiment confirmed the non-zero value of θ_{13} at a 90% confidence level and the MINOS [4] experiment disfavored the $\theta_{13} = 0$ at a 89% confidence level. To accommodate the experimental data, the TB mixing pattern was modified [5, 6]. The modified matrix is called the trimaximal (TM) mixing matrix.

The β decay experiments give direct information on the absolute neutrino mass scale. Recently, the Karlsruhe Tritium Neutrino (KATRIN) experiment reported the upper bound of effective electron anti-neutrino mass m_ν at a 90% confidence level

I. A. Mazumder (✉) · R. Dutta
National Institute of Technology Silchar, Silchar 788010, India
e-mail: iffat_rs@phy.nits.ac.in

R. Dutta
e-mail: rupak@phy.nits.ac.in

J. W. F. Valle et al. (eds.), *Proceedings of the International Conference on Future Prospects in Neutrino and Astroparticle Physics*, Springer Proceedings in Physics 361,
https://doi.org/10.1007/978-981-97-7441-8_8

to be $m_\nu < 0.8\,\text{eV}$ [7]. Similarly, the Project 8 experiment [8] aims to achieve the absolute mass scale of the neutrino up to 40 meV/c^2 sensitivity. Also, the cosmological observations provide impressive constraint on the absolute neutrino mass. The Planck collaboration reported the current upper bound on the sum of neutrino masses to be $\sum m_i < 0.12\,\text{eV}$ at a 95% confidence level [9]. The effective Majorana mass term M_{ee} determines the nature of the neutrinos. The upper bound of M_{ee} by using ^{48}Ca isotope reported by GERDA experiment to be $m_{ee} < (0.079 - 0.18)\,\text{eV}$ [10].

8.2 Theoretical Framework and Two Minor Zeros in Neutrino Mass Matrix

The trimaximal neutrino mixing is parameterized in terms of two unknown parameters ϕ and θ as

$$U_{TM} = \begin{pmatrix} \sqrt{\frac{2}{3}}\cos\theta & \frac{1}{\sqrt{3}} & \sqrt{\frac{2}{3}}\sin\theta \\ -\frac{\cos\theta}{\sqrt{6}} + \frac{e^{-i\phi}\sin\theta}{\sqrt{2}} & \frac{1}{\sqrt{3}} & -\frac{\sin\theta}{\sqrt{6}} - \frac{e^{-i\phi}\cos\theta}{\sqrt{2}} \\ -\frac{\cos\theta}{\sqrt{6}} - \frac{e^{-i\phi}\sin\theta}{\sqrt{2}} & \frac{1}{\sqrt{3}} & -\frac{\sin\theta}{\sqrt{6}} + \frac{e^{-i\phi}\cos\theta}{\sqrt{2}} \end{pmatrix}. \tag{8.1}$$

The neutrino mass matrix corresponding to this mixing matrix can be written as

$$M_{\psi\omega} = (U M_{diag} U^T)_{\psi\omega} \text{ where } \psi, \omega = e, \mu, \tau, \tag{8.2}$$

where $U = U_{TM}P$, $P = \text{diag}(1, e^{i\alpha}, e^{i\beta})$ is the phase matrix and $M_{diag} = \text{diag}(m_1, m_2, m_3)$ is the matrix contains three mass state.

In terms of ϕ and θ, we can write the three mixing angles, Jarlskog rephrasing parameter J and the Dirac CP-violating phase δ as

$$\sin^2\theta_{12} = \frac{1}{3 - 2\sin^2\theta}, \sin^2\theta_{13} = \frac{2}{3}\sin^2\theta, \sin^2\theta_{23} = \frac{1}{2}\Big(1 + \frac{\sqrt{3}\sin 2\theta\cos\phi}{3 - 2\sin^2\theta}\Big),$$
$$J = \frac{1}{6\sqrt{3}}\sin 2\theta\sin\phi, \csc^2\delta = \csc^2\phi - \frac{3\sin^2 2\theta\cot^2\phi}{(3 - 2\sin^2\theta)^2}. \tag{8.3}$$

The effective Majorana mass term $|M_{ee}|$ and the effective electron anti-neutrino mass M_ν can be expressed as

$$|M_{ee}| = \left|\frac{1}{3}(2m_1\cos^2\theta + m_2 e^{2i\alpha} + 2m_3\sin^2\theta e^{2i\beta})\right|, \tag{8.4}$$

$$M_\nu^2 = \frac{1}{3}(2m_1^2 + m_2^2\cos^2\theta + 2m_3^2\sin^2\theta). \tag{8.5}$$

Table 8.1 Two minor zeros

Class	Constraining equations	Class	Constraining equations	Class	Constraining equations
A_1	$C_{33}=0, C_{32}=0$	A_2	$C_{22}=0, C_{32}=0$	B_3	$C_{33}=0, C_{31}=0$
B_4	$C_{22}=0, C_{21}=0$	B_5	$C_{33}=0, C_{12}=0$	B_6	$C_{22}=0, C_{13}=0$
D	$C_{33}=0, C_{22}=0$	S_1	$C_{31}=0, C_{11}=0$	S_2	$C_{21}=0, C_{11}=0$
S_3	$C_{13}=0, C_{12}=0$	F_1	$C_{33}=0, C_{11}=0$	F_2	$C_{22}=0, C_{11}=0$
F_3	$C_{32}=0, C_{11}=0$	F_4	$C_{31}=0, C_{32}=0$	F_5	$C_{21}=0, C_{32}=0$

Fifteen cases of two minor zeros are possible in the neutrino mass matrix. We denote the minor corresponding to the ijth element of M_{ij} as C_{ij}. The fifteen possible patterns of two minor zeros in the neutrino mass matrix are listed in Table 8.1.

The conditions for two minor zeros can be written as

$$
\begin{aligned}
M_{pq}M_{rs} - M_{uv}M_{xy} &= 0\,, \\
M_{p'q'}M_{r's'} - M_{u'v'}M_{x'y'} &= 0\,.
\end{aligned} \tag{8.6}
$$

We can write Eq. 8.6 in terms of complex equations as

$$
\begin{aligned}
m_1m_2X_3e^{2i\alpha} + m_2m_3X_1e^{2i(\alpha+\beta)} + m_3m_1X_2e^{2i\beta} &= 0, \\
m_1m_2Y_3e^{2i\alpha} + m_2m_3Y_1e^{2i(\alpha+\beta)} + m_3m_1Y_2e^{2i\beta} &= 0,
\end{aligned} \tag{8.7}
$$

where X_k and Y_k are associated to the TM mixing matrix elements [11]. Using Eq. 8.7, we can write the mass ratios as

$$
\begin{aligned}
\frac{m_1}{m_2}e^{-2i\alpha} &= \frac{X_3Y_1 - X_1Y_3}{X_2Y_3 - X_3Y_2}, \\
\frac{m_1}{m_3}e^{-2i\beta} &= \frac{X_1Y_2 - X_2Y_1}{X_2Y_3 - X_3Y_2}, \\
\frac{m_3}{m_2}e^{-2i(\alpha-\beta)} &= \frac{X_3Y_1 - X_1Y_3}{X_1Y_2 - X_2Y_1}.
\end{aligned} \tag{8.8}
$$

The CP-violating Majorana phases can be written as

$$
\alpha = -\frac{1}{2}\arg\left(\frac{X_3Y_1 - X_1Y_3}{X_2Y_3 - X_3Y_2}\right), \beta = -\frac{1}{2}\arg\left(\frac{X_1Y_2 - X_2Y_1}{X_2Y_3 - X_3Y_2}\right). \tag{8.9}
$$

The values of m_1, m_2, and m_3 can be calculated using Eq. 8.8 and mass square differences (Δm^2_{21}, Δm^2_{32}). We can write as

$$m_1 = \sqrt{\Delta m_{21}^2} \sqrt{\frac{|\frac{m_1}{m_2}|^2}{|1 - |\frac{m_1}{m_2}|^2|}},$$

$$m_2 = \sqrt{|\Delta m_{32}^2|} \sqrt{\frac{1}{||\frac{m_3}{m_2}|^2 - 1|}},$$

$$m_3 = \sqrt{|\Delta m_{32}^2|} \sqrt{\frac{1}{|1 - |\frac{m_2}{m_3}|^2|}}. \tag{8.10}$$

The ratio of squared mass difference can be written as

$$r \equiv \left| \frac{\Delta m_{21}^2}{\Delta m_{32}^2} \right| = \left| \frac{1 - (\frac{m_1}{m_2})^2}{(\frac{m_3}{m_2})^2 - 1} \right|, \tag{8.11}$$

where Δm_{21}^2 and Δm_{32}^2 represent solar and atmospheric mass squared difference, respectively.

8.3 Results and Discussion

We use the oscillation parameters from NuFIT [12] for numerical analysis. Out of fifteen possible two minor zeros classes in the neutrino mass matrix, only class A_1 and class A_2 are allowed. The classes B_3, B_4, B_5, B_6, S_1, S_2, F_1, F_2, and F_3 are not allowed as the value of r for these classes are not within the experimental range. The classes S_3, F_4, and F_5 predict $m_1 = m_2$. So, these classes are not allowed. The class D is also not allowed as this class predicts $m_1 = m_3$. We will discuss the results of only two classes: class A_1 and class A_2.

It is evident from Eq. 8.9, the CP-violating Majorana phases (α, β) depend on ϕ and θ. Also from Eq. 8.10, it is clear that the neutrino mass m_i depends on ϕ, θ and the mass squared differences Δm_{32}^2 and Δm_{21}^2. We use the 3σ allowed range of ϕ and θ of that are obtained from the measured values of the mixing angles θ_{13}, θ_{12} and θ_{23}. We also use the 3σ allowed ranges of Δm_{32}^2, Δm_{21}^2 and r for constraining the values of the neutrino masses. We carry out a χ^2 analysis to find the best-fit values of mixing parameters. The χ^2 is defined as

$$\chi^2 = \sum_{k=1}^{3} \frac{\left(\theta_k^{mea} - \theta_k^{exp}\right)^2}{(\sigma_k^{exp})^2} + \sum_{l=32,21} \frac{\left(\Delta m_l^{mea} - \Delta m_l^{exp}\right)^2}{(\sigma_l^{exp})^2}, \tag{8.12}$$

where $\theta_k = (\theta_{12}, \theta_{23}, \theta_{13})$ and $\Delta m_l = (\Delta m^2_{32},\ \Delta m^2_{21})$. The θ_k^{mea} and Δm_l^{mea} are the measured value of θ_k and Δm_l, respectively, whereas, θ_k^{exp} and Δm_l^{exp} are the experimental central values of θ_k and Δm_l, respectively. The σ_k^{exp} and σ_l^{exp} are the errors corresponding to the experimental value of θ_k and Δm_l, respectively. Now we will analyze the class A_1 and class A_2 individually.

8.3.1 *Class* A_1

The class A_1 corresponds to minor zeros of the (3, 3) and (3, 2) elements of M_{ij} matrix. For this class, we obtained the 3σ allowed range of θ to be $(10.03° - 10.99°)$ using Eq. 8.3 for class A_1. The 3σ allowed range of θ_{12} and θ_{23} are $(35.68° - 35.75°)$ and $(39.00° - 50.99°)$, respectively. The allowed range of θ_{12} is consistent only with the 3σ values. This is because the value of θ_{12} will be greater than or equal to the value obtained for the TB mixing matrix. The 3σ allowed ranges of J and δ are $[0, \pm 3.39 \times 10^{-2}]$ and $[(0, 90)°,\ (270, 360)°]$, respectively. Figure 8.1a represent the variation of neutrino mass states m_1, m_2 and m_3 with respect to ϕ for class A_1. This class shows the normal mass ordering. We show the dependence of $\sum m_i$ with respect to ϕ in Fig. 8.1b. The variation of α and β with ϕ are shown in Fig. 8.1c and d, respectively. The correlation of M_{ee} and $\sum m_i$ is shown in Fig. 8.1e. In Fig. 8.1f, we have shown the correlation of M_ν with $\sum m_i$. The best-fit values and the corresponding 3σ allowed ranges of $\sum m_i$, α, β, M_{ee}, and M_ν, for this class are listed in Table 8.2.

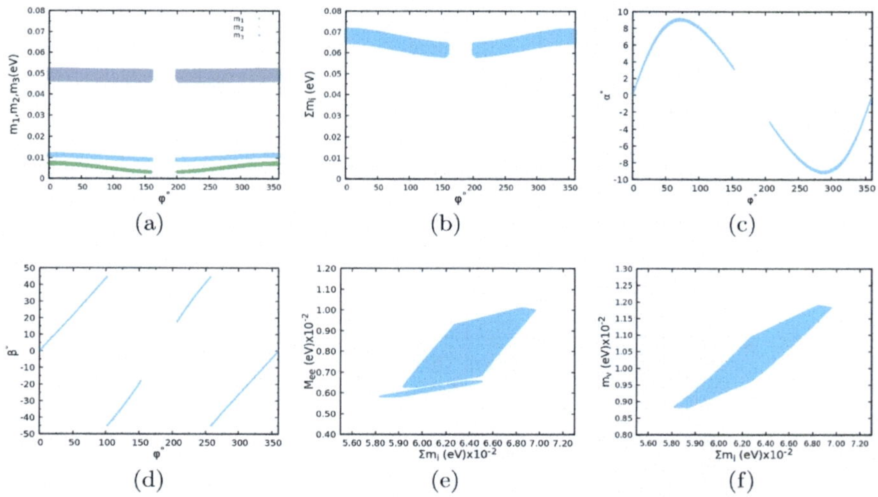

Fig. 8.1 The dependence of ϕ with the neutrino mass states (m_1, m_2, m_3), the total neutrino mass $(\sum m_i)$, the Majorana phase (α), and the Majorana phase (β) are shown by **(a)**, **(b)**, **(c)** and **(d)**, respectively. The correlation of $\sum m_i$ with respect to M_{ee} and the correlation of $\sum m_i$ with respect to M_ν are shown in **(e)** and **(f)**, respectively

Table 8.2 Various neutrino oscillation parameters

Class	Values	$\sum m_i$ (eV)	α°	β°	M_{ee} (eV)	m_ν (eV)
A_1	3σ range	[0.058, 0.071]	[0, ±9.18]	[0, ±44.99]	[0.005, 0.010]	[0.009, 0.012]
	Best fit	0.069	±6.62	±16.46	0.009	0.011

8.3.2 *Class A_2*

The class A_2 corresponds to minor zeros of the (2, 2) and (3, 2) elements of M_{ij} matrix. We obtained the 3σ allowed range of θ to be (10.03°–10.99°) using Eq. 8.3 for class A_2. The 3σ allowed range of θ_{23} and θ_{12} are (39.00°–50.99°) and (35.68° – 35.75°), respectively. The 3σ allowed ranges of J and δ are $[0, \pm 3.39 \times 10^{-2}]$ and [(0, 90)°, (270, 360)°], respectively. Figure 8.2a represent the plot of neutrino masses m_1, m_2, and m_3 versus ϕ. The plot of $\sum m_i$ versus ϕ is shown in Fig. 8.2b. In Fig. 8.2c, we have shown the dependence of α and β with respect to ϕ. In Fig. 8.2e, we have shown the variation of M_{ee} with $\sum m_i$. We have shown the variation of M_ν with $\sum m_i$ in Fig. 8.2f. The phenomenology of this class is similar to class A_1. The best-fit values and the corresponding 3σ allowed ranges of $\sum m_i$, α, β, M_{ee}, and M_ν for this class are listed in Table 8.3.

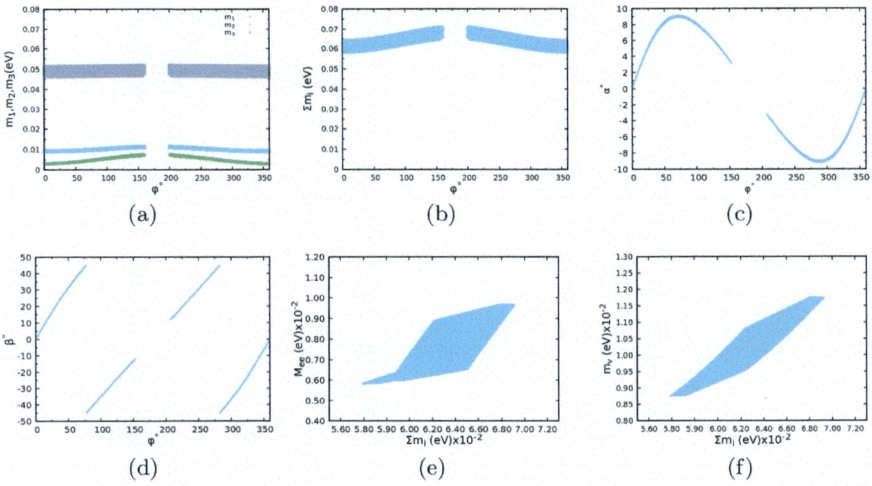

Fig. 8.2 The dependence of ϕ with the neutrino mass states (m_1, m_2, m_3), the total neutrino mass ($\sum m_i$), the Majorana phase (α) and the Majorana phase (β) are shown by **(a)**, **(b)**, **(c)** and **(d)**, respectively. The correlation of $\sum m_i$ with respect to M_{ee} and the correlation of $\sum m_i$ with respect to M_ν are depicated by **(e)** and **(f)**, respectively

Table 8.3 Various neutrino oscillation parameters

Class	Values	$\sum m_i$ (eV)	α°	β°	M_{ee} (eV)	m_ν (eV)
A_2	3σ Range	[0.057, 0.071]	[0, ±9.18]	[0, ±44.99]	[0.006, 0.009]	[0.008, 0.012]
	Best fit	0.064	±7.84	±29.21	0.006	0.009

8.4 Symmetry Realization

The symmetry of two minor zeros in the neutrino mass matrix can be realized through the type-I seesaw mechanism along with Abelian symmetry. For this, we need three left-handed lepton doublets D_{Lp}, three right-handed neutrinos ν_{Rp} $(p = 1, 2, 3)$, and three right-handed charged lepton l_{Rp}. Also, we require Higgs doublet Φ_r $(r = 1, 2, 3)$ for non-zero elements $(M_l)_{pq}$ or $(M_D)_{pq}$, and scalar singlet for non-zero elements $(M_R)_{pq}$ where $q = 1, 2, 3$. Higgs doublets get vacuum expectation values (vevs) at the electroweak scale, while the scalar singlets get the vevs at the seesaw scale. Under Z_8 symmetry, leptons of the first family remain invariant, leptons of the second family change the sign and leptons of the third family get multiplied by $\omega = \exp(\frac{i\pi}{4})$.

The leptonic fields under Z_8 transform as

$$\begin{aligned}
&\bar{\nu}_{R1} \to \bar{\nu}_{R1}, && \bar{l}_{R1} \to \bar{l}_{R1}, && D_{L1} \to \bar{D}_{L1},\\
&\bar{\nu}_{R2} \to \omega^4 \bar{\nu}_{R2}, && \bar{l}_{R2} \to \omega^4 \bar{l}_{R2}, && D_{L2} \to \omega^4 \bar{D}_{L2},\\
&\bar{\nu}_{R3} \to \omega \bar{\nu}_{R3}, && \bar{l}_{R3} \to \omega \bar{l}_{R3}, && D_{L3} \to \omega^7 \bar{D}_{L3}.
\end{aligned} \tag{8.13}$$

The bilinears $\bar{l}_{Rp} D_{Lq}$ and $\bar{\nu}_{Rp} D_{Lq}$ relevant for $(M_l)_{pq}$ and $(M_D)_{pq}$ transform as

$$\bar{l}_{Rp} D_{Lq} \cong \bar{\nu}_{Rp} D_{Lq} \cong \begin{pmatrix} 1 & \omega^4 & \omega^7 \\ \omega^4 & 1 & \omega^3 \\ \omega & \omega^5 & 1 \end{pmatrix} \tag{8.14}$$

and the bilinears $\bar{\nu}_{Rp}\bar{\nu}_{Rq}$ relevant for $(M_R)_{pq}$ transform as

$$\bar{\nu}_{Rp}\bar{\nu}_{Rq} \cong \begin{pmatrix} 1 & \omega^4 & \omega \\ \omega^4 & 1 & \omega^5 \\ \omega & \omega^5 & \omega^2 \end{pmatrix}. \tag{8.15}$$

These transformations automatically generate the diagonal Dirac mass matrices for charged leptons and neutrinos. The real scalar field χ_{12} gives non-zero element $(1, 2)$ and changes sign under Z_8 transformation. Additionally, under Z_8 transformation, the elements $(1, 1)$ and $(2, 2)$ of Majorana mass matrix M_R remain invariant. Complex scalar field χ_{13} generates $(1, 3)$ element which under Z_8 transformation gets multiplied by ω^7 for M_R. Other elements of M_R remain zero without scalar singlets. The Majorana mass matrix M_R can be written as

Table 8.4 Symmetry realization for class A_1 and class A_2 under Z_8 transformation. The index l_{1F} represents the lepton first family and so on

Class	l_{1F}	l_{2F}	l_{3F}	χ_{11}	χ_{12}	χ_{13}	χ_{22}	χ_{23}	χ_{33}
A_1	1	−1	ω	Absent	−1	ω^7	Absent	Absent	Absent
A_2	1	ω	−1	Absent	ω^7	−1	Absent	Absent	Absent

$$M_R = \begin{pmatrix} 1 & b & c \\ b & 1 & 0 \\ c & 0 & 0 \end{pmatrix}. \tag{8.16}$$

This provides two minor zeros class A_1 in the neutrino mass matrix. The Z_8 invariant Yukawa Lagrangian for this class can be written as

$$\begin{aligned} -\mathcal{L} = {} & Y^l_{11} \bar{D}_{eL} \Phi_2 \, e_R + Y^l_{22} \bar{D}_{\mu L} \Phi_3 \, \mu_R + Y^l_{33} \bar{D}_{\tau L} \Phi_1 \, \tau_R + Y^D_{11} \bar{D}_{eL} \tilde{\Phi}_3 \, \nu_{eR} + Y^D_{22} \bar{D}_{\mu L} \tilde{\Phi}_1 \\ & \nu_{\mu R} + Y^D_{33} \bar{D}_{\tau L} \tilde{\Phi}_2 \, \nu_{\tau R} + \frac{M^M_{11}}{2} \bar{\nu}^T_{eR} C^{-1} \nu_{eR} + \frac{Y^M_{12}}{2} \bar{\nu}^T_{eR} C^{-1} \nu_{\mu R} \, \chi_{12} + \frac{Y^M_{13}}{2} \bar{\nu}^T_{eR} C^{-1} \\ & \nu_{\tau R} \, \chi_{13} + \frac{M^M_{22}}{2} \bar{\nu}^T_{\mu R} C^{-1} \nu_{\mu R} + h.c. \end{aligned} \tag{8.17}$$

Here $\tilde{\Phi} = i\tau_2 \Phi^*$. Similarly, class A_2 can also be realized for different M_R. For completeness, the field transformations for both the classes under Z_8 transformation are listed in the Table. 8.4.

8.5 Conclusion

Using the trimaximal mixing matrix, we explore the two minor zeros in the neutrino mass matrix. There are 15 two minor zero cases possible in the neutrino mass matrix. Only class A_1 and class A_2 are phenomenologically valid. We calculated the unknown parameter θ for classes A_1 and A_2. We also got the values of θ_{12} and θ_{23} for both classes. Class A_1 and class A_2 show normal mass ordering. We predict the absolute neutrino mass scale, the range of the effective Majorana mass, and the effective electron anti-neutrino mass range. Also, we determine the values of δ, α, β and J. These values can be testable in future experiments. Moreover, we explain the symmetry realization of class A_1 in the framework of the type-I seesaw model using Abelian symmetry group Z_8.

References

1. P.F. Harrison, D.H. Perkins, W.G. Scott, Phys. Lett. B **530**, 167 (2002). https://doi.org/10.1016/S0370-2693(02)01336-9. ([arXiv:hep-ph/0202074 [hep-ph]].)
2. Z.Z. Xing, Phys. Lett. B **533**, 85–93 (2002). https://doi.org/10.1016/S0370-2693(02)01649-0 [arXiv:hep-ph/0204049 [hep-ph]]
3. K. Abe, et al., [T2K], Phys. Rev. Lett. **107**, 041801 (2011) https://doi.org/10.1103/PhysRevLett.107.041801 [arXiv:1106.2822 [hep-ex]]
4. P. Adamson, et al., [MINOS], Phys. Rev. Lett. **107**, 181802 (2011). https://doi.org/10.1103/PhysRevLett.107.181802 [arXiv:1108.0015 [hep-ex]]
5. S. Kumar, Phys. Rev. D **82**, 013010 (2010). [erratum: Phys. Rev. D **85**, 079904 (2012)]. https://doi.org/10.1103/PhysRevD.82.013010 [arXiv:1007.0808 [hep-ph]]
6. W. Grimus, L. Lavoura, JHEP **09**, 106 (2008). https://doi.org/10.1088/1126-6708/2008/09/106. ([arXiv:0809.0226 [hep-ph]].)
7. M. Aker, et al., [KATRIN], Nat. Phys. **18**(2), 160–166 (2022). https://doi.org/10.1038/s41567-021-01463-1 [arXiv:2105.08533 [hep-ex]]
8. A. A. Esfahani *et al.* [Project 8],[arXiv:2203.07349 [nucl-ex]]
9. M. Zhang, J.F. Zhang, X. Zhang, Commun. Theor. Phys. **72**(12), 125402 (2020) [arXiv:2005.04647 [astro-ph.CO]]
10. M. Agostini, et al., [GERDA], Phys. Rev. Lett. **111**(12), 122503 (2013). https://doi.org/10.1103/PhysRevLett.111.122503 [arXiv:1307.4720 [nucl-ex]]
11. I.A. Mazumder, R. Dutta, [arXiv:2309.04394 [hep-ph]]
12. I. Esteban, M.C. Gonzalez-Garcia, M. Maltoni, T. Schwetz, A. Zhou, JHEP **09**, 178 (2020). https://doi.org/10.1007/JHEP09(2020)178 [arXiv:2007.14792 [hep-ph]]. NuFIT 5.0 (2020), http://www.nu-t.org

Chapter 9
Leptogenesis and Dark Matter Through Relativistic Bubble Walls

Indrajit Saha, Debasish Borah, and Arnab Dasgupta

Abstract We study a scenario where both dark matter (DM) and heavy right-handed neutrino (RHN) responsible for leptogenesis acquire masses by crossing the relativistic bubble walls formed as a result of a supercooled first-order phase transition above electroweak scale. This leads to a large out-of-equilibrium abundance of RHN inside the bubble sufficient to produce the required lepton asymmetry. A minimal scenario with three RHN, one inert scalar doublet and one singlet scalar as additional fields beyond the standard model is sufficient to realise this possibility which also favours inert RHN DM over inert scalar doublet. While low-scale leptogenesis scenario can be probed at future gravitational wave detectors like LISA, a sufficiently high-scale leptogenesis scenario can be constrained from LIGO-VIRGO data as well.

9.1 Introduction

The visible matter content of the present Universe is highly asymmetric with negligible trace of antimatter. The observer baryon asymmetry is often quantified in terms of the baryon to photon ratio [1]

$$\eta_B = \frac{n_B - n_{\bar{B}}}{n_\gamma} = 6.1 \times 10^{-10}. \tag{9.1}$$

Starting with a baryon symmetric Universe, one can generate such an asymmetry dynamically if Sakharov's conditions [12] are satisfied. While the standard model (SM) fails to satisfy these criteria in the required amount, several beyond standard model (BSM) have been proposed out of which leptogenesis [8] has been a very

I. Saha (✉) · D. Borah
Department of Physics, Indian Institute of Technology Guwahati, Assam 781039, India
e-mail: s.indrajit@iitg.ac.in

A. Dasgupta
Pittsburgh Particle Physics, Astrophysics, and Cosmology Center, Department of Physics and Astronomy, University of Pittsburgh, Pittsburgh, PA 15206, USA

J. W. F. Valle et al. (eds.), *Proceedings of the International Conference on Future Prospects in Neutrino and Astroparticle Physics*, Springer Proceedings in Physics 361,
https://doi.org/10.1007/978-981-97-7441-8_9

popular one due to its connection to the origin of light neutrino masses, another observed phenomena which the SM fails to address. Apart from the asymmetric nature of the visible sector, the present Universe also has a large amount of non-luminous, non-baryonic form of matter, known as dark matter (DM). As none of the SM particles satisfy the criteria for DM, several BSM proposals have been put forward out of which the weakly interacting massive particle (WIMP) and lately, the feebly interacting massive particle (FIMP) paradigms have been the most widely studied ones.

Typically, leptogenesis is a high-scale phenomena which makes its direct probe challenging. While the scale of leptogenesis can be brought down in the framework of resonant leptogenesis [11], the corresponding coupling of the heavy decaying particle, like a right-handed neutrino (RHN), with the SM bath gets suppressed. In view of these, some complementary detection prospects may help in probing leptogenesis indirectly. Detection of stochastic gravitational wave (GW) background in present and future experiments may provide such an opportunity. In view of this, we consider a specific leptogenesis scenario where RHN acquires mass in a first-order phase transition (FOPT) by crossing the walls of bubbles formed. Such a mass-gain mechanism, originally proposed in [3] followed by leptogenesis implementation in [4, 5, 7, 9] can lead to stochastic GW whose peak frequency has tight correlation with the scale of FOPT or leptogenesis. We apply this mechanism to generate DM and lepton asymmetry simultaneously in a minimal seesaw model and study the GW detection prospects at present and near future experiments.

This paper is organised as follows. In Sect. 9.2, we discuss the particle physics model followed by the discussion of FOPT and GW signatures in Sects. 9.3, 9.4 respectively. We show the results related to leptogenesis and DM in Sect. 9.5 and finally conclude in Sect. 9.6.

9.2 The Model

We consider a minimal setup where the SM is extended by three gauge singlet right-handed neutrinos N_i (i = 1, 2, 3), one extra scalar doublet η and a scalar singlet S. A classical conformal invariance is imposed which help in realising a supercooled FOPT with enhanced GW amplitude. An additional Z_2 symmetry is imposed under which the lightest RHN namely N_1 and η are odd while all other particles including the SM ones are even.

The relevant part of the Yukawa Lagrangian is [4]

$$-\mathcal{L} \supset \frac{1}{2} Y'_{ij} S N_i N_j + \left(Y_{i1}\, \bar{L}_i \tilde{\eta} N_1 + \sum_{k=2,3} (y_D)_{ik} \bar{L}_i \tilde{\Phi}_1 N_k + \text{h.c.} \right), \qquad (9.2)$$

where Φ_1 is the SM Higgs doublet. Due to two different types of Dirac Yukawa coupling of neutrinos, their masses arise from a combination of type-I and scotogenic [10] contributions. The scalar potential of the model can be written as

$$\begin{aligned} V(\Phi_1, \eta, S) &= \frac{\lambda_1}{4}|\Phi_1|^4 + \frac{\lambda_2}{4}|\eta|^4 + \lambda_3|\Phi_1|^2|\eta|^2 + \frac{1}{4}\lambda_S S^4 + \lambda_4|\Phi_1^\dagger\eta|^2 \\ &+ \left[\frac{\lambda_5}{2}(\Phi_1^\dagger\eta)^2 + \text{h.c.}\right] + \lambda_6|\Phi_1|^2 S^2 + \lambda_7|\eta|^2 S^2 . \end{aligned} \tag{9.3}$$

Thus, the Lagrangian does not have any mass scales and the singlet scalar S driven FOPT is responsible for generating the mass scales of RHN as well as electroweak symmetry breaking.

9.3 First-Order Phase Transition

In order to realise leptogenesis, the phase transition is driven by singlet scalar field at a scale above the electroweak scale. To study the first-order phase transition, the effective potential considering one-loop correction can be written as

$$V_{\rm tot} = V_{\rm tree} + V_{\rm CW} + V_{\rm th}, \tag{9.4}$$

where $V_{\rm tree}$, $V_{\rm CW}$ and $V_{\rm th}$ denote the tree level scalar potential, the one-loop Coleman–Weinberg potential, the thermal effective potential, respectively. The Coleman–Weinberg (CW) potential is given by

$$V_{\rm CW} = \sum_i (-1)^{n_f} \frac{n_i}{64\pi^2} m_i^4(\phi) \left(\log\left(\frac{m_i^2(\phi)}{\mu^2}\right) - \frac{3}{2}\right), \tag{9.5}$$

where suffix i represents particle species, and m_i, $n_i(\phi)$ are field dependent masses and the degrees of freedom (dof) of i'th particle, μ is the renormalisation scale and $(-)^{n_f}$ is $+1$ for bosons and -1 for fermions, respectively. Here, we consider vacuum expectation value (VEV) of the singlet scalar, denoted by M as the renormalisation scale, $\mu = M = \langle S \rangle$. The singlet scalar can be written as $S = (\phi + M)/\sqrt{2}$. In the above correction, the contributing field dependent masses along with their dof are

$$m_\eta^2 = \lambda_7\phi^2/2 \; (n_\eta = 4), m_s^2 = 3\lambda_s\phi^2 \; (n_s = 1), m_{y_i}^2 = 2y_i^2\phi^2 \; (n_y = 2). \tag{9.6}$$

Considering the finite temperature loop correction to effective potential, the temperature dependent contribution calculated as

$$V_{\rm th} = \sum_i \left(\frac{n_{\rm B_i}}{2\pi^2} T^4 J_B\left[\frac{m_{\rm B_i}}{T}\right] - \frac{n_{\rm F_i}}{2\pi^2} T^4 J_F\left[\frac{m_{\rm F_i}}{T}\right]\right), \tag{9.7}$$

where n_{B_i} and n_{F_i} denote the dof of the bosonic and fermionic particles, respectively and

$$J_B(x) = \int_0^\infty dz z^2 \log\left[1 - e^{-\sqrt{z^2+x^2}}\right], \tag{9.8}$$

$$J_F(x) = \int_0^\infty dz z^2 \log\left[1 + e^{-\sqrt{z^2+x^2}}\right]. \tag{9.9}$$

In our calculation of the thermal potential, we also consider the correction from the daisy diagrams. The thermal part of the potential with the daisy contributions is given by

$$V_T(\phi, T) = V_{\text{th}} + V_{\text{daisy}}(\phi, T), \tag{9.10}$$

$$V_{\text{daisy}}(\phi, T) = -\sum_i \frac{g_i T}{12\pi}\left[m_i^3(\phi, T) - m_i^3(\phi)\right],$$

where, V_{th} is the thermal correction and V_{daisy} is the daisy subtraction. The relevant masses with thermal contribution can be written as $m_i^2(\phi, T) = m_i^2(\phi) + \Pi_i(T)$, where

$$\Pi_\eta(T) = \left(\frac{g_2^2}{8} + \frac{g_1^2 + g_2^2}{16} + \frac{\lambda_2}{2} + \frac{\lambda_3 + \lambda_4}{12}\right) T^2$$

$$\Pi_s(T) = \left(\frac{\lambda_s}{4} + \frac{\lambda_6}{3} + \frac{\lambda_7}{3} + \frac{y_1^2}{8} + \frac{y_2^2}{8} + \frac{y_2^2}{8}\right) T^2.$$

The FOPT proceeds via tunnelling and the rate of tunnelling per unit time per unit volume is given by $\Gamma(T) = \mathcal{A}(T)e^{-S_3(T)/T}$, where $\mathcal{A}(T) \sim T^4$ and $S_3(T)$ are determined by the dimensional analysis and bounce solution configuration, respectively. The bubbles are nucleated at some specific temperature below the critical temperature and called as the nucleation temperature T_n. This is calculated by comparing the above tunnelling rate to the Hubble expansion rate as $\Gamma(T_n) = \mathbf{H}^4(T_n)$. Here, considering usual radiation dominated universe, the Hubble parameter is given by $\mathbf{H}(T) \simeq 1.66\sqrt{g_*}T^2/M_{\text{Pl}}$ with g_* being the dof of the radiation component.

We write the zero temperature one-loop effective potential as

$$V_0 = V_{\text{tree}} + V_{\text{CW}} = \frac{1}{4}\lambda_S(t)G^4(t)\phi^4, \tag{9.11}$$

where $t = \log(\phi/\mu)$ with $\mu = M$ being the scale of renormalisation. $G(t)$ is given by

$$G(t) = e^{-\int_0^t dt'\gamma(t')}, \quad \gamma(t) = \frac{1}{32\pi^2}\text{Tr}[Y'^\dagger Y']. \tag{9.12}$$

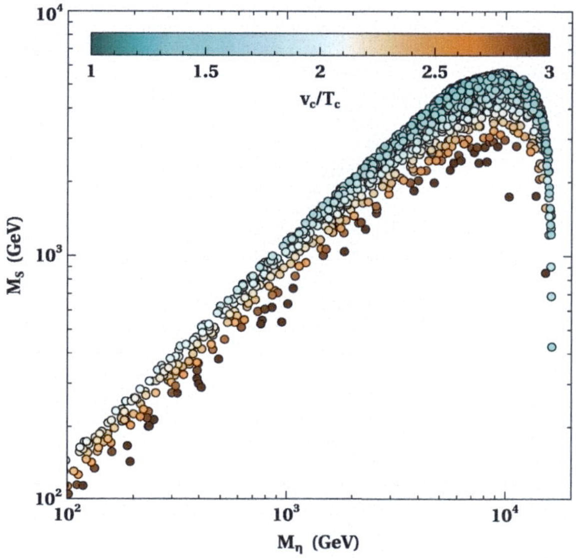

Fig. 9.1 Parameter space in inert doublet Mass M_η versus scalar singlet mass M_S plane satisfying FOPT and the colour code indicates the strength of the FOPT

After solving the renormalisation group evolution (RGE) equations of the model, the quartic coupling and Yukawa couplings at the renormalisation scale are estimated. The minimization condition $\frac{dV}{d\phi}|_{\phi=M} = 0$ gives us,

$$10\lambda_s^2(0) + 32\pi^2\lambda_s(0) + 3Y'^2(0)\lambda_s(0) - 3Y'^4(0) + \lambda_6^2(0) + \lambda_7^2(0) = 0 \quad (9.13)$$

and we assumed all three Yukawa couplings to be identical.

We have implemented the model in `PhaseTracer` [2] to find the parameter space satisfying FOPT. We have done a numerical scan for parameter space in new scalar masses in Fig. 9.1. From the colour code, it is cleared that the strength of the FOPT can be large such as $v_c/T_c \geq 3$ for some region of the parameter space.

9.4 Stochastic Gravitational Waves

A FOPT can sourced to the formation of stochastic gravitational waves (GW) background mainly due to three mechanisms: the bubble collisions during the expansion phase generate a quadrupole contribution to the stress-energy tensor, which act as source of GW; the sound wave in the plasma medium and the turbulence of the plasma medium created from the decay of magneto-hydrodynamic turbulence produced by bulk motions of bubbles. The last two mechanism can continue to work for longer time and the contribution is also more. The energy released during the FOPT can also be estimated as $\alpha_* = \frac{\epsilon_*}{\rho_{\rm rad}}$, where, $\epsilon_* = \left[\Delta V_{\rm tot} - \frac{T}{4}\frac{\partial \Delta V_{\rm tot}}{\partial T}\right]_{T=T_*}$, $\rho_{\rm rad} = g_*\pi^2 T^4/30$ and

$\Delta V_{\text{tot}} \equiv V_{\text{tot}}(\phi_{\text{false}}, T) - V_{\text{tot}}(\phi_{\text{true}}, T)$. The duration of the FOPT is related to the parameter β, defined as $\frac{\beta}{\mathbf{H}(T)} \simeq T\frac{d}{dT}\left(\frac{S_3}{T}\right)$. Here, α_* and $\beta/\mathbf{H}(T)$ are calculated at the nucleation temperature $T = T_n$.

Now, the total GW power spectrum can be expressed as [6]

$$\Omega_{\text{GW}}(f) = \Omega_{\phi}(f) + \Omega_{\text{sw}}(f) + \Omega_{\text{turb}}(f). \tag{9.14}$$

Different contributions are also characterised by its own peak frequency and each GW spectrum can be expressed in the parametric form as

$$h^2\Omega(f) = \mathcal{R}\Delta(v_w)\left(\frac{\kappa\alpha_*}{1+\alpha_*}\right)^p\left(\frac{\mathbf{H}_*}{\beta}\right)^q \mathcal{S}(f/f_{\text{peak}}). \tag{9.15}$$

Here, the factor $\mathcal{R} \simeq 7.69 \times 10^{-5} \boldsymbol{g}_*^{-1/3}$ is due to the red-shift of the GW energy density, $\mathcal{S}(f/f_{\text{peak}})$ tells about the shape of the spectrum and $\Delta(v_w)$ is the normalisation factor. The Hubble parameter at the nucleation temperature $T = T_n$ is denoted by $\mathbf{H}_*$. Figure 9.2 shows the combined GW spectrum for four different benchmark points shown in Table 9.1.

If the gravitational waves come within the sensitivity of the detector, it will be detected otherwise it is considered as noise. The signal-to-noise ratio (SNR) ρ can be written by integrating over the experiment's total observing time t_{obs} and accessible frequency range $[f_{\text{min}}, f_{\text{max}}]$

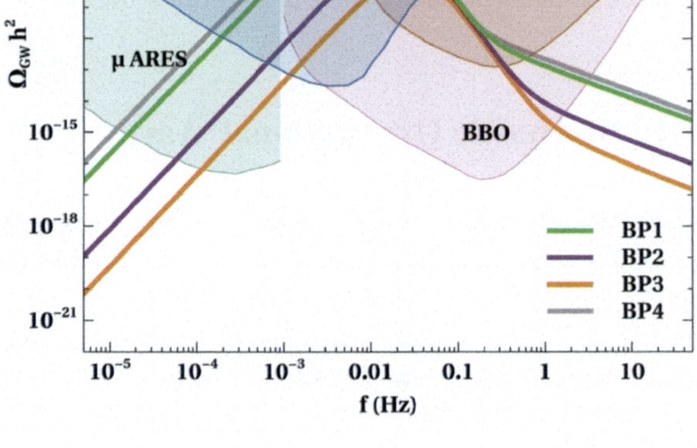

Fig. 9.2 Gravitational wave spectrum due to FOPT in the model for four different benchmark points given in Table 9.1

Table 9.1 Benchmark parameters and other details required for the GW spectrum calculation of the model

	v_c (GeV)	T_c (GeV)	v (GeV)	v_c/T_c	$\lambda_7(0)$	$Y_2'(0) \approx Y_3'(0)$	$\lambda_s(0)$	T_n (GeV)	T_p (GeV)	$(\beta/\mathbf{H}_*)$	v_J	α_*
BP1	9634	2521	9934	3.82	1.5	0.5	0.02	988	974	151	0.89	0.46
BP2	9553	2416	9757	3.95	1.6	0.7	0.02	896	887	110	0.91	0.66
BP3	9698	2370	9988	4.09	1.2	0.3	0.02	779	770	103	0.92	0.80
BP4	9692	2391	9978	4.05	1.3	0.4	0.02	1207	1190	204.51	0.84	0.24

$$\rho = \left[n_{\text{det}} t_{\text{obs}} \int_{f_{\text{min}}}^{f_{\text{max}}} df \left(\frac{\Omega_{\text{signal}}(f)}{\Omega_{\text{noise}}(f)} \right)^2 \right]^{1/2}, \tag{9.16}$$

where n_{det} distinguishes between experiments that aim at detecting the stochastic gravitational wave background by means of an auto-correlation ($n_{\text{det}} = 1$) measurement.

Since a high-scale phase transition generates stochastic gravitational waves which fall in LIGO-VIRGO ballpark, there is a possibility to detect those signals in ground based experiments. For such ongoing experiment, if the SNR is more than 1, a null result can be used to constrain the corresponding parameter space. In the left panel of Fig. 9.2, we show the peak amplitudes of the GW spectrum with the colour code denoting the SNR. In the right panel of the same figure, we show the masses of heavier RHNs $N_{2,3}$ and η with the SNR being indicated by shapes of the points. We also

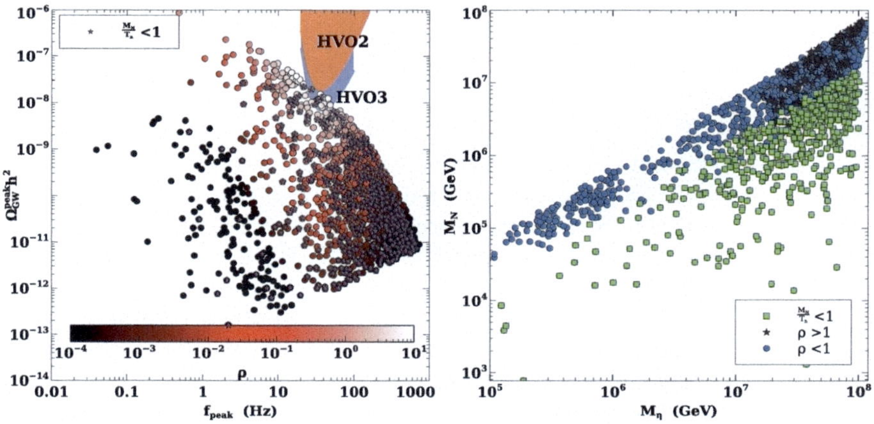

Fig. 9.3 Left panel: GW spectrum peak vs its corresponding Peak frequency with the colour code showing the SNR. Right panel: Parameter space in inert doublet Mass M_η versus RHN mass M_N plane satisfying FOPT in the model with the SNR for LIGO-VIRGO O3

identify the region of parameter space where $M_N < T_n$ and one cannot distinguish the leptogenesis from thermal leptogenesis.

9.5 Leptogenesis and Dark Matter

As noticed above, the scalar doublet η needs to be heavier compared to the RHNs in order to realise the singlet driven FOPT. Therefore, only the Z_2-odd RHN namely, N_1 can be DM in this model. The heavier RHN $N_{2,3}$ can have CP violating out-of-equilibrium decay into Φ_1, L generating the lepton asymmetry.

A first-order phase transition in the singlet (S) sector will create bubbles. As particles such as N_i and η enter these bubbles, they gain mass due to $\langle S \rangle \neq 0$ within the bubble. Consequently, the subsequent decays of $N_{2,3}$ into leptons and SM Higgs generate the leptonic asymmetry. Meanwhile, N_1, being lighter than η, remains stable due to the unbroken Z_2 symmetry and consequently serves as a Dark Matter (DM) candidate. As supercooled phase transition leads to the relativistic nature of bubble walls ensuring all the particles penetrate to the true vacuum following equilibrium abundance. Although $T_n < M_{2,3}$, one can still achieve a large abundance of $N_{2,3}$ without Boltzmann suppression. The abundance of RHN after entering the bubble is

$$Y_N = \frac{135}{8\pi^4}\xi(3)\frac{g_N}{g_*}, \tag{9.17}$$

where g_* and g_N are the total relativistic degrees of freedom in the energy density of the universe and the degrees of freedom of RHN N, respectively. The final baryonic asymmetry can be written as

$$Y_B = \epsilon_N \kappa_{\rm sph} Y_N \left(\frac{T_n}{T_{RH}}\right)^3. \tag{9.18}$$

where ϵ_N is the CP-asymmetry, $\kappa_{\rm Sph} = 8/23$ is the sphaleron conversion factor for two Higgs doublets and $T_{\rm RH}$ is the reheating temperature after the FOPT. Although the inverse decay washout processes are sub-dominant at $T = T_n$, the sudden release of latent heat after the FOPT can lead to a larger temperature $T_{\rm RH}$ of the bath, resulting in a dilution of the asymmetry by a factor of $(T_n/T_{\rm RH})^3$. However, for the chosen benchmark points in the analysis, we do not have such dilution.

In Table 9.2, we show the required CP-asymmetry with Dirac Yukawa couplings among others for chosen benchmark points such that successful leptogenesis via mass-gain mechanism can be realised. Figure 9.4 shows the parameter space in terms of particle masses consistent with dark matter relic and FOPT. The part of the parameter space where leptogenesis becomes purely thermal is also identified. Due to the constraints from light neutrino masses on Yukawa coupling of DM, we consider DM to be produced from non-thermal framework. In non-thermal or feebly interacting massive particle (FIMP) dark matter scenario, the required abundance is produced

Table 9.2 The required CP-asymmetry along with other relevant details involved in leptogenesis for the model

	ϵ_N	$T_{\rm RH}$ (GeV)	T_n (GeV)	$M_{N_2} \approx M_{N_3}$ (GeV)	y_D	$\Delta V_{\rm tot}$ (GeV)^4
BP1	6.22×10^{-8}	988	988.32	4966	5.12×10^{-8}	1.7546×10^{13}
BP2	6.22×10^{-8}	896	896.89	6826	6.00×10^{-8}	1.6973×10^{13}
BP3	6.22×10^{-8}	779	779.49	2996	3.98×10^{-8}	1.1569×10^{13}
BP4	6.22×10^{-8}	1207	1207.58	3991	4.59×10^{-8}	2.1549×10^{13}

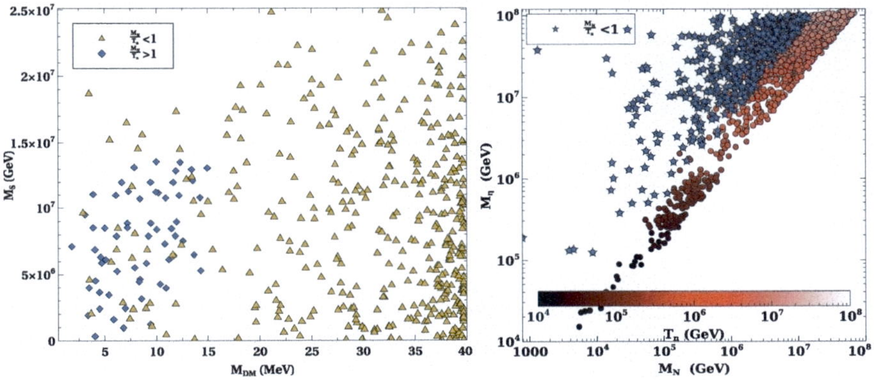

Fig. 9.4 Light panel: Parameter space in dark matter mass M_{DM} versus singlet scalar mass M_S plane with correct relic, Right panel: Parameter space in inert doublet Mass M_η versus RHN mass M_N plane satisfying FOPT in the model

from the decay of singlet scalar into pair of DM, Higgs and $N_{2,3}$. Also, a purely thermal DM will be overproduced in our framework as a large abundance of such particles is captured inside the bubble, in a way similar to $N_{2,3}$ discussed before.

9.6 Conclusion

We have studied the possibility of realising leptogenesis and dark matter via relativistic bubble walls arising from a strong first-order phase transition. We adopt a minimal model with conformal invariance where light neutrino masses arise from a combination of type-I and scotogenic contributions. A singlet scalar field responsible for generating the scale of seesaw as well as electroweak scale dynamically drives the FOPT. We show that, depending upon the scale of the FOPT, the parameter space can be probed at present or near future experiments. While successful leptogenesis is possible with or without resonant enhancement, dark matter is preferred to be in the non-thermal ballpark.

References

1. Aghanim, N., et al., Planck 2018 results. VI. Cosmological parameters. Astron. Astrophys. **641**, A6 (2020), [Erratum: Astron. Astrophys. **652**, C4 (2021)]
2. P. Athron, C. Balázs, A. Fowlie, Y. Zhang, PhaseTracer: tracing cosmological phases and calculating transition properties. Eur. Phys. J. C **80**(6), 567 (2020)
3. I. Baldes, S. Blasi, A. Mariotti, A. Sevrin, K. Turbang, Baryogenesis via relativistic bubble expansion. Phys. Rev. D **104**(11), 115029 (2021)
4. D. Borah, A. Dasgupta, I. Saha, Leptogenesis and dark matter through relativistic bubble walls with observable gravitational waves. JHEP **11**, 136 (2022)
5. D. Borah, A. Dasgupta, I. Saha, LIGO-VIRGO constraints on dark matter and leptogenesis triggered by a first order phase transition at high scale (2023)
6. C. Caprini et al., Science with the space-based interferometer eLISA. II: gravitational waves from cosmological phase transitions. JCAP **04**, 001 (2016)
7. A. Dasgupta, P.S.B. Dev, A. Ghoshal, A. Mazumdar, Gravitational wave pathway to testable leptogenesis. Phys. Rev. D **106**(7), 075027 (2022)
8. M. Fukugita, T. Yanagida, Baryogenesis without grand unification. Phys. Lett. B **174**, 45–47 (1986)
9. P. Huang, K.P. Xie, Leptogenesis triggered by a first-order phase transition. JHEP **09**, 052 (2022)
10. E. Ma, Verifiable radiative seesaw mechanism of neutrino mass and dark matter. Phys. Rev. D **73**, 077301 (2006)
11. A. Pilaftsis, Heavy Majorana neutrinos and baryogenesis. Int. J. Mod. Phys. A **14**, 1811–1858 (1999)
12. A.D. Sakharov, Violation of CP Invariance, C asymmetry, and baryon asymmetry of the universe. Pisma Zh. Eksp. Teor. Fiz. **5**, 32–35 (1967)

Chapter 10
Exploring New Physics Effects on Leggett-Garg-Type Inequalities for NOνA and T2K

Juhi Vardani, Bhavna Yadav, and Lekhashri Konwar

Abstract The recent anomalies observed in NOνA and T2K experiments may serve as indications of physics extending beyond the standard model (SM). In this work, we analyzed the violation of Leggett-Garg-type Inequalities (LGtI) for NOνA and T2K experiments focusing on the Non-Standard Interaction (NSI) parameters that allow such anomalies. We find that LGtI violation enhanced in the presence of NSI in normal ordering (NO) instead of inverted ordering (IO) in NOνA and T2K experiments. Moreover, LGtI parameter gives a clear distinction between SM and NSI for NO.

10.1 Introduction

Neutrino oscillations, a groundbreaking discovery in the realm of particle physics, are a consequence of the superposition principle that demands that the flavor eigenstate (ν_α) mix with mass eigenstate (ν_i). A neutrino characterized by a specific flavor (ν_α) travels, and it can undergo change in proportional of its mass state when detected these results change in the neutrino flavor state. The oscillation of the flavor states of neutrinos demonstrates that neutrinos do indeed have a tiny mass. Neutrinos also have non-standard interactions with matter, which can be characterized by effective couplings ($\epsilon_{\alpha\beta}$) that can be real or complex. Recently, two long baseline accelerator experiments NOνA and T2K released data for δ_{CP} [1, 2], which depict a mismatch in CP-phase for normal ordering while results for inverse ordering are matched. To address this discrepancy, we consider the existence of flavor off-diagonal ($\epsilon_{e\mu}$ and $\epsilon_{e\tau}$) complex neutral current NSI [3–5]. The most effective method for studying $\epsilon_{\mu\tau}$ is

J. Vardani (✉) · B. Yadav · L. Konwar
Indian Institute of Technology Jodhpur, Jodhpur 342037, India
e-mail: vardani.1@iitj.ac.in

B. Yadav
e-mail: yadav.18@iitj.ac.in

L. Konwar
e-mail: konwar.3@iitj.ac.in

J. W. F. Valle et al. (eds.), *Proceedings of the International Conference on Future Prospects in Neutrino and Astroparticle Physics*, Springer Proceedings in Physics 361,
https://doi.org/10.1007/978-981-97-7441-8_10

the use of atmospheric neutrinos [6], whereas $\epsilon_{e\mu}$ and $\epsilon_{e\tau}$ have been explored through long baseline accelerator experiments [7]. NSI has the potential to influence spatial and temporal quantum correlations in neutrino oscillations. The spatial quantum correlations can be quantified in terms of Bell's inequality, flavor entropy [8]. On the other hand, temporal quantum correlation can be quantified in terms of Leggett-Garg Inequalities (LGI), which is based on macrorealism and noninvasive measurement (NIM) [9].

The violation of the LGI signifies that either the system cannot be accurately described within a realistic framework or measurements cannot be taken without disrupting it. However, quantum mechanic systems unequivocally diverge from these [9, 10]. A weaker criterion known as stationarity is used to formulate Leggett-Garg-type Inequalities (LGtI) in place of the NIM assumption [11]. Since neutrino measurements undermine NIM assumptions, LGtI is preferred to study temporal quantum correlation in neutrinos [12, 13]. Moreover, LGtI can be expressed in terms of neutrino survival and oscillation probability [13, 14]. In this work, we explore the impact of complex NSI parameters on LGtI in NOνA, and T2K experiments to address observed anomalies in NOνA, and T2K experiments.

10.2 Formalism

Neutrino Oscillation in matter: The flavor state (ν_α) can be expressed in terms of mass eigenstate (ν_i) as follows:

$$|\nu_\alpha\rangle = \sum_{i=1}^{3} U_{\alpha i} |\nu_i\rangle \,. \tag{10.1}$$

Where $U_{\alpha i}$ is 3×3 unitary mixing matrix. At time t, the time-evolved flavor eigenstate is written as follows:

$$|\nu_\alpha(t)\rangle = e^{-\iota H_f t} \left|\nu_\beta\right\rangle = U_f(t) \left|\nu_\beta\right\rangle \tag{10.2}$$

In the flavor basis, the neutrino's Hamiltonian H_f, as it travels through matter, is expressed as:

$$\mathcal{H}_f = U \begin{pmatrix} E_1 & 0 & 0 \\ 0 & E_2 & 0 \\ 0 & 0 & E_3 \end{pmatrix} U^+ + \begin{pmatrix} A & 0 & 0 \\ 0 & 0 & 0 \\ 0 & 0 & 0 \end{pmatrix} . \tag{10.3}$$

Here, $A = \pm\sqrt{2} G_f N_e$, A represents standard matter potential, N_e is the electron number density in matter and G_f is the Fermi constant, E_i ($i = 1, 2, 3$) are the eigenvalues corresponding to $|\nu_i\rangle$. Neutrinos have positive matter potential while it is negative for antineutrinos. The flavor evolution operator in the ultrarelativistic limit ($t = L$) is given as [15]:

$$U_f(L) = e^{-\iota \mathcal{H}_f L} = U e^{-\iota \mathcal{H}_m L} U^{-1} \tag{10.4}$$
$$= \phi \sum_{a=1}^{3} e^{-\iota L \lambda_a} \frac{1}{3\lambda_a^2 + c_1} [(\lambda_a^2 + c_1) I + \lambda_a \tilde{T} + \tilde{T}^2],$$

where $\tilde{T} = UTU^{-1}$, $T = \mathcal{H}_m - (tr\mathcal{H}_m)\frac{I}{3}$, $H_f = UH_mU^{+}$, $\lambda_a (a = 1, 2, 3)$ are eigen values of T matrix, and $c_1 = det(T)Tr(T^{-1)}$.

Non-Standard Interaction(NSI) in Neutrino Oscillation: In addition to standard interaction, neutrino oscillation undergoes NSI with matter [16]. NSI can be categorized into two types: Charged Current (CC) and Neutral Current (NC) NSI. The dimension-six fermion operators can be used to describe the Lagrangian for both CC and NC-NSI as follows:

$$\mathcal{L}_{CC-NSI} = 2\sqrt{2} G_F \epsilon_{\alpha\beta}^{ff',L} (\bar{\nu_\alpha} \gamma^\mu P_L l_\beta)(\bar{f}' \gamma_\mu P_L f) + 2\sqrt{2} G_F \epsilon_{\alpha\beta}^{ff',R} (\bar{\nu_\alpha} \gamma^\mu P_L l_\beta)(\bar{f}' \gamma_\mu P_R f)$$

$$\mathcal{L}_{NC-NSI} = 2\sqrt{2} G_F \epsilon_{\alpha\beta}^{f,L} (\bar{\nu_\alpha} \gamma^\mu P_L \nu_\beta)(\bar{f} \gamma_\mu P_L f) + 2\sqrt{2} G_F \epsilon_{\alpha\beta}^{f,R} (\bar{\nu_\alpha} \gamma^\mu P_L \nu_\beta)(\bar{f} \gamma_\mu P_R f) \tag{10.5}$$

Here, $P_{L/R} = \frac{1 \mp \gamma_5}{2}$ is chirality operators, The dimensionless coefficient $(\epsilon_{\alpha\beta}^{f,f'}, \epsilon_{\alpha\beta}^{f})$ quantifies the strength of NSI. For CC interactions f', f are first-generation quarks while for NC interactions, f can be e, u, d fermions, α and β represent different lepton flavors $\{\alpha, \beta\} \in \{e, \mu, \tau\}$.

The flavor basis interaction Hamiltonian in the presence of NSI is expressed as follows:

$$\mathcal{H}_f = U \begin{pmatrix} E_1 & 0 & 0 \\ 0 & E_2 & 0 \\ 0 & 0 & E_3 \end{pmatrix} U^{+} + A \begin{pmatrix} 1 + \epsilon_{ee}(x) & \epsilon_{e\mu}(x) & \epsilon_{e\tau}(x) \\ \epsilon_{\mu e}(x) & \epsilon_{\mu\mu}(x) & \epsilon_{\mu\tau}(x) \\ \epsilon_{\tau e}(x) & \epsilon_{\tau\mu}(x) & \epsilon_{\tau\tau}(x) \end{pmatrix}. \tag{10.6}$$

The diagonal terms are considered to be real by the hermiticity condition, while off-diagonal terms may be complex.

Leggett Garg-type inequalities (LGtI): Leggett-Garg inequalities (LGI) serve as a temporal quantum correlation measure, quantifying the system's correlations at different instances of time. LGI is based on [9] (1) Macro-realism (MR), i.e.., a macroscopic system possessing two or more macroscopically distinct states will invariably exist in one of those states. (2) Noninvasive measurement (NIM), i.e., a measurement can be carried out on a system without affecting its dynamics. LGI reveals the extent to which quantum mechanics applies to many particle systems undergoing decoherence. LGI offers the notion of hidden variable theory. The breach of LGI implies that relying on such a hidden variable theory is not a viable alternative for explaining the time evolution of quantum mechanics. Neutrino measurements destroy the NIM assumptions. To overcome this, stationarity condition is applied. For three-flavor neutrino oscillation in ultrarelativistic, LGtI can be expressed in terms of neutrino oscillation probability as [13, 14]:

$$K_3 = 1 + 2P_{\alpha\beta}(2L, E) - 4P_{\alpha\beta}(L, E). \quad (10.7)$$

This manifests the experimental feasibility of the LG function by employing the condition $P_{\alpha,\beta}(2L, E) = P_{\alpha,\beta}(L, \tilde{E})$ with a suitable choice of E and $\tilde{E}$ [14].

10.3 Results

The impact of NSI scenarios concerning $\epsilon_{e\mu}$ and $\epsilon_{e\tau}$ parameters on the temporal measurement of quantum correlations for the neutrino system in NOνA and T2K experiment setups has been investigated in this section. The values of the standard oscillation parameters have been taken from Ref. [17]. Additionally, the NSI parameters, $\epsilon_{e\mu}$ and $\epsilon_{e\tau}$, have been derived from a comprehensive global analysis of both appearance and disappearance channels using the combined NOνA and T2K data, as stipulated in Table 10.1 [3]. Figure 10.1 illustrates that the parameter K_3 exhibits sensitivity to both SM and NSI interactions for normal as well as for inverted mass ordering. It can be seen that for NO, K_3 exceeds its classical limit for energy above 1.25 GeV in T2K whereas for NOνA experiment violation of K_3 consistently occurs at energies exceeding 3.5 GeV. Similarly, for IO, it shows violation for energy above 1.45 GeV for T2K, and it consistently exhibits violation for energy above 5 GeV for NOνA. It can be seen from Fig. 10.1 that for NO we can readily distinguish between SM and NSI scenarios in T2K as well as in NOνA. Whereas for IO, it is hard to distinguish SM and NSI scenarios for T2K while for NOνA SM and $\epsilon_{e\mu}$ exhibit almost identical characteristics and $\epsilon_{e\tau}$ scenario stands out as distinguishable.

10.4 Conclusion

In this paper, we explored the influence of the $\epsilon_{e\mu}$ and $\epsilon_{e\tau}$ NSI scenario on temporal correlation, assessed through LGtI within the framework of three-flavor neutrino oscillation. The outcomes are then compared with SM predictions for two distinct setups of accelerator experiments. The emergence of tension in the NOνA and T2K data could potentially be alleviated by these two NSI scenarios. The inclusion of

Table 10.1 Best fit values of NSI parameters used in our analysis [3]

Mass ordering	NSI	$\epsilon_{\alpha\beta}$	$\phi_{\alpha\beta}/\pi$	δ_{CP}/π
NO	$\epsilon_{e\mu}$	0.15	1.38	1.48
IO	$\epsilon_{e\mu}$	0.02	0.96	1.50
NO	$\epsilon_{e\tau}$	0.27	1.62	1.46
IO	$\epsilon_{e\tau}$	0.15	1.58	1.52

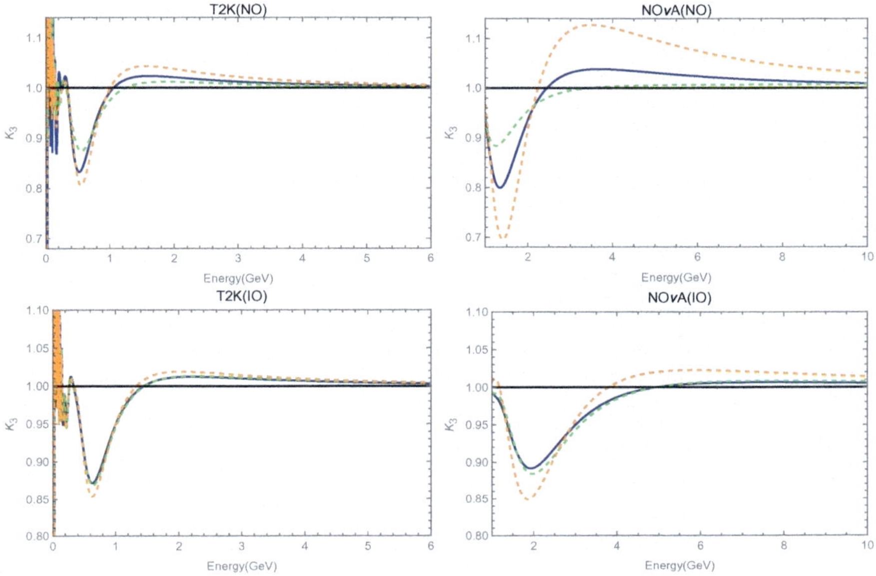

Fig. 10.1 K_3 ($\nu_\mu \rightarrow \nu_e$) is plotted with respect to neutrino energy (GeV) for NO and IO for SM interaction (blue solid curve), SM+NSI ($\epsilon_{e\mu}$) (green dashed), and SM+NSI ($\epsilon_{e\tau}$) (orange dashed) in T2K and NOνA experiments

NSI with $\epsilon_{e\tau}$ results in a substantial enhancement of LGtI-violation compared to the standard model, and there is a clear distinction between SM and NSI scenario for NO. The level of enhancement of K_3 in NOνA stands out than in T2K.

References

1. A. Himmel, New oscillation results from the NOνA experiment (Talk presented at Neutrino 2020), 22 Jun 2020–2 Jul 2020, Virtual Meeting (2020)
2. P. Dunne, Latest neutrino oscillation results from T2K (Talk presented at Neutrino 2020), 22 Jun 2020–2 Jul 2020, Virtual Meeting (2020)
3. S.S. Chatterjee, A. Palazzo, Nonstandard neutrino interactions as a solution to the $NO\nu A$ and T2K discrepancy. Phys. Rev. Lett. **126**, 051802 (2021). https://doi.org/10.1103/PhysRevLett.126.051802. arXiv:2008.04161 [hep-ph]
4. P.B. Denton, J. Gehrlein, R. Pestes, CP -violating neutrino nonstandard interactions in long-baseline-accelerator data. Phys. Rev. Lett. **126**(5), 051801 (2021). https://doi.org/10.1103/PhysRevLett.126.051801. arXiv:2008.01110 [hep-ph]
5. A. Cherchiglia, P. Pasquini, O.L.G. Peres, F.F. Rodrigues, R.R. Rossi, E.S. Souza, arXiv:2310.18401 [hep-ph]
6. F. Fukuda et al., Super-Kamiokande collaboration. Phys. Lett. B **539** 179–187 (2002)
7. K. Abe et al., T2K collaboration. Phys. Rev. Lett. **107**, 041801 (2011)
8. J.S. Bell, On the Einstein Podolsky Rosen paradox. Phys. Phys. Fizika **1**(3), 195 (1964)

9. A.J. Leggett, A. Garg, Quantum mechanics versus macroscopic realism: is the flux there when nobody looks? Phys. Rev. Lett. **54**, 857 (1985). https://doi.org/10.1103/PhysRevLett.54.857
10. C. Emary, N. Lambert, F. Nori, Leggett-Garg inequalities. Rep. Prog. Phys. **77**, 1361–6633 (2013). https://doi.org/10.1088/0034-4885/77/1/016001. arXiv:1304.5133 [quant-ph]
11. S.F. Huelga, W.M. Trevor, E. Santos, Phys. Rev. A **52** (1995)
12. J. Naikoo, A.K. Alok, S. Banerjee, S. Uma Sankar, G. Guarnieri, C. Schultze, B.C. Hiesmayr, A quantum information theoretic quantity sensitive to the neutrino mass-hierarchy. Nucl. Phys. B **951**, 114872 (2020). :https://doi.org/10.1016/j.nuclphysb.2019.114872. arXiv:1710.05562 [hep-ph]
13. J. Naikoo, A.K. Alok, S. Banerjee, S. Uma Sankar, Leggett-Garg inequality in the context of three-flavor neutrino oscillation. Phys. Rev. D **99**(9), 095001 (2019). https://doi.org/10.1103/PhysRevD.99.095001. arXiv:1901.10859 [hep-ph]
14. J.A. Formaggio, D.I. Kaiser, M.M. Murskyj, T.E. Weiss, Violation of the leggett-garg inequality in neutrino oscillations. Phys. Rev. Lett. **117**(5), 050402 (2016). https://doi.org/10.1103/PhysRevLett.117.050402. arXiv:1602.00041 [quant-ph]
15. T. Ohlsson, H. Snellman, Three flavor neutrino oscillations in matter. J. Math. Phys. **41**, 2768–2788 (2000). https://doi.org/10.1063/1.533270. arXiv:hep-ph/9910546 [hep-ph]
16. L. Wolfenstein, Neutrino oscillations in matter. Phys. Rev. D **17**, 2369 (1978)
17. F. Capozzi, E. Di Valentino, E. Lisi, A. Marrone, A. Melchiorri, A. Palazzo, Global constraints on absolute neutrino masses and their ordering. Phys. Rev. D **95**, 096014 (2017). https://doi.org/10.1103/PhysRevD.95.096014. arXiv:1703.04471 [hep-ph]

Chapter 11
TeV Scale Leptogenesis in Extended Scotogenic Model

Labh Singh, Devabrat Mahanta, and Surender Verma

Abstract We explore the potential for low-scale leptogenesis in the singlet-triplet scotogenic model where dark matter participates in neutrino mass generation radiatively. Our investigation focuses on two possible scenarios. The first scenario introduces two heavy right-handed (RH) fermion fields, F and Σ, with moderate hierarchy. In this case, the out-of-equilibrium decay of these particles results in the emergence of a lepton asymmetry. Our analysis reveals that the scale of leptogenesis in this scenario resembles that of standard thermal leptogenesis with $M_{F,\Sigma} \sim 10^9$ GeV. In the second scenario, we propose a framework to lower the scale of leptogenesis. In particular, we have considered two singlet fermions (F_1, F_2) and one triplet fermion (Σ), which results in non-zero lightest active neutrino mass, i.e. $m_1 \neq 0$. With the mass hierarchy $M_{F_1} < M_\Sigma \ll M_{F_2}$, this scenario yields a lower bound on heavy RH fermion, significantly reducing the scale of leptogenesis up to the TeV range. Also, real component of the neutral part of inert scalar (η) is the dark matter candidate in the model.

11.1 Introduction

Scientific evidence strongly indicates an asymmetry in the abundance of baryonic matter and antimatter in the Universe [1–3]. This baryon asymmetry (BA) is quantified by the ratio of baryon to photon $\eta_B = 6.1 \times 10^{-10}$ [4, 5]. To dynamically generate this asymmetry, Sakharov's three conditions must be satisfied [6]. Leptogenesis, proposed by Fukugita and Yanagida [7], stands as the predominant theory explaining

L. Singh (✉) · S. Verma
Department of Physics and Astronomical Science, Central University of Himachal Pradesh, Dharamshala, India
e-mail: sainilabh5@gmail.com

S. Verma
e-mail: s_7verma@hpcu.ac.in

D. Mahanta
Department of Physics, Abhayapuri College, Bongaigaon, Assam, India

J. W. F. Valle et al. (eds.), *Proceedings of the International Conference on Future Prospects in Neutrino and Astroparticle Physics*, Springer Proceedings in Physics 361,
https://doi.org/10.1007/978-981-97-7441-8_11

the origin of the BA. This theory involves decays of right-handed neutrinos (RHN) in a way it satisfies Sakhrov's conditions, resulting in a lepton asymmetry subsequently transformed into the BA through electroweak sphaleron interactions [8]. In the conventional thermal leptogenesis based on Type-I seesaw, a notably high scale for the RHN is required [9–11]. For successful leptogenesis, a minimal RHN mass, known as the Davidson-Ibarra bound, is determined to be $M_{\rm min} = 10^9$ GeV [12]. However, the experimental verification of such high-scale phenomena surpasses the capabilities of current and foreseeable collider experiments. This challenge prompts the exploration of experimentally accessible alternative approaches.

We investigate the feasibility of achieving low-scale leptogenesis within the singlet-triplet scotogenic model (STScoM) [13–22]. This model extends the minimal scotogenic model [23] by introducing a scalar and fermion triplet with zero hypercharges. The STScoM offers increased phenomenological richness, particularly with the hyperchargeless real scalar triplet contributing to the W boson's mass, as observed by the CDF-II experiment [24]. This effect has been explored within the STScoM framework in a previous study [17]. Additionally, the scalar triplet induces mixing between the singlet (F) and triplet (Σ) fermions, leading to a one-loop generation of neutrino masses. The model's dark matter candidate is the real part of an inert scalar doublet.

The minimal scotogenic model explores TeV-scale leptogenesis through the CP-violating and out-of-equilibrium decay of a singlet fermion, contributing to leptogenesis without any degeneracy in the RHN mass spectrum [25, 34]. In the present work, we investigate the potential for achieving TeV-scale leptogenesis within the STScoM. Initially, we examine a scenario with two moderately hierarchical heavy Majorana fermions, resulting in leptogenesis in the robust washout domain due to large Yukawa couplings. Consequently, the leptogenesis scale aligns with that of thermal leptogenesis in Type-I seesaw mode ($M_{F,\Sigma} \geq 10^9$ GeV). To further lower the scale of leptogenesis, we introduce two additional singlet right-handed Majorana fermions, leading to a non-zero value of the lightest active neutrino eigenstate. By selecting a specific mass hierarchy among the RHNs, both F_1 and Σ contribute to the overall lepton asymmetry. In this case, the Yukawa couplings of the RHNs are sufficiently small to maintain the leptogenesis in the weak washout domain, thereby reducing the leptogenesis scale to the TeV energy range.

11.2 Model and Formalism

The field content of the STScoM is presented in Table 11.1. The relevant invariant Yukawa Lagrangian is expressed as follows:

$$\mathcal{L} = Y_{\ell_L}^{\alpha\beta}\overline{\ell_{L_\alpha}}Hl_\beta + Y_F^\alpha\overline{\ell_{L_\alpha}}i\sigma_2\eta N + Y_\Sigma^\alpha\overline{\ell_{L_\alpha}}C\Sigma^\dagger i\sigma_2\eta + Y_\Omega Tr(\overline{\Sigma}\Omega)F$$
$$+\frac{1}{2}M_F\overline{F^c}F + \frac{1}{2}M_\Sigma Tr(\overline{\Sigma^c}\Sigma) + h.c., \tag{11.1}$$

Table 11.1 Symmetry groups and field content of the STScoM model

Symmetry	ℓ_L	l	H	F	Σ	η	Ω
Generation	3	3	1	1	1	1	1
$SU(2)_L$	2	1	2	1	3	2	3
$U(1)_Y$	−1/2	−1	1/2	0	0	$\frac{1}{2}$	0
$\mathbb{Z}_2$	+	+	+	−	−	−	+

where $\alpha, \beta = 1, 2, 3$ represents the three flavors.

The invariant scalar potential of the model is given by

$$\begin{aligned} V = &- m_H^2 H^\dagger H + m_\eta^2 \eta^\dagger \eta + \frac{\lambda_1}{2}\left(H^\dagger H\right)^2 + \frac{\lambda_2}{2}\left(\eta^\dagger \eta\right)^2 + \frac{m_\Omega^2}{2}\mathrm{Tr}\left(\Omega^\dagger \Omega\right) + \lambda_3\left(H^\dagger H\right)\left(\eta^\dagger \eta\right) \\ &+ \lambda_4\left(H^\dagger \eta\right)\left(\eta^\dagger H\right) + \frac{\lambda_5}{2}\left[\left(H^\dagger \eta\right)^2 + h.c.\right] + \frac{\lambda^\eta}{2}\left(\eta^\dagger \eta\right)\mathrm{Tr}\left(\Omega^\dagger \Omega\right) + \frac{\lambda_1^\Omega}{2}\left(H^\dagger H\right)\mathrm{Tr}\left(\Omega^\dagger \Omega\right) \\ &+ \frac{\lambda_2^\Omega}{4}\mathrm{Tr}\left(\Omega^\dagger \Omega\right)^2 + \mu_1 H^\dagger \Omega H + \mu_2 \eta^\dagger \Omega \eta. \end{aligned} \tag{11.2}$$

It is important to emphasize that $\lambda_i \leq 1$ to uphold the perturbative nature of the theory [26]. Due to the odd charge of η field under $\mathbb{Z}_2$ cyclic symmetry, the process of spontaneous electroweak symmetry breaking is caused mainly by the H^0 and Ω^0, given by

$$\langle H^0 \rangle = \frac{v_H}{\sqrt{2}}, \quad \langle \Omega^0 \rangle = v_\Omega.$$

The scalar part is divided into two parts: $\mathbb{Z}_2$-even (H^0, $H^\pm$, Ω^0, $\Omega^\pm$) and $\mathbb{Z}_2$-odd (η^0, $\eta^\pm$) scalars. The neutral even scalars experience mixing characterized by a 2×2 mixing matrix given by

$$\begin{pmatrix} \mathcal{S}_1 \\ \mathcal{S}_2 \end{pmatrix} = \begin{pmatrix} \cos\Theta & \sin\Theta \\ -\sin\Theta & \cos\Theta \end{pmatrix} \begin{pmatrix} H^0 \\ \Omega^0 \end{pmatrix}, \text{ where } \tan(2\Theta) = \frac{4 v_H \left(\sqrt{2}\mu_1 - 2\lambda_1^\Omega v_\Omega\right)}{8\lambda_2^\Omega v_\Omega^2 - 4\lambda_1 v_H^2 + \sqrt{2}\mu_1 \frac{v_H^2}{v_\Omega}}. \tag{11.3}$$

Here $\mathcal{S}_1$ is recognized as the SM Higgs boson (HB), whereas $\mathcal{S}_2$ is the new HB in the theory. Also, the charged even scalars undergo mixing through a 2×2 mixing matrix given by

$$\begin{pmatrix} \mathcal{S}_1^\pm \\ \mathcal{S}_2^\pm \end{pmatrix} = \begin{pmatrix} \cos\delta & \sin\delta \\ -\sin\delta & \cos\delta \end{pmatrix} \begin{pmatrix} H^\pm \\ \Omega^\pm \end{pmatrix}, \text{ with } \tan(2\delta) = -\frac{4 v_\Omega v_H}{v_H^2 - 4 v_\Omega^2}. \tag{11.4}$$

The light charged scalar $\mathcal{S}_1^\pm$ is the Goldstone boson, whereas, $\mathcal{S}_2^\pm$ is the novel charged scalar. Further, in the $\mathbb{Z}_2$-odd scalars, masses of η^0 and $\eta^\pm$, are written as

$$m^2_{\eta^\pm} = m^2_\eta + \frac{1}{2}\lambda_3 v_H^2 + \frac{1}{2}\lambda^\eta v_\Omega^2 + \frac{1}{\sqrt{2}} v_\Omega \mu_2, \tag{11.5}$$

$$m^2_{\eta^R(\eta^I)} = m^2_\eta + \frac{1}{2}\lambda_3 v_H^2 + \frac{1}{2}\left(\lambda_4 \pm \lambda_5\right) v_H^2 - \frac{1}{\sqrt{2}} v_\Omega \mu_2 + \frac{1}{2}\lambda^\eta v_\Omega^2. \tag{11.6}$$

In our model, η^R can be a weakly interacting massive particle type DM candidate, which has been explored in the Refs. [17–22] in context of STScoM.

In the fermionic sector, the Lagrangian term $Y_\Omega \mathrm{Tr}(\overline{\Sigma}\Omega)F$ in Eq. (11.1) reveals that F and Σ^0 undergo mixing subsequent to the acquisition of a non-zero *vev* by the scalar triplet Ω. Consequently, the Majorana mass matrix is expressed as

$$M_\zeta = \begin{pmatrix} M_\Sigma & v_\Omega Y_\Omega \\ v_\Omega Y_\Omega & M_F \end{pmatrix}, \tag{11.7}$$

and it is diagonalized by a 2×2 matrix

$$\begin{pmatrix} \zeta_1^0 \\ \zeta_2^0 \end{pmatrix} = V \begin{pmatrix} \Sigma^0 \\ F \end{pmatrix}, \quad \text{where} \quad V = \begin{pmatrix} \cos\alpha & \sin\alpha \\ -\sin\alpha & \cos\alpha \end{pmatrix} \quad \text{so as} \quad V.M_\zeta.V = \mathrm{diag}(m_{\zeta_1^0}, m_{\zeta_2^0}) \tag{11.8}$$

Subsequently, the masses of the $\zeta^\pm$ and ζ_i^0 eigenstates at tree-level are given by

$$m_{\zeta^\pm} = M_\Sigma \quad \text{and} \quad m_{\zeta^0_{2,1}} = \frac{1}{2}\left(M_F + M_\Sigma \pm \sqrt{(M_\Sigma - M_F)^2 + 4(v_\Omega Y_\Omega)^2}\right), \tag{11.9}$$

with the mixing angle α satisfying the relation

$$\tan\left(2\alpha\right) = \frac{2 v_\Omega Y_\Omega}{M_\Sigma - M_F}. \tag{11.10}$$

Further, the neutral components of the new fermionic eigenstates generate neutrino mass radiatively, as shown in Fig. 11.1. The neutrino mass matrix is given by

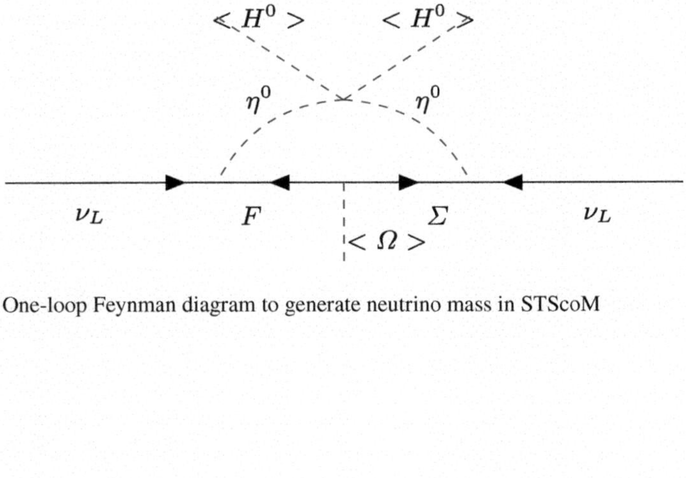

Fig. 11.1 One-loop Feynman diagram to generate neutrino mass in STScoM

$$(\mathcal{M}_\nu)_{\alpha\beta} = \sum_{i=1}^{2} \frac{h_{\alpha i} h_{\beta i} m_{\zeta_i^0}}{2(4\pi)^2} \left[\frac{\ln\left(\frac{m^2_{\zeta_i^0}}{m^2_{\eta R}}\right)}{\left(\frac{m^2_{\zeta_i^0}}{m^2_{\eta R}}\right)-1} - \frac{\ln\left(\frac{m^2_{\zeta_i^0}}{m^2_{\eta I}}\right)}{\left(\frac{m^2_{\zeta_i^0}}{m^2_{\eta I}}\right)-1} \right] = \left(h\kappa h^T\right)_{\alpha\beta} \quad (11.11)$$

where matrices h and κ are given by

$$h = \begin{pmatrix} Y_\Sigma^1 & Y_F^1 \\ Y_\Sigma^2 & Y_F^2 \\ Y_\Sigma^3 & Y_F^3 \end{pmatrix} V^T, \quad \kappa = \begin{pmatrix} \kappa_1 & 0 \\ 0 & \kappa_2 \end{pmatrix}, \text{ with } \kappa_i = \frac{m_{\zeta_i^0}}{2(4\pi)^2} \left[\frac{\ln\left(\frac{m^2_{\zeta_i^0}}{m^2_{\eta R}}\right)}{\left(\frac{m^2_{\zeta_i^0}}{m^2_{\eta R}}\right)-1} - \frac{\ln\left(\frac{m^2_{\zeta_i^0}}{m^2_{\eta I}}\right)}{\left(\frac{m^2_{\zeta_i^0}}{m^2_{\eta I}}\right)-1} \right]. \quad (11.12)$$

The matrix h can be expressed in terms of other parameters through the utilization of the Casas-Ibarra (CI) parameterization [27, 28], which can be written as

$$h = \mathcal{U}^* \sqrt{\widetilde{\mathcal{M}}} R \sqrt{\kappa}^{-1}, \text{ with } \widetilde{\mathcal{M}} = \text{diag}\,(m_1, m_2, m_3) \text{ and } R = \begin{pmatrix} 0 & 0 \\ \cos\theta & \sin\theta \\ -\sin\theta & \cos\theta \end{pmatrix} \quad (11.13)$$

where $\mathcal{U}$ is Pontecorvo-Maki-Nakagawa-Sakata matrix and θ is a complex angle.

11.3 Dark Matter Phenomenology

In STScoM, we considered the neutral component of scalar doublet (η_R) as our DM candidate. To study the DM phenomenology, we execute the model in the SARAH [29] and SPheno [30] to extract the model file and numerically solve the Boltzmann equations in the micrOMEGAs-5.0.8 [31], which yields relic density (Ωh^2) as shown in the Fig. 11.2. The black horizontal band represents the 3σ experimental range of Ωh^2 from Planck observation [5].

11.4 Leptogenesis in STScoM

In this section, we explore leptogenesis within the framework of STScoM by considering two scenarios: one involving two heavy RHN and the other involving three. The goal is to achieve leptogenesis up to the TeV scale.

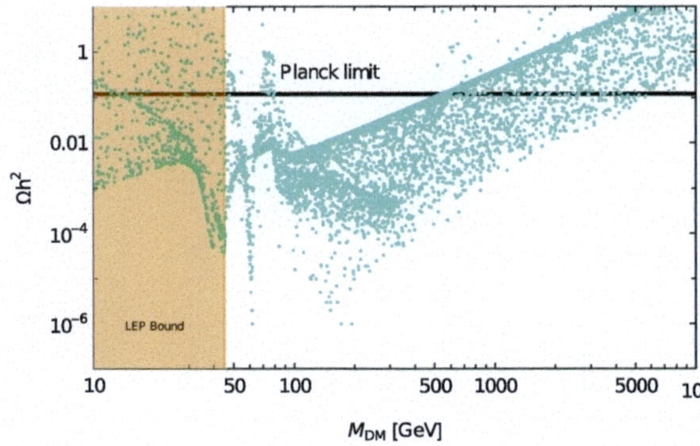

Fig. 11.2 Correlation plot between dark matter mass (M_{DM}) and dark matter relic density

11.4.1 Two Heavy RHN Case

In two heavy RHN scenario, the lepton asymmetry (η_{B-L}) is generated by CP-violating decay of heavy RHNs (F, Σ). The CP asymmetries (ϵ) arising from the decays of F and Σ are illustrated in Fig. 11.3, can be written as

$$\epsilon^{F(\Sigma)} = \frac{1}{8\pi\left(Y^{\dagger}_{F(\Sigma)}Y_{F(\Sigma)}\right)}\sum_{j\neq i}\text{Im}\left[\left(Y^{\dagger}_{F(\Sigma)}Y_{\Sigma(F)}\right)^{2}_{ij}\right]\mathcal{F}\left(\frac{M^2_{\Sigma(F)}}{M^2_{F(\Sigma)}}\right). \quad (11.14)$$

The Boltzmann equation (BE) can be written as

$$\frac{dn_F}{dz} = -D_F(n_F - n_F^{eq}), \quad \frac{dn_\Sigma}{dz} = -D_\Sigma(n_\Sigma - n_\Sigma^{eq}) - s_a(n_\Sigma^2 - (n_\Sigma^{eq})^2), \quad (11.15)$$

$$\frac{dn_{B-L}}{dz} = -\epsilon^{\Sigma}D_\Sigma(n_\Sigma - n_\Sigma^{eq}) - \epsilon^{F}D_F(n_F - n_F^{eq}) - (W_{\text{ID}}^F + W_{\text{ID}}^{\Sigma} + W_{\Delta L})n_{B-L}, \quad (11.16)$$

In this context, we denote the comoving number densities (n) associated with $B-L$ asymmetry, F, and Σ as n_{B-L}, n_N, and n_Σ, respectively. The equilibrium number densities for F and Σ are expressed as n_F^{eq} and n_Σ^{eq}. The decay terms for F and Σ are represented by D_F and D_Σ, respectively. The washout term results from the combined influence of inverse decays and washout processes that violate lepton

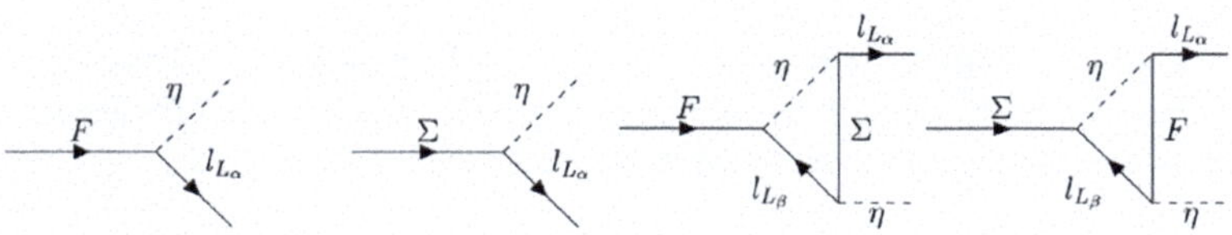

Fig. 11.3 Feynman diagrams depicting lepton number violating decays of heavy RHN (F, Σ) at tree level and one-loop vertex corrections

number (L). In Eq. (11.16), $W_{\rm ID}^{F}$ and $W_{\rm ID}^{\Sigma}$ denote the inverse decay terms for F and Σ, respectively. The gauge boson-mediated scattering term for Σ is denoted as s_a [32]. The lepton asymmetry obtained from solving the BEs is subsequently transformed into the baryon asymmetry using the relation:

$$\eta_B \approx 0.0092 \times n_{B-L} \tag{11.17}$$

11.4.2 Three Heavy RHN Case

In this case, we consider two singlet heavy RHN alongside a triplet fermion (F_1, F_2,Σ). The Yukawa matrix h takes a specific structure owing to the inclusion of an extra singlet RHN, which is given by

$$h = \begin{pmatrix} Y_{F_1}^1 & Y_{\Sigma}^1 & Y_{F_2}^1 \\ Y_{F_1}^2 & Y_{\Sigma}^2 & Y_{F_2}^2 \\ Y_{F_1}^3 & Y_{\Sigma}^3 & Y_{F_2}^3 \end{pmatrix} V^T. \tag{11.18}$$

We chose a specific mass hierarchy among the heavy RHNs ($M_{F_1} \lesssim M_\Sigma \ll M_{F_2}$). As F_2 is significantly heavier than the other two fermions, the η_{B-L} produced through the F_2 decay is washed out by the ID of F_1 and Σ. Therefore, only F_1 and Σ contribute to the generation of η_{B-L}. The ϵ parameter due to the decays of F_1 and Σ is found to be

$$\epsilon^{F_1} = \frac{1}{8\pi\left(Y_{F_1}^\dagger Y_{F_1}\right)}\left[\mathrm{Im}\left[\left(Y_{F_1}^\dagger Y_\Sigma\right)^2\right]\mathcal{F}\left(\frac{M_\Sigma^2}{M_{F_1}^2}\right) + \left(Y_{F_1}^\dagger Y_\Sigma\right)^2 \mathcal{F}\left(\frac{M_\Sigma^2}{M_{F_1}^2}\right)\right], \tag{11.19}$$

$$\epsilon^{\Sigma} = \frac{1}{8\pi\left(Y_{\Sigma}^\dagger Y_{\Sigma}\right)}\left[\mathrm{Im}\left[\left(Y_{\Sigma}^\dagger Y_{F_1}\right)^2\right]\mathcal{F}\left(\frac{M_{F_1}^2}{M_{\Sigma}^2}\right) + \left(Y_{\Sigma}^\dagger Y_{F_2}\right)^2 \mathcal{F}\left(\frac{M_{F_2}^2}{M_{\Sigma}^2}\right)\right]. \tag{11.20}$$

The relevant BEs for three RHN case are

$$\frac{dn_\Sigma}{dz} = -D_\Sigma(n_\Sigma - n_\Sigma^{eq}) - s_a(n_\Sigma^2 - (n_\Sigma^{eq})^2), \quad \frac{dn_{F_1}}{dz} = -D_F\left(n_{F_1} - n_{F_1}^{eq}\right), \tag{11.21}$$

$$\frac{dn_{B-L}}{dz} = -\epsilon^{F_1} D_{F_1}(n_{F_1} - n_{F_1}^{eq}) - \epsilon^{\Sigma} D_\Sigma(n_\Sigma - n_\Sigma^{eq}) - (W_{\rm ID}^{F_1} + W_{\rm ID}^{\Sigma} + W_{\Delta L})n_{B-L}. \tag{11.22}$$

11.5 Numerical Analysis and Results

In the case of two heavy RHN, we chose a moderate hierarchy ($M_F \lesssim M_\Sigma$) among the heavy RHN so that the out-of-equilibrium and CP-violating decay of both F and Σ contribute to the leptogenesis process. For $\lambda_5 = -1$ and $M_\Sigma = 10^{0.5} M_F$, the upper

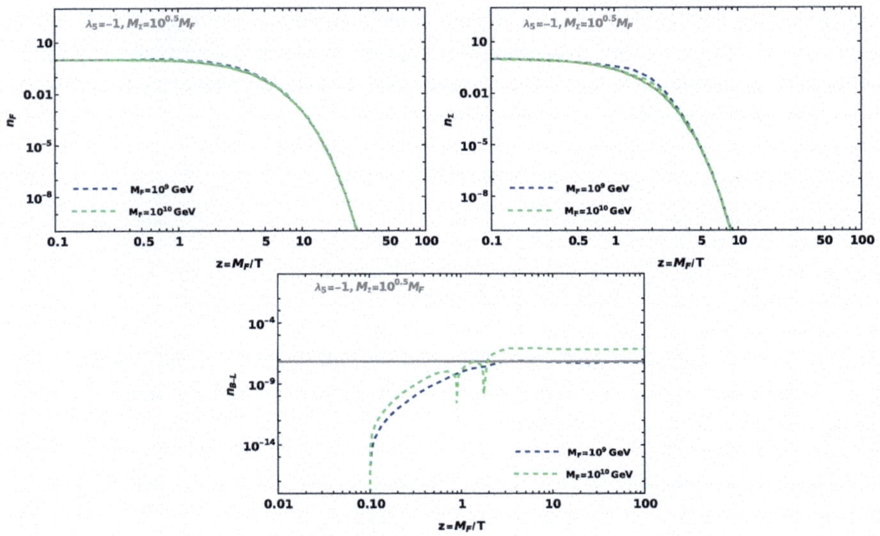

Fig. 11.4 The n of F (Σ) is depicted in the upper left (right) panel as functions of $z = M_F/T$ for various values of M_F in a scenario involving two heavy RHNs. The lepton asymmetry is shown in the lower panel. The grey horizontal line indicates the required experimental value of η_{B-L}.

panel of Fig. 11.4 illustrates that n_F and n_Σ closely match n_F^{eq} and n_Σ^{eq}. With only two heavy fermions contributing to neutrino masses, the large Yukawa couplings (Y_F,Y_Σ) ensure that decay and inverse decay maintain abundances close to equilibrium. Due to the same reason, the leptogenesis remains in the robust washout region. This leads to a high scale of leptogenesis with $M_{(F,\Sigma)} \geq 10^9$ GeV, as shown in the lower panel of Fig. 11.4. It can be observed from Fig. 11.4 that an increase in $M_{F,\Sigma}$ is correlated with a higher $B - L$ asymmetry. This correlation can be attributed to the mitigation of inverse decay washouts, which is facilitated by increasing values of $M_{(F,\Sigma)}$. It should be noted that we do not consider influence of flavor effects on the leptogenesis as their overall significance remains limited as discussed in Refs. [33, 34].

In the three heavy RHN case (F_1, F_2, Σ), the lightest neutrino eigenstate has a non-zero mass ($m \neq 0$). A specific mass hierarchy is chosen for heavy fermions to enable the CP-violating decay of both F_1 and Σ to contribute to the η_{B-L} ($M_{F_1} \lesssim M_\Sigma \ll M_{F_2}$). From the CI parameterization, the R matrix is chosen in a way that Y_{F_1} is proportional to m_1 within the h matrix. As m_1 approaches very small values, the Yukawa coupling for the lightest fermion F_1 decreases, leading to weak washout leptogenesis. The washout effects become negligible when $m_1 \lesssim 10^{-5}$ eV reduces the leptogenesis scale to the TeV range. In Fig. 11.5, the progression of n_{F_1}, n_Σ, and n_{B-L} with respect to the $z = M_{F_1}/T$ is depicted for various M_{F_1} values for fixed λ_5 in the scenario involving three RHNs. Analogous trends in the behavior of n_{F_1}, n_Σ, and n_{B-L} are observed upon varying lightest heavy RHN, akin to the patterns observed in the case of two heavy RHNs. Moreover, prior research has

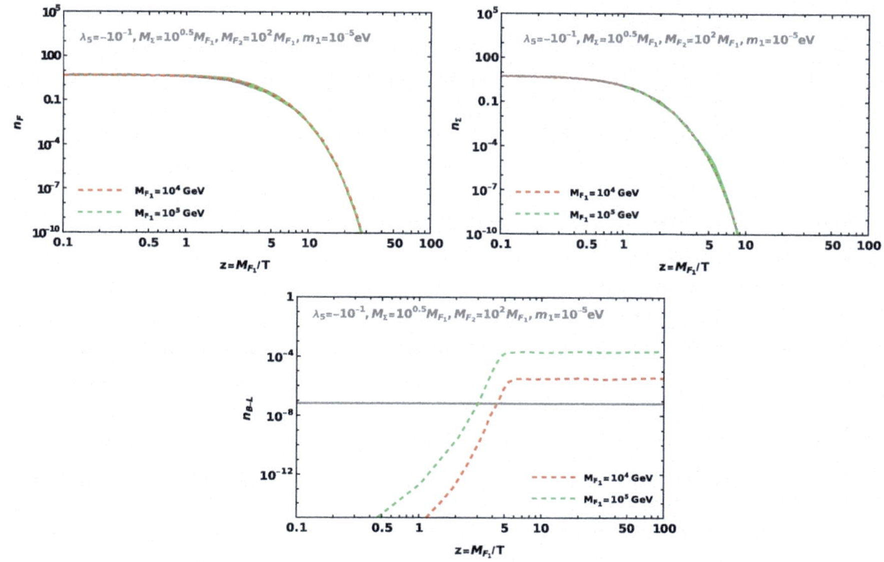

Fig. 11.5 The comoving number density of F_1 (Σ) is depicted in the upper left (right) panel as functions of $z = M_{F_1}/T$ for various values of M_{F_1} in a scenario involving three heavy RHNs. The lepton asymmetry is shown in the lower panel. The grey horizontal line indicates the required experimental value of η_{B-L}.

demonstrated that the lightest RHN exclusively contributes to the η_{B-L}, with heavier fermions being dismissed owing to their considerable washout effects. Nevertheless, our investigation posits that the heavier RHN can substantially contribute to the η_{B-L} by choosing moderately hierarchical heavy RHN masses. In Fig. 11.6, the variation of the η_{B-L} with $z = M_{F_1}/T$ is shown for two scenarios. The first scenario involves the combined system of $F_1 + \Sigma$, depicted by the red dashed line, while the second scenario involves the F_1 only, represented by the blue dashed line. In the

Fig. 11.6 The lepton asymmetry (n_{B-L}) for F_1(blue dashed line) and $F_1 + \Sigma$(red dashed line) as a function of $z = M_{F_1}/T$. The required experimental value of n_{B-L} is indicated by grey horizontal line

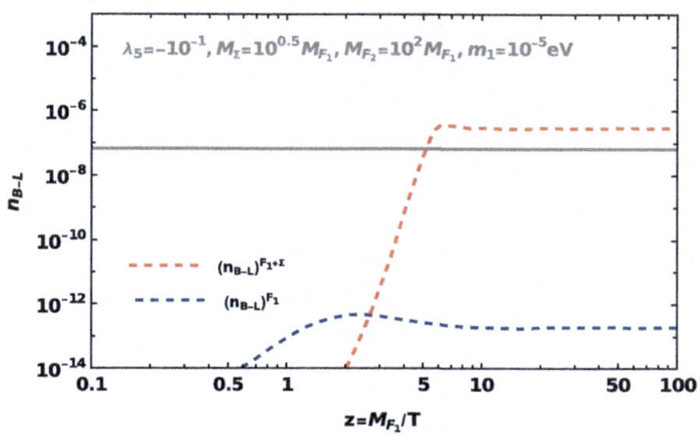

latter scenario, η_{B-L} remains significantly lower compared to the equilibrium value of n_{B-L}. However, when we consider the moderately hierarchical arrangement of heavy RHNs ($M_{F_1} \lesssim M_\Sigma < M_{F_2}$), the contribution of the triplet fermion to achieving the final lepton asymmetry becomes substantial, as evident from Fig. 11.6. Also, it is possible to detect a TeV scale triplet fermion in future collider experiments [35, 36].

11.6 Conclusion

This study investigated the potential for TeV-scale leptogenesis within the context of STScoM. The TeV-scale leptogenesis has been explored in the minimal version of scotogeic model where CP-violating decay of the lightest RHN (with hierarchical mass spectrum) take part in the process of leptogenesis. However, in the present work, we consider the moderately hierarchical masses of RHNs so that heavier particles can also contribute to the $B-L$ asymmetry. We considered two scenarios: one with two heavy RHNs (F, Σ) and the other with three (F_1, F_2, Σ). We explicitly differentiated between these cases. In the two RHN case, the lightest active neutrino remains massless. We chose both heavy RHNs to be moderately hierarchical so that the out-of-equilibrium decay of both F and Σ contribute to the $B-L$ asymmetry. The Casas-Ibarra parameterization necessitates large Yukawa couplings (Y_F,Y_Σ) to generate light neutrino masses. Consequently, the scale of leptogenesis is estimated to be around 10^{10} GeV. To lower the leptogenesis scale, we consider three heavy RHN, where we chose the specific hierarchy among the heavy fermions ($M_{F_1} \lesssim M_\Sigma \ll M_{F_2}$). For $m_1 \lesssim 10^{-4}$ eV, the impact of washouts becomes negligible and, therefore, significantly reduces the scale of leptogenesis up to the TeV scale. It is worth noting that a fermion triplet with a TeV scale could potentially be detected in upcoming collider experiments.

References

1. G. Steigman, Int. J. Mod. Phys. E **15**, 1–36 (2006)
2. R.H. Cyburt, B.D. Fields, K.A. Olive, E. Skillman, Astropart. Phys. **23**, 313–323 (2005)
3. W. Hu, S. Dodelson, Ann. Rev. Astron. Astrophys. **40**, 171–216 (2002)
4. R.L. Workman et al., [Particle Data Group], PTEP **2022**, 083C01 (2022)
5. N. Aghanim et al., Planck. Astron. Astrophys. **641**, A6 (2020)
6. A.D. Sakharov, Pisma. Zh. Eksp. Teor. Fiz. **5**, 32–35 (1967)
7. M. Fukugita, T. Yanagida, Phys. Lett. B **174**, 45–47 (1986)
8. V.A. Kuzmin, V.A. Rubakov, M.E. Shaposhnikov, Phys. Lett. B **155**, 36 (1985)
9. M.A. Luty, Phys. Rev. D **45**, 455–465 (1992)
10. M. Plumacher, Z. Phys. C **74**, 549–559 (1997)
11. L. Covi, E. Roulet, F. Vissani, Phys. Lett. B **384**, 169–174 (1996)
12. S. Davidson, A. Ibarra, Phys. Lett. B **535**, 25–32 (2002)
13. M. Hirsch, R.A. Lineros, S. Morisi, J. Palacio, N. Rojas, J.W.F. Valle, JHEP **10**, 149 (2013)
14. L. Singh, D. Mahanta, S. Verma, arXiv:2309.12755 [hep-ph]

15. J. Fiaschi, M. Klasen, S. May, JHEP **05**, 015 (2019)
16. S. Choubey, S. Khan, M. Mitra, S. Mondal, Eur. Phys. J. C **78**(4), 302 (2018)
17. A. Batra, S.K.A., S. Mandal, R. Srivastava, arXiv:2204.09376 [hep-ph]
18. M.A. Díaz, N. Rojas, S. Urrutia-Quiroga, J.W.F. Valle, JHEP **08**, 017 (2017)
19. G.J. Ding, J.N. Lu, J.W.F. Valle, Phys. Lett. B **815**, 136122 (2021)
20. D. Restrepo, A. Rivera, JHEP **04**, 134 (2020)
21. I.M. Ávila, V. De Romeri, L. Duarte, J.W.F. Valle, Eur. Phys. J. C **80**(10), 908 (2020)
22. G. Bélanger, S. Choubey, R.M. Godbole, S. Khan, M. Mitra, A. Roy, JHEP **11**, 133 (2022)
23. E. Ma, Phys. Rev. D **73**, 077301 (2006)
24. T. Aaltonen et al., CDF. Science **376**(6589), 170–176 (2022)
25. T. Hugle, M. Platscher, K. Schmitz, Phys. Rev. D **98**(2), 023020 (2018)
26. K. Kannike, Eur. Phys. J. C **72**, 2093 (2012)
27. J.A. Casas, A. Ibarra, Nucl. Phys. B **618**, 171–204 (2001)
28. A. Ibarra, G.G. Ross, Phys. Lett. B **591**, 285–296 (2004)
29. F. Staub, Adv. High Energy Phys. **2015**, 840780 (2015)
30. W. Porod, F. Staub, Comput. Phys. Commun. **183**, 2458–2469 (2012)
31. G. Bélanger, F. Boudjema, A. Pukhov, A. Semenov, Comput. Phys. Commun. **192**, 322–329 (2015)
32. A. Biswas, M. Chakraborty, S. Khan, JCAP **08**, 026 (2023)
33. P.S.B. Dev, P. Di Bari, B. Garbrecht, S. Lavignac, P. Millington, D. Teresi, Int. J. Mod. Phys. A **33**, 1842001 (2018)
34. D. Mahanta, D. Borah, JCAP **11**, 021 (2019)
35. S. Ashanujjaman, K. Ghosh, Phys. Lett. B **825**, 136889 (2022)
36. T. Li, H. Qin, C.Y. Yao, M. Yuan, Phys. Rev. D **106**(3), 035021 (2022)

Chapter 12
A Spotlight on the μ-τ *Mixed Symmetry*

Manash Dey, Pralay Chakraborty, and Subhankar Roy

Abstract We study the elegant neutrino mass matrix texture referred to as the μ-τ *mixed symmetry*, emphasizing two straightforward correlations among its elements. A comprehensive analysis is conducted to explore its phenomenological implications. The said texture is motivated within the framework of Seesaw mechanism and A_4 symmetry.

12.1 Introduction

Neutrino oscillation is a peculiar phenomenon in Particle Physics which involves the flavour transformation of neutrinos when they travel from source to detector. This phenomenon was first conceptualized by B. Pontecorvo in [1]. In the recent past, various experiments, such as SNO [2], KamLAND [3] and Super-Kamiokande [4], witnessed the phenomenon of neutrino oscillation. The latter shares the fact that the three flavour neutrino states ($\nu_{l=e,\mu,\tau}$) are an admixture of three neutrino mass eigenstates ($\nu_{i=1,2,3}$) having mass eigenvalues ($m_{i=1,2,3}$). The neutrino oscillation experiments aim to measure six observational parameters: solar mixing angle (θ_{12}), reactor mixing angle (θ_{13}), atmospheric mixing angle (θ_{23}), and two mass squared differences, $\Delta m^2_{21} = m^2_2 - m^2_1$ and $\Delta m^2_{31} = |m^2_3 - m^2_1|$. Apart from these, the Dirac CP-violating phase (δ) is also a crucial parameter under examination. Beyond these, there are other two observational parameters termed Majorana phases (α and β) on which the oscillation experiments stay silent. It is noteworthy that the experiments have not yet determined the octant of θ_{23}, neutrino mass ordering, and the precise

M. Dey (✉) · P. Chakraborty · S. Roy
Department of Physics, Gauhati University, Gopinath Bordoloi Nagar, Jalukbari, Guwahati 781014, Assam, India
e-mail: manashdey@gauhati.ac.in

P. Chakraborty
e-mail: pralay@gauhati.ac.in

S. Roy
e-mail: subhankar@gauhati.ac.in

J. W. F. Valle et al. (eds.), *Proceedings of the International Conference on Future Prospects in Neutrino and Astroparticle Physics*, Springer Proceedings in Physics 361,
https://doi.org/10.1007/978-981-97-7441-8_12

range of δ. To address these problems in neutrino physics phenomenology, we need to move beyond the Standard Model (SM). The extension of SM is evident through the neutrino mass matrix (M_ν) present in the Yukawa Lagrangian ($\mathcal{L}_Y$). The M_ν contains crucial information related to all the physical parameters mentioned above, and the theorists in general try to predict a promising neutrino mass matrix starting from different discrete symmetries [5–11]. Several phenomenological ideas such as $\mu - \tau$ symmetry [6, 12, 13], texture zeroes [14], hybrid textures [15, 16], and vanishing minors [17, 18] are ascribed to the neutrino mass matrix and are studied thoroughly from both *bottom-up* and *top-down* viewpoints. It is to be underlined that all these methodologies aim to reduce the number of independent parameters within the mass matrix. This is achieved either by introducing correlations among the mass matrix elements or by restricting the mass matrix elements to assume specific values. We declare that our work is based on the assumption that neutrinos are Majorana particles [19, 20] and this assumption forces M_ν to take a complex symmetric structure.

12.2 The Neutrino Mass Matrix and Unphysical Phases

Apart from M_ν, the fundamental aspects of neutrino physics phenomenology revolve around a crucial quantity, V, which refers to the Pontecorvo-Maki-Nakagawa-Sakata matrix (PMNS matrix) [21]. In this context, we consider two basis systems: the *flavour basis*, where the charged lepton mass matrix M_l is diagonal, and the *symmetry basis*, where it is not. Generally, the PMNS matrix is defined in the flavour basis and serves to diagonalize M_ν: $V^T.M_\nu.V = diag(m_1, m_2, m_3)$. The PMNS matrix V is a 3×3 unitary matrix, encompassing nine free parameters. The parameterization of V according to the PDG (Particle Data Group) is presented below [22].

$$V = P_\phi.U.P_M, \tag{12.1}$$

where $P_\phi = diag(e^{i\phi_1}, e^{i\phi_2}, e^{i\phi_3})$ and $P_M = diag(e^{i\alpha}, e^{i\beta}, 1)$. The matrix U is represented as shown in the following,

$$U = \begin{pmatrix} c_{12}c_{13} & c_{13}s_{12} & e^{-i\delta}s_{13} \\ -c_{23}s_{12} - e^{i\delta}c_{12}s_{13}s_{23} & c_{12}c_{23} - e^{i\delta}s_{12}s_{13}s_{23} & c_{13}s_{23} \\ s_{12}s_{23} - e^{i\delta}c_{12}c_{23}s_{13} & -c_{12}s_{23} - e^{i\delta}c_{23}s_{12}s_{13} & c_{13}c_{23} \end{pmatrix} \tag{12.2}$$

where $s_{ij} = \sin\theta_{ij}$ and $c_{ij} = \cos\theta_{ij}$.

In principle, the three arbitrary phases ($\phi_{i=1,2,3}$) can be eliminated from the picture by redefining the charged lepton fields in terms of these phases. However, it is to be noted that such a redefinition of the charged lepton fields is purely a matter of choice. The neutrino mass matrix M_ν is a 3×3 complex symmetric matrix; therefore, in general, it comprises twelve parameters. Hence, there exists a possibility for M_ν to incorporate the three unphysical phases. When the PMNS matrix V is derived from M_ν, one needs to structure V in the standard form, as specified in (12.1). This process allows the isolation of the unphysical phases, thereby aiding in extracting the information about mixing angles and other physical phases. However, in the case of M_ν, details about the unphysical phases are intrinsically embedded within it. To simplify the analysis, we distinguish between the general neutrino mass matrix M_ν and $\tilde{M}_\nu$, the latter of which excludes the unphysical phases. These two matrices are connected through the following transformation,

$$M_\nu = P_\phi^*.\tilde{M}_\nu.P_\phi^*. \tag{12.3}$$

Hence, the scenarios may differ depending on whether we are dealing with $M\nu$ or $\tilde{M}_\nu$. Because, for the former case, the involvement of unphysical phases is indisputable. To understand the importance [23, 24] of the unphysical phases while defining a neutrino mass matrix texture, we take the example of the μ-τ reflection symmetry [23, 25–27] which says

$$M = \begin{bmatrix} A & B & B^* \\ B & C & D \\ B^* & D & C^* \end{bmatrix}, \tag{12.4}$$

where, except A and D, all other elements are complex quantities. If the said symmetry is true for $\tilde{M}_\nu$, then following relations must be valid,

$$Arg\,[(\tilde{M}_\nu)_{e\tau}] - Arg\,[(\tilde{M}_\nu)_{e\mu}] = 0, \tag{12.5}$$

$$Arg\,[(\tilde{M}_\nu)_{\mu\mu}] - Arg\,[(\tilde{M}_\nu)_{\tau\tau}] = 0, \tag{12.6}$$

$$Arg\,[(\tilde{M}_\nu)_{ee}] = 0, \tag{12.7}$$

$$Arg\,[(\tilde{M}_\nu)_{\mu\tau}] = 0, \tag{12.8}$$

whereas, for the same symmetry when imposed on M_ν, the following two relations appear,

$$\begin{aligned} Arg\,[(\tilde{M})_{\mu\tau}] &= Arg\,[(\tilde{M})_{e\mu}] + Arg\,[(\tilde{M})_{e\tau}] - Arg\,[(\tilde{M})_{ee}], \\ &= \frac{1}{2}\,[Arg\,[(\tilde{M})_{\mu\mu}] + Arg\,[(\tilde{M})_{\tau\tau}]. \end{aligned} \tag{12.9}$$

Needless to mention the following two relations,

$$|(\tilde{M})_{e\mu}| = |(\tilde{M})_{e\tau}|, \tag{12.10}$$

$$|(\tilde{M})_{\mu\mu}| = |(\tilde{M})_{\tau\tau}|, \tag{12.11}$$

hold good for the above discussion. In our further discussion, we shall stick to M_ν. The general μ-τ symmetry focuses on correlations: $(M_\nu)_{e\tau} = (M\nu)_{e\mu}$ and $(M\nu)_{\tau\tau} = (M\nu)_{\mu\mu}$. Yet, these connections are partially successful because they don't align with the reality that $\theta 13 \neq 0$. So, looking for afresh correlations motivated by any underlying symmetry is desirable. In this regard, a unique texture of the neutrino mass matrix in the flavour basis was put forward in Ref. [24] as follows:

$$M_\nu = \begin{bmatrix} A & -B & B \\ -B & C & D \\ B & D & C^* \end{bmatrix}. \tag{12.12}$$

The texture highlights the following correlations:$(M_\nu)_{e\tau} = -(M_\nu)_{e\mu}$ and $(M_\nu)_{\tau\tau} = (M_\nu)^*_{\mu\mu}$. Here, the elements A, B, C, and D are in general complex quantities. The above texture was coined as μ-τ *mixed symmetry* [24], to demarcate it from the $\mu - \tau$ symmetry [13, 26, 28], $\mu - \tau$ reflection symmetry [23, 25–27], and $\mu - \tau$ antisymmetry [25, 26]. For the μ-τ *mixed symmetry*, the following two relations hold good,

$$\begin{aligned} \phi_2 = & -\frac{\pi}{2} + \frac{1}{4}\left(Arg[(\tilde{M}_\nu)_{\mu\mu}] + Arg[(\tilde{M}_\nu)_{\tau\tau}]\right) \\ & -\frac{1}{2}\left(Arg[(\tilde{M}_\nu)_{e\tau}] - Arg[(\tilde{M}_\nu)_{e\mu}]\right), \end{aligned} \tag{12.13}$$

$$\begin{aligned} \phi_3 = & \frac{\pi}{2} + \frac{1}{4}\left(Arg[(\tilde{M}_\nu)_{\mu\mu}] + Arg[(\tilde{M}_\nu)_{\tau\tau}]\right) \\ & +\frac{1}{2}\left(Arg[(\tilde{M}_\nu)_{e\tau}] - Arg[(\tilde{M}_\nu)_{e\mu}]\right). \end{aligned} \tag{12.14}$$

First, we shall evaluate how well the posited texture aligns with the experiment and later, we will delve into understanding the first principle associated with this symmetry.

12.3 Numerical Analysis and Results

The two constraints, appearing in the mass matrix in (12.12), tell about four equations as shown below,

$$Re\,[(M_\nu)_{e\mu}] + Re\,[(M_\nu)_{e\tau}] = 0, \tag{12.15}$$

$$Im\,[(M_\nu)_{e\mu}] + Im\,[(M_\nu)_{e\tau}] = 0, \tag{12.16}$$

$$Re\,[(M_\nu)_{\mu\mu}] - Re\,[(M_\nu)_{\tau\tau}] = 0, \tag{12.17}$$

$$Im\,[(M_\nu)_{\mu\mu}] + Im\,[(M_\nu)_{\tau\tau}] = 0, \tag{12.18}$$

and hence, one expects the prediction of four observable parameters. In the light of normal ordering and based on the oscillation data [29] and cosmological upper bound on the sum of three neutrino masses which is $0.12\,eV$ [30], we see that the upper and lower bounds of m_3 are approximately $0.06eV$ and $0.05eV$, respectively. For our analysis, we set m_3 to $0.06eV$. We solve the four equations namely (12.15)–(12.18) to generate a sufficiently large set of random numbers for θ_{12}, θ_{13}, θ_{23}, and δ. The associated correlation plots are highlighted in Fig. 12.1 a, b, and c.

In addition, we extract the information of the effective Majorana neutrino mass ($m_{\beta\beta}$),

$$m_{\beta\beta} = |V_{ei}^2 m_i|, \quad \text{where, } i = 1, 2, 3. \tag{12.19}$$

The $m_{\beta\beta}$ is an important observable parameter in neutrinoless double beta ($0\nu\beta\beta$)-decay [31, 32]. So far, numerous experiments have set the upper bounds for this parameter, some of them are SuperNEMO(Se^{82}): $67-131$ meV, GERDA(Ge^{76}): $104-228$ meV, EXO-200(Xe^{136}): $111-477$ meV, CUORE(Te^{130}) : $75-350$ meV, and KamLAND-Zen(Xe^{136}): $61-165$ meV [32–38]. It is interesting to note that the predictions of the θ_{ij}'s and δ are in good agreement with the experimental 3σ bounds [29]. We obtain sharp bounds on θ_{23} and δ; specifically, $45.08° \le \theta_{23} \le 45.10°$ (θ_{23} lies in the upper octant) and $395.02° \le \delta \le 395.75°$, respectively. In addition to these, the bound on $m_{\beta\beta}$ is obtained as $18.82\,\text{meV} \le m_{\beta\beta} \le 59.77\,\text{meV}$. The correlation plots for $m_{\beta\beta}$ vs the other mixing angles are highlighted in Fig. 12.1 d, e, and f. To achieve these predictions, the bounds for $\Delta m_{21}^2 = m_2^2 - m_1^2$ and $\Delta m_{31}^2 = |m_3^2 - m_1^2|$ are tuned within the 3σ ranges as per Ref. [29], and the other input arbitrary phases are assigned certain numerical bounds (see Table 12.1). In Table 12.2, we tabulate the predictions of the work in the light of normal ordering. Similarly, the analysis is extended for an inverted ordering of neutrino masses; for this, the numerical values of the input parameters are shown in Table 12.3. The predictions are consistent with the experimental data [29], and they are tabulated in Table 12.4.

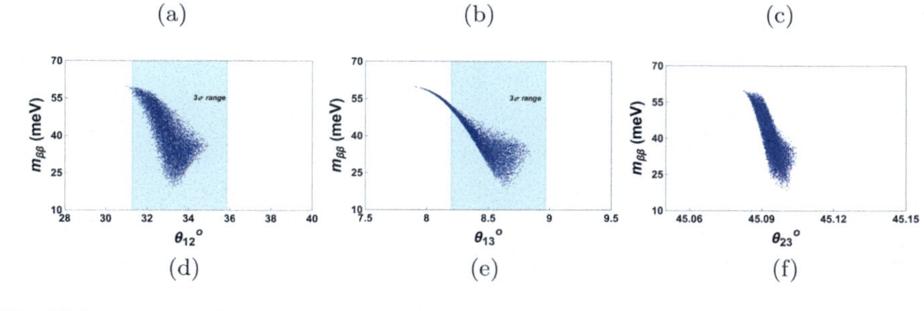

Fig. 12.1 Represents the correlation plots between **a** δ and θ_{23} **b** θ_{12} and θ_{23} **c** θ_{13} and θ_{12} **d** $m_{\beta\beta}$ and θ_{12} **e** $m_{\beta\beta}$ and θ_{13} **f** $m_{\beta\beta}$ and θ_{23}, for normal ordering of neutrino masses with $m_3 = 0.06\,eV$

Table 12.1 Displays the numerical values of the inputs parameters for normal ordering of neutrino masses

Input parameters	Numerical values
m_3/eV	0.060
$\frac{\Delta m_{21}^2}{10^{-5} eV^2}$	7.00–7.45
$\frac{\Delta m_{31}^2}{10^{-3} eV^2}$	2.49–2.53
$\phi_1/°$	0–360
$\phi_2/°$	72.80–73
$\phi_3/°$	128.90–129.10
$\alpha/°$	159–159.20
$\beta/°$	142.70–142.90

Table 12.2 Displays the maximum and minimum values of the predicted parameters for normal ordering of neutrino masses

Predictions	Minimum value	Maximum value
$\theta_{12}/°$	30.92	34.94
$\theta_{13}/°$	7.90	8.84
$\theta_{23}/°$	45.08	45.10
$\delta/°$	395.02	395.75
$m_{\beta\beta}/\text{meV}$	18.82	59.77

Table 12.3 Displays the numerical values of the inputs parameters for inverted ordering of neutrino masses

Input parameters	Numerical values
m_3/eV	0.019
$\frac{\Delta m_{21}^2}{10^{-5} eV^2}$	7.41–7.44
$\frac{\Delta m_{31}^2}{10^{-3} eV^2}$	–2.50–(–2.49)
$\phi_1/°$	44.43–44.6
$\phi_2/°$	29.42–29.58
$\phi_3/°$	171.94–172.10
$\alpha/°$	85.30–85.97
$\beta/°$	81.30–81.70

The associated correlation plots are highlighted in Fig. 12.2 a, b, c, d, e, and f. It is to be noted that for the inverted ordering of neutrino masses, θ_{23} is also in the upper octant. The next section examines whether the proposed texture is driven by any underlying symmetry.

Table 12.4 Displays the maximum and minimum values of the predicted parameters for inverted ordering of neutrino masses

Predictions	Mininum value	Maximum value
$\theta_{12}/°$	30.60	37.32
$\theta_{13}/°$	7.22	9.11
$\theta_{23}/°$	45.09	45.13
$\delta/°$	181.09	183.47
$m_{\beta\beta}$/meV	0.22	49.18

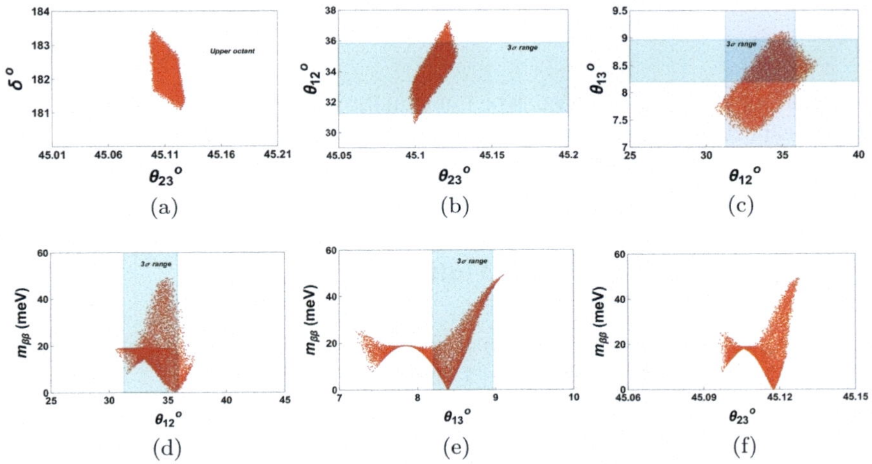

Fig. 12.2 Represents the correlation plots between **a** δ and θ_{23} **b** θ_{12} and θ_{23} **c** θ_{13} and θ_{12} **d** $m_{\beta\beta}$ and θ_{12} **e** $m_{\beta\beta}$ and θ_{13} **f** $m_{\beta\beta}$ and θ_{23}, for inverted ordering of neutrino masses with $m_3 = 0.019\,eV$

Table 12.5 The transformation properties of the field contents under $A_4 \times Z_3 \times SU(2)_L$

Fields	D_{l_L}	l_R	ν_{l_R}	ϕ	ψ	ζ	Δ
A_4	3	$(1, 1'', 1')$	$(1, 1', 1'')$	3	3	1	3
Z_3	1	$(\omega^2, \omega^2, \omega^2)$	(ω, ω, ω)	ω	ω^2	ω	1
$SU(2)_L$	2	1	1	2	2	1	3

12.4 Symmetry Realization

We try to obtain the texture appearing in (12.12) by extending the field content of SM. In this regard, we introduce right-handed neutrinos (ν_{l_R}), complex scalar field doublets (ψ), a scalar singlet (ζ), and scalar triplets (Δ), alongside the existing left-handed lepton doublets (D_{l_L}), right-handed lepton singlets (l_R), and complex Higgs doublets (ϕ). Further, we associate the discrete flavour symmetry A_4 [5, 39] and define the transformation properties of the field contents as shown in Table 12.5. In addition, we supplement the framework with Z_3 symmetry to cut short some undesired terms in $\mathcal{L}_Y$ which are invariant under A_4. The $A_4 \times Z_3 \times SU(2)_L$ invariant Yukawa Lagrangian is formulated in the S basis [5, 39] and it is presented as shown below,

$$-\mathcal{L}_Y = y_e(\bar{D}_{l_L}\phi)_1\, e_R + y_\mu(\bar{D}_{l_L}\phi)_{1'}\,\mu_{R_{1''}} + y_\tau\,(\bar{D}_{l_L}\phi)_{1''}\tau_{R_{1'}} + y_d\,[(\bar{D}_{l_L}\psi)_1\,\nu_{eR} + (\bar{D}_{l_L}\psi)_{1'}\,\nu_{\tau_{R_{1''}}} + (\bar{D}_{l_L}\psi)_{1''}\,\nu_{\mu_{R_{1'}}}] + \frac{1}{2}\,y_{r_1}(\nu^{\bar{c}}_{eR}\,\nu_{eR})_1\,\zeta + \frac{1}{2}\,y_{r_2}\,[(\nu^{\bar{c}}_{\mu R}\,\nu_{\tau R}) + (\nu^{\bar{c}}_{\tau R}\,\nu_{\mu R})]\,\zeta + y_{t_2}\,(\bar{D}_{l_L}\,D^c_{l_L})_{3_S}\,\Delta + h.c. \tag{12.20}$$

The details of the A_4 symmetry can be found in the Refs. [5, 39, 40]. We choose the vacuum expectation values (vev) in the following way, $\langle\phi\rangle_\circ = v_\phi\,(1, 1, 1)^T$, $\langle\psi\rangle_\circ = v_\psi\,(1, 1, 1)^T$, $\langle\zeta\rangle_\circ = v_\zeta$, and obtain the charged lepton mass matrix, Dirac neutrino mass matrix (M_D), and right-handed neutrino mass matrix (M_R) as shown below,

$$M_l = v_\phi \begin{bmatrix} y_e & y_\mu & y_\tau \\ y_e & \omega^2\, y_\mu & \omega\, y_\tau \\ y_e & \omega\, y_\mu & \omega^2\, y_\tau \end{bmatrix}, \quad M_D = y_d v_\psi \begin{bmatrix} 1 & 1 & 1 \\ 1 & \omega & \omega^2 \\ 1 & \omega^2 & \omega \end{bmatrix}, \quad M_R = \begin{bmatrix} a & 0 & 0 \\ 0 & 0 & b \\ 0 & b & 0 \end{bmatrix} \tag{12.21}$$

where $a = y_{r_1} v_\zeta$, $b = y_{r_2} v_\zeta$, $\omega = e^{i\,2\pi/3}$, and $\omega^2 = \omega^*$. We fetch the contribution from Type-I seesaw mechanism [41, 42] as $M_{T_1} = -M_D M_R^{-1} M_D^T$, and on choosing the vev, $\langle\Delta\rangle_\circ = v_\Delta(0, 1, -1)^T$, we obtain the contribution from Type-II mechanism [43] as shown below,

$$M_{T_2} = y_{t_2} v_\Delta \begin{bmatrix} 0 & -1 & 1 \\ -1 & 0 & 0 \\ 1 & 0 & 0 \end{bmatrix}. \tag{12.22}$$

The neutrino mass matrix is constructed in the symmetry basis in the following way: $M_{\nu s} = M_{T_1} + M_{T_2}$. To switch to the flavour basis, we diagonalize M_l appearing in (12.21): $M_l^{diag} = U_{l_L}^\dagger M_l U_{l_R}$, where $M_l^{diag} = \sqrt{3}\,v_\phi\,diag\,(y_e, y_\mu, y_\tau)$. The U_{l_L} and U_{l_R} represent the left-handed and right-handed charged lepton diagonalization matrices, respectively. They can be expressed as

$$U_{l_L} = \frac{1}{\sqrt{3}} \begin{bmatrix} e^{i\eta} & e^{-i\kappa} & e^{-i\xi} \\ e^{i\eta} & \omega^2\, e^{-i\kappa} & \omega\, e^{-i\xi} \\ e^{i\eta} & \omega\, e^{-i\kappa} & \omega^2\, e^{-i\xi} \end{bmatrix}, \tag{12.23}$$

$$U_{l_R} = diag(e^{i\eta}, e^{-i\kappa}, e^{-i\xi}). \tag{12.24}$$

The presence of arbitrary phases, η, κ, and ξ, in U_{l_L} arises from the non-uniqueness in the choice of eigenvectors or diagonalizing matrices, and this form of the charged lepton diagonalization matrix is more general. The presence of these phases makes U_{l_L} special, and proper choices of them help obtain different correlations within the final neutrino mass matrix. By keeping η arbitrary and setting $\kappa = \xi = \frac{\pi}{2}$, the final neutrino mass matrix in the flavour basis, after charged lepton correction and simplification, takes a form similar to that of the posited texture in (12.12),

$$M_\nu = U_{l_L}^T M_{\nu_s} U_{l_L} = \begin{bmatrix} W & -X & X \\ -X & Y & Z \\ X & Z & Y^* \end{bmatrix}. \tag{12.25}$$

12.5 Summary and Discussions

We have studied a promising texture for the neutrino mass matrix entitled $\mu - \tau$ mixed symmetry. For our analysis, we adhere to both normal and inverted ordering of neutrino masses such that $m_1 + m_2 + m_3$ is consistent with the cosmological upper bound. The said texture not only constrains the value of δ but also successfully predicts mixing angles and $m_{\beta\beta}$. These predictions are in good agreement with the experimental data, and it is to be emphasized that θ_{23} is in the upper octant for both the ordering of neutrino masses. We have tried to obtain the said texture in the framework of $A_4 \times Z_3 \times SU(2)_L$ by introducing additional scalar fields and incorporating Type-I and Type-II seesaw mechanisms. Additionally, we've highlighted how the arbitrary phases within the charged lepton diagonalizing matrix play a crucial role in achieving the desired structure of the neutrino mass matrix.

Acknowledgements The research work of MD is supported by Council of Scientific and Industrial Research (CSIR), Government of India, through a NET Junior Research Fellowship vide grant No. 09/0059(15346)/2022-EMR-I. PC acknowledges support from Innovation in Science Pursuit for Inspired Research (INSPIRE), Department of Science and Technology, Government of India, New Delhi, vide grant No. IF190651.

References

1. B. Pontecorvo, Sov. Phys. JETP **6**, 429 (1957)
2. Q.R. Ahmad et al., Phys. Rev. Lett. **89**, 011301 (2002). https://doi.org/10.1103/PhysRevLett.89.011301
3. K. Eguchi et al., Phys. Rev. Lett. **90**, 021802 (2003). https://doi.org/10.1103/PhysRevLett.90.021802
4. Y. Fukuda et al., Phys. Rev. Lett. **82**, 2644 (1999). https://doi.org/10.1103/PhysRevLett.82.2644
5. G. Altarelli, F. Feruglio, Rev. Mod. Phys. **82**, 2701 (2010). https://doi.org/10.1103/RevModPhys.82.2701
6. P.F. Harrison, D.H. Perkins, W.G. Scott, Phys. Lett. B **530**, 167 (2002). https://doi.org/10.1016/S0370-2693(02)01336-9

7. E. Ma, Mod. Phys. Lett. A **21**, 1917 (2006). https://doi.org/10.1142/S0217732306021190
8. P. Chakraborty, M. Dey, S. Roy, J. Phys. G **51**(1), 015003 (2024). https://doi.org/10.1088/1361-6471/ad074c
9. M. Dey, S. Roy, A realistic neutrino mixing scheme arising from A4 symmetry. EPL **147**, 14002 (2023). https://inspirehep.net/literature/2651457
10. G. Altarelli, F. Feruglio, L. Merlo, JHEP **05**, 020 (2009). https://doi.org/10.1088/1126-6708/2009/05/020
11. M. Dey, S. Roy, (2023)
12. S.F. King, C.C. Nishi, Phys. Lett. B **785**, 391 (2018). https://doi.org/10.1016/j.physletb.2018.08.056
13. Y. Lin, Nucl. Phys. B **824**, 95 (2010). https://doi.org/10.1016/j.nuclphysb.2009.08.018
14. P.O. Ludl, W. Grimus, JHEP **07**, 090 (2014). https://doi.org/10.1007/JHEP07(2014)090. [Erratum: JHEP 10, 126 (2014)]
15. R. Kalita, D. Borah, Int. J. Mod. Phys. A **31**(06), 1650008 (2016). https://doi.org/10.1142/S0217751X16500081
16. M. Singh, Adv. High Energy Phys. **2018**, 5806743 (2018). https://doi.org/10.1155/2018/5806743
17. E.I. Lashin, N. Chamoun, Phys. Rev. D **80**, 093004 (2009). https://doi.org/10.1103/PhysRevD.80.093004
18. T. Araki, J. Heeck, J. Kubo, JHEP **07**, 083 (2012). https://doi.org/10.1007/JHEP07(2012)083
19. S.M. Bilenky, Universe **6**(9), 134 (2020). https://doi.org/10.3390/universe6090134
20. J.B. Albert et al., Nature **510**, 229 (2014). https://doi.org/10.1038/nature13432
21. Z. Maki, M. Nakagawa, S. Sakata, Prog. Theor. Phys. **28**, 870 (1962). https://doi.org/10.1143/PTP.28.870
22. P.A. Zyla, et al., PTEP **2020**(8), 083C01 (2020). https://doi.org/10.1093/ptep/ptaa104
23. Z.z. Xing, Rept. Prog. Phys. **86**(7), 076201 (2023). https://doi.org/10.1088/1361-6633/acd8ce
24. M. Dey, P. Chakraborty, S. Roy, Phys. Lett. B **839**, 137767 (2023). https://doi.org/10.1016/j.physletb.2023.137767
25. Z.z. Xing, Z.h. Zhao, Rept. Prog. Phys. **79**(7), 076201 (2016). https://doi.org/10.1088/0034-4885/79/7/076201
26. Z.Z. Xing, Phys. Rept. **854**, 1 (2020). https://doi.org/10.1016/j.physrep.2020.02.001
27. Z.C. Liu, C.X. Yue, Z.h. Zhao, JHEP **10**, 102 (2017). https://doi.org/10.1007/JHEP10(2017)102
28. E. Ma, Phys. Rev. D **73**, 057304 (2006). https://doi.org/10.1103/PhysRevD.73.057304
29. M.C. Gonzalez-Garcia, M. Maltoni, T. Schwetz, Universe **7**(12), 459 (2021). https://doi.org/10.3390/universe7120459
30. N. Aghanim et al., Astron. Astrophys. **641**, A6 (2020). https://doi.org/10.1051/0004-6361/201833910. [Erratum: Astron. Astrophys. 652, C4 (2021)]
31. A. Barabash, Universe **9**(6), 290 (2023). https://doi.org/10.3390/universe9060290
32. M. Agostini, G. Benato, J.A. Detwiler, J. Menéndez, F. Vissani, Rev. Mod. Phys. **95**(2), 025002 (2023). https://doi.org/10.1103/RevModPhys.95.025002
33. D.Q. Adams et al., Phys. Rev. Lett. **124**(12), 122501 (2020). https://doi.org/10.1103/PhysRevLett.124.122501
34. A. Gando et al., Phys. Rev. Lett. **117**(8), 082503 (2016). https://doi.org/10.1103/PhysRevLett.117.082503. [Addendum: Phys. Rev. Lett. 117, 109903 (2016)]
35. R. Arnold et al., JINST **16**(07), T07012 (2021). https://doi.org/10.1088/1748-0221/16/07/T07012
36. A.S. Barabash et al., Nucl. Instrum. Meth. A **868**, 98 (2017). https://doi.org/10.1016/j.nima.2017.06.044
37. C. Alduino et al., Eur. Phys. J. C **79**(9), 795 (2019). https://doi.org/10.1140/epjc/s10052-019-7275-5
38. M. Agostini et al., Science **365**, 1445 (2019). https://doi.org/10.1126/science.aav8613
39. S.F. King, M. Malinsky, Phys. Lett. B **645**, 351 (2007). https://doi.org/10.1016/j.physletb.2006.12.006

40. S.F. King, C. Luhn, Rept. Prog. Phys. **76**, 056201 (2013). https://doi.org/10.1088/0034-4885/76/5/056201
41. V. Brdar, A.J. Helmboldt, S. Iwamoto, K. Schmitz, Phys. Rev. D **100**, 075029 (2019). https://doi.org/10.1103/PhysRevD.100.075029
42. R.N. Mohapatra, in *SEESAW25: International Conference on the Seesaw Mechanism and the Neutrino Mass* (2004), pp. 29–44. https://doi.org/10.1142/9789812702210_0003
43. A. Melfo, M. Nemevsek, F. Nesti, G. Senjanovic, Y. Zhang, Phys. Rev. D **85**, 055018 (2012). https://doi.org/10.1103/PhysRevD.85.055018

Chapter 13
NSI Effects on Entanglement of Formation in Three-Flavor Neutrino Oscillations

Bhavna Yadav

Abstract Quantum correlation measures are extensively studied in neutrino systems as some measures show nonclassical features in neutrino oscillation. Entanglement is also a splendid measure to study in the neutrino system. Entanglement of formation is a fundamental concept in the realm of quantum physics, and it is a widely used measure for quantifying entanglement. We analyze the effects of non-standard interaction on the entanglement of formation in the context of various experimental setups for a three-flavor neutrino oscillation scenario. We find that the impact of NSI can be maximum in the DUNE, i.e., the long-baseline neutrino experiment.

13.1 Introduction

Quantum correlations are a multifaceted concept within the domain of quantum mechanics, and they cannot be fully described by a single quantity or measure. Instead, there are various aspects used to characterize these correlations. One of the most famous and intriguing facets of quantum correlations is entanglement. Entanglement serves as a fundamental resource in various quantum information processing tasks, including quantum teleportation [1] and quantum cryptography [2]. Importantly, entanglement is not a single-valued quantity; rather, there are numerous methods to quantify it.

Entanglement of Formation (EOF) is a fundamental concept in the realm of quantum physics and is a widely used measure for quantifying entanglement. It is defined as the von Neumann entropy of the reduced density matrix as perceived by any of the three parties involved, namely A, B, or C. In simpler terms, it quantifies the minimum level of entanglement that must be shared among three parties to create a specified quantum state. An intriguing property of EOF is that its value is only zero when the quantum state is separable, underlining its usefulness in identifying entanglement [3–6].

B. Yadav (✉)
Indian Institute of Technology Jodhpur, Jodhpur 342037, India
e-mail: yadav.18@iitj.ac.in

J. W. F. Valle et al. (eds.), *Proceedings of the International Conference on Future Prospects in Neutrino and Astroparticle Physics*, Springer Proceedings in Physics 361,
https://doi.org/10.1007/978-981-97-7441-8_13

The phenomenon of neutrino oscillation, validated through a multitude of experimental observations, incontrovertibly signifies the presence of nonzero neutrino masses. Within the confines of the Standard Model (SM) of electroweak interactions, the three-flavor neutrino oscillations elegantly account for the comprehensive dataset stemming from a variety of neutrino oscillation experiments. Nonetheless, acknowledging the limitations of the SM as the definitive theory governing fundamental interactions, it becomes imperative to incorporate the potential influence of new physics into the domain of neutrino oscillations. Such deviations from the standard framework, often referred to as non-standard interactions (NSI), are anticipated to manifest as subtle perturbations within the landscape of neutrino oscillations, leading to modifications in oscillation probabilities.

The study of quantum correlations in the oscillating neutrino system can also serve as an alternative platform to probe NSI effects. Numerous studies have delved into the investigation of quantum correlations within oscillating neutrino systems, as evidenced by various works such as [7–14]. However, it is noteworthy that most of these studies operate under the assumption that the dynamics of oscillating neutrino systems adhere to the SM. In this work, we study NSI effects on EOF in the context of three-flavor neutrino oscillations. We explore these NSI effects across a range of experimental scenarios, encompassing various experimental setups, including reactors and accelerators such as Daya Bay [15], JUNO [16], KamLAND [17], T2K [18], MINOS [19], and DUNE [20], all within the framework of three-flavor neutrino oscillations.

The work is structured as follows: Sect. 13.2 outlines the formalization of our study in conjunction with matter and NSI. In the same section, we delve into the discussion of EOF. Subsequently, Sect. 13.3 presents our findings, and Sect. 13.4 encapsulates our concluding remarks.

13.2 Formalism

Neutrino oscillations in the presence of NSI: In the formalism of three-flavor neutrino oscillations, the flavor states ν_e, ν_μ, and ν_τ are expressed as a combination of the mass eigenstates ν_1, ν_2, and ν_3. This relationship is represented by the equation

$$|\nu_\alpha\rangle = \sum_i U_{\alpha i}|\nu_i\rangle. \tag{13.1}$$

This formalism reflects the fact that the initial flavor states at $t = 0$ are composed of the underlying mass eigenstates due to the phenomenon of mixing. The 3×3, Pontecorvo-Maki-Nakagawa-Sakata (PMNS) mixing matrix is responsible for this transformation. In this work, for the sake of simplicity, we will disregard the CP-violating phase.

Neutrinos, as they traverse through matter, can engage in interactions through both charged currents (CC) and neutral currents (NC). These matter interactions can have consequences for the characteristic flavor oscillation pattern.

$$\mathcal{H}_m = \mathcal{H}_{vac} + \mathcal{H}_{mat} \tag{13.2}$$

Here, the Hamiltonian for neutrino oscillations in vacuum is denoted as $\mathcal{H}vac$, while $\mathcal{H}mat$ represents the Hamiltonian accounting for oscillations in the presence of matter.

The impact of physics beyond the Standard Model (SM), specifically non-standard interactions (NSI), is expected to play a subheading role in neutrino flavor oscillations. These NSI effects can be incorporated by employing four-fermion, dimension-6 operators. For our current analysis, we will exclude CC-NSI effects due to these tight constraints. The total Hamiltonian in the presence of NSI can be written as

$$\begin{aligned}\mathcal{H}_{tot} &= \mathcal{H}_{vac} + \mathcal{H}_{mat} + \mathcal{H}_{NSI} = \begin{pmatrix} E_1 & 0 & 0 \\ 0 & E_2 & 0 \\ 0 & 0 & E_3 \end{pmatrix} \\ &+ U^\dagger \begin{pmatrix} A & 0 & 0 \\ 0 & 0 & 0 \\ 0 & 0 & 0 \end{pmatrix} U + U^\dagger A \begin{pmatrix} \epsilon_{ee}(x) & \epsilon_{e\mu}(x) & \epsilon_{e\tau}(x) \\ \epsilon_{\mu e}(x) & \epsilon_{\mu\mu}(x) & \epsilon_{\mu\tau}(x) \\ \epsilon_{\tau e}(x) & \epsilon_{\tau\mu}(x) & \epsilon_{\tau\tau}(x) \end{pmatrix} U.\end{aligned} \tag{13.3}$$

Here, $A = \pm\sqrt{2}G_F N_e$ stands for the matter potential, where G_F is the Fermi constant and N_e corresponds to the number density of electrons within the matter. Notably, the sign of the matter potential hinges on the type of neutrinos being considered. Specifically, for neutrinos, the matter potential is positive, whereas for antineutrinos, it is negative. In our analysis, the approximate value of the matter potential, denoted as A, is around 10^{-13} eV, corresponding to Earth's matter density of $\rho = 2.8$ gm/cc. The NSI parameters $\epsilon_{\alpha\beta}(x)$, where $\alpha, \beta = e, \mu, \tau$, can be expressed as follows:

$$\epsilon_{\alpha\beta}(x) = \sum_{f=e,u,d} \frac{N_f(x)}{N_e(x)} \epsilon_{\alpha\beta}^f, \tag{13.4}$$

where x denotes the position or location within the medium, f is summed over the fermions $f \in e, u, d$, and $N_f(x)$ is the matter fermion density. The strength of NC NSI is measured by $\epsilon_{\alpha\beta}^f$, which is the dimensionless coefficient. It measures the strength as compared to weak interaction coupling constant G_F , $i.e.$, $\epsilon_{\alpha\beta}^{f,P} \sim \mathcal{O}(G_x/G_F)$.

The constraints on NSI parameters are derived from a comprehensive analysis of data from a range of experiments, including both oscillation and non-oscillation studies. References such as [21–24] are instrumental in providing these constraints based on a global evaluation of experimental data.

In the ultra-relativistic limit, the time evolution operator for a neutrino mass eigenstate is given by $U_m(L) = e^{-i\mathcal{H}_{tot}L}$. The evolution operator $U_f(L)$, which operates in the flavor basis, can be computed as $U_f(L) = U^\dagger U_m(L)U$ [25]. This evolution operator in the flavor basis is essential for calculating the probabilities of neutrino oscillations. To calculate the probabilities, we focus on the matrix elements of the evolution operator on the flavor basis. By considering the absolute values of these elements and squaring them, we obtain the oscillation probabilities for the transition from ν_α to ν_β. These probabilities are expressed as follows:

$$\tilde{P}_{\alpha\beta} \equiv \left|A_{\alpha\beta}\right|^2 = \left|\langle\beta|U_f(L)|\alpha\rangle\right|^2. \tag{13.5}$$

In this context, $\tilde{P}\alpha\beta$ represents the probability of neutrino oscillation in the presence of non-standard interactions (NSI). In the formalism described above, the mass evolution operator $U_m(L)$ depends on the Hamiltonian of the system. When considering vacuum oscillations, the operator $U_m(L)$ is constructed using the vacuum Hamiltonian $\mathcal{H}_{vac}$. In the presence of NSI, the operator $U_m(L)$ is based on the total Hamiltonian $\mathcal{H}_{tot}$, which encompasses contributions from vacuum, matter, and non-standard interactions. The choice of the appropriate Hamiltonian depends on the specific context being considered, whether it's vacuum, matter, or the inclusion of NSI effects.

In the context of the relativistic limit, neutrino flavor states are treated as individual modes. When two modes become entangled, it signifies a form of interconnection between their quantum states. Neutrino mode entanglement refers to a specific type of entanglement that emerges among distinct modes within a neutrino field. As a consequence, the time-evolved flavor state can be conceived as an entangled superposition of flavor modes. The density matrix allows us to compute the tripartite entanglement. For any state, at time t, where the initial neutrino state is $|\nu_\alpha\rangle$, denoted as $\rho^\alpha_{ABC}(t) = |\nu_\alpha(t)\rangle\langle\nu_\alpha(t)|$, the density matrix is expressed as [27]

$$\rho^\alpha_{ABC}(t) = \begin{pmatrix} 0 & 0 & 0 & 0 & 0 & 0 & 0 & 0 \\ 0 & \rho^\alpha_{22} & \rho^\alpha_{23} & 0 & \rho^\alpha_{25} & 0 & 0 & 0 \\ 0 & \rho^\alpha_{32} & \rho^\alpha_{33} & 0 & \rho^\alpha_{35} & 0 & 0 & 0 \\ 0 & 0 & 0 & 0 & 0 & 0 & 0 & 0 \\ 0 & \rho^\alpha_{52} & \rho^\alpha_{53} & 0 & \rho^\alpha_{55} & 0 & 0 & 0 \\ 0 & 0 & 0 & 0 & 0 & 0 & 0 & 0 \\ 0 & 0 & 0 & 0 & 0 & 0 & 0 & 0 \\ 0 & 0 & 0 & 0 & 0 & 0 & 0 & 0 \end{pmatrix}, \tag{13.6}$$

where the elements of this matrix can be written as

$$\rho^\alpha_{22} = \left|\bar{U}_{\alpha\tau}(t)\right|^2; \; \rho^\alpha_{23} = \bar{U}_{\alpha\tau}(t)\bar{U}^*_{\alpha\mu}(t); \; \rho^\alpha_{25} = \bar{U}_{\alpha\tau}(t)\bar{U}^*_{\alpha e}(t);$$
$$\rho^\alpha_{32} = \bar{U}_{\alpha\mu}(t)\bar{U}^*_{\alpha\tau}(t) \; \rho^\alpha_{33} = \left|\bar{U}_{\alpha\mu}(t)\right|^2; \; \rho^\alpha_{35} = \bar{U}_{\alpha\mu}(t)\bar{U}^*_{\alpha e}(t);$$

$$\rho_{52}^{\alpha} = \bar{U}_{\alpha e}(t)\bar{U}_{\alpha\tau}^{*}(t);\ \rho_{53}^{\alpha} = \bar{U}_{\alpha e}(t)\bar{U}_{\alpha\mu}^{*}(t);\ \rho_{55}^{\alpha} = \left|\bar{U}_{\alpha e}(t)\right|^{2}. \tag{13.7}$$

The associated probabilities are as follows: $P_{\alpha e}(t) = \left|\bar{U}\alpha e(t)\right|^2$ and $P\alpha\mu(t) = \left|\bar{U}_{\alpha\mu}(t)\right|^2$.

In the upcoming section, we delve into the description of tripartite entanglement measures, both in terms of probabilities and through the lens of reduced density matrices. It's evident that these measures can be computed by leveraging the formalism elucidated in the preceding section.

Entanglement of Formation: Let's consider a density matrix representing two quantum systems, A and B. We can decompose this density matrix into all possible pure-state ensembles $|\psi_i\rangle$ and their associated probabilities p_i. Mathematically, this can be expressed as

$$\rho = \sum_i p_i |\psi_i\rangle\langle\psi_i|, \tag{13.8}$$

where ρ is the density matrix for the combined systems A and B, and $|\psi_i\rangle$ are the pure states in the ensemble, while p_i represents the corresponding probabilities. The sum of probabilities over all pure states should be equal to one, ensuring the completeness of the ensemble. This decomposition allows us to represent the density matrix in terms of its constituent pure states and their probabilities, enabling a more detailed understanding of the quantum correlations present in the system.

The entropy of subsystem A or B is employed to define the entanglement of formation for each pure state within this ensemble as

$$EOF(|\psi_i\rangle) = -Tr(\rho_A \log_2 \rho_A) = -Tr(\rho_B \log_2 \rho_B). \tag{13.9}$$

Here, $\rho_{A(B)}$ denotes the partial trace of the density matrix ρ, yielding the reduced density matrix of subsystem A (B). The entanglement characterized by the EOF can then be generalized to encompass an arbitrary tripartite pure state $\rho_{ABC}(t)$. In this context, the EOF for such a tripartite state is often defined as follows:

$$EOF(\rho_{ABC}(t)) = \frac{1}{2}[S(\rho_A) + S(\rho_B) + S(\rho_C)], \tag{13.10}$$

where ρ_A, ρ_B, and ρ_C are reduced density matrices which are $\rho_A = Tr_{BC}(\rho_{ABC}(t))$, $\rho_B = Tr_{AC}(\rho_{ABC}(t))$, and $\rho_C = Tr_{AB}(\rho_{ABC}(t))$. Further $S_(\rho_A)$, $S_(\rho_B)$, and $S_(\rho_C)$ are von Neumann entropies defined as $S_(\rho_A) = -Tr(\rho_A \log \rho_A)$ and same with $S_(\rho_B)$ and $S_(\rho_C)$.

EOF in terms of survival and oscillation probabilities for vacuum can be presented as follows [27]:

$$EOF^{\alpha} = -\frac{1}{2}[P_{\alpha e}\log_2 P_{\alpha e} + P_{\alpha\mu}\log_2 P_{\alpha\mu} + P_{\alpha\tau}\log_2 P_{\alpha\tau} + (P_{\alpha\mu} + P_{\alpha\tau})\log_2(P_{\alpha\mu} + P_{\alpha\tau}) + (P_{\alpha e} + P_{\alpha\tau})\log_2(P_{\alpha e} + P_{\alpha\tau}) + (P_{\alpha\mu} + P_{\alpha e})\log_2(P_{\alpha\mu} + P_{\alpha e})]. \tag{13.11}$$

Here $\alpha = e, \mu, \tau$, representing the initial flavor state of the neutrino.

13.3 Results and Discussion

In this section, we delve into an exploration of how NSI affects EoF, which is a tripartite entanglement measure within a three-flavor neutrino system. Our examination encompasses various experimental setups, specifically those involving reactors and accelerators. Through detailed analysis, we aim to discern the impact of NSI on the entanglement characteristics of neutrinos, shedding light on the intricate interplay between quantum correlations and particle interactions. We consider the following reactor and accelerator experimental setups: (1) Daya Bay ($L \approx 2$ km, $E \approx 0.8 - 6$ MeV) [15], (2) JUNO ($L \approx 53$ km, $E \approx 1 - 8$ MeV) [16], (3) KamLAND ($L \approx 180$ km, $E \approx 1 - 16$ MeV) [17], (4) T2K ($L = 295$ km, $E \approx 0 - 6$ GeV) [18], (5) MINOS ($L = 735$ km, $E \approx 1 - 10$ GeV) [19], (6) DUNE ($L \approx 1300$ km, $E \approx 1 - 14$ GeV) [20].

The standard neutrino oscillation parameters are presented in Table 13.1, while the NSI parameters are provided in Table 13.2. These tables pertain to the framework of three-flavor neutrino oscillations. In the context of reactor experiments, the initial

Table 13.1 Standard neutrino oscillation parameters [26]

Parameters	θ_{12}^o	θ_{13}^o	θ_{23}^o	$\Delta m_{21}^2 \times 10^{-5}$ (eV^2)	$\Delta m_{31}^2 \times 10^{-3}$ (eV^2)
Best fit$\pm 1\sigma$	34.3 ± 1.0	$8.58^{+0.11}_{-0.15}$	$48.79^{+0.93}_{-1.25}$	$7.5^{+0.22}_{-0.20}$	$2.56^{+0.03}_{-0.04}$

Table 13.2 NSI parameters [24] are given here and these parameters are assumed to be real

NSI Parameters	Range (1σ)	Range (2σ)
$\epsilon_{ee}^{\oplus}$	[–0.30, 0.20] $\oplus$ [0.95, 1.3]	[–1.00, 1.4]
$\epsilon_{e\mu}^{\oplus}$	[–0.12, 0.011]	[–0.20, 0.09]
$\epsilon_{e\tau}^{\oplus}$	[–0.16, 0.083]	[–0.24, 0.30]
$\epsilon_{\mu\mu}^{\oplus}$	[–0.43, 0.14] $\oplus$ [0.91, 1.3]	[–0.80, 1.4]
$\epsilon_{\mu\tau}^{\oplus}$	[–0.047, 0.012]	[–0.021,0.021]
$\epsilon_{\tau\tau}^{\oplus}$	[–0.43, 0.21] $\oplus$ [0.83, 1.3]	[–0.85, 1.4]

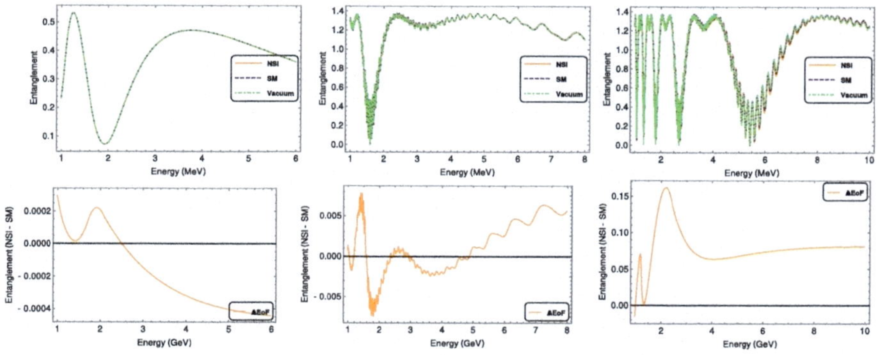

Fig. 13.1 The three panels depict the prediction of EOF for various reactor neutrino oscillation experimental setups, i.e., Daya Bay (left), JUNO (center), and KamLAND (right). Standard neutrino oscillation parameters and NSI parameters are given in Tables 13.1 and 13.2, respectively. The vacuum, SM, and NSI in the upper panels are represented by green dot-dashed, purple-dashed, and orange solid lines, respectively

neutrino is considered to be the antineutrino $\bar{\nu}_e$, whereas for accelerator experiments, the initial neutrino is denoted as ν_μ.

Figure 13.1 illustrates the projected results for EoF in the context of different reactor neutrino oscillation experiments, namely Daya Bay, JUNO, and KamLAND. The lower panel of each experiment showcases the difference in EoF, denoted as ΔEoF, between the scenario with NSI and the SM scenario. This visual representation serves to emphasize the discernible distinctions between the two scenarios, facilitating a clearer understanding of the impact of NSI on the entanglement measure.

In the lower panel of Fig. 13.1, it is noteworthy that the impact of NSI is most subdued for the Daya Bay setup, whereas it is most pronounced for the KamLAND setup. This observation can be attributed to the fact that the baseline distance is the greatest for the KamLAND configuration compared to the other two setups. The extended baseline enhances the sensitivity of the KamLAND setup to NSI effects, leading to a more pronounced influence on the entanglement measure.

Figure 13.2 illustrates the analysis of the entanglement measure within accelerator experiments, specifically T2K, MINOS, and DUNE. In the upper panel of the figure, it becomes evident that in the low-energy range, in the DUNE experimental setup, EoF can discriminate among all three scenarios, i.e., vacuum, SM, and NSI. However, in the high-energy domain, MINOS and DUNE experiments can distinguish between SM and NSI.

The lower panel of Fig. 13.2 provides a closer look at the impact of NSI, characterized by the difference between values in the presence of NSI and SM interactions. Notably, the NSI effects are least pronounced for the T2K setup among all three configurations, while they are most prominent for the DUNE experiment. The MINOS experiment experiences moderate NSI effects. This discrepancy can be attributed to differences in the baseline length.

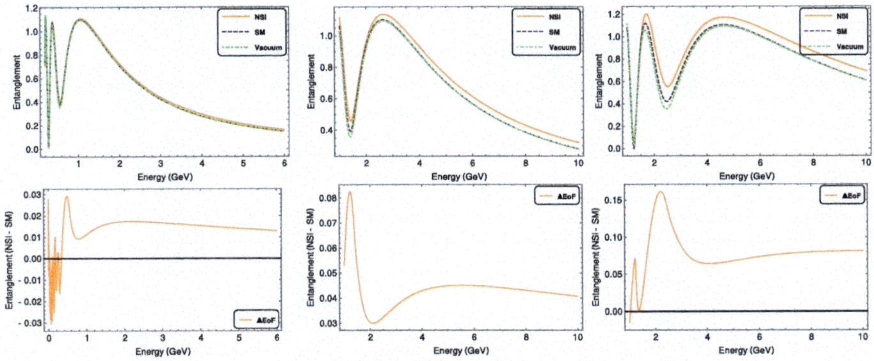

Fig. 13.2 The three panels depict the prediction of EOF for various accelerator neutrino oscillation experimental setups, i.e., T2K (left), MINOS (center), and DUNE (right). Green dot-dashed, purple-dashed, and orange solid lines represent the vacuum, SM, and NSI, respectively, in the upper panels. Figure nomenclature is the same as given in Fig. 13.1

13.4 Conclusion

The impact of NSI on the entanglement measure for the three-flavor neutrino oscillation scenarios is studied in this paper. We studied the entanglement measure, EoF, for six neutrino experiments, i.e., T2K, MINOS, DUNE, Daya Bay, JUNO, and KamLAND. DUNE has the longest baseline among the six experiments. Therefore, the DUNE experiment exhibits a greater deviation toward NSI in comparison to MINOS and T2K. For reactor experiments, EoF showed less sensitivity toward NSI.

Acknowledgements We would like to acknowledge that this conference paper serves as a precursor to our comprehensive work presented in the main paper [28], which has been recently uploaded on arXiv. The conference paper encapsulates a condensed version of our findings, while the main paper delves deeper, providing a more extensive analysis and discussion.

References

1. C.H. Bennett, G. Brassard, C. Crépeau, R. Jozsa, A. Peres, W.K. Wootters, Teleporting an unknown quantum state via dual classical and Einstein-Podolsky-Rosen channels. Phys. Rev. Lett **70**, 1895 (1993)
2. D. Deutsch, A. Ekert, R. Jozsa, C. Macchiavello, S. Popescu, A. Sanpera, Quantum privacy amplification and the security of quantum cryptography over noisy channels. Phys. Rev. Lett. **67**, 661–663 (1991)
3. C.H. Bennett, D.P. DiVincenzo, J.A. Smolin, W.K. Wootters, Mixed-state entanglement and quantum error correction. Phys. Rev. A **54**(5), 3824 (1996)
4. C. Sabín, G. García-Alcaine, A classification of entanglement in three-qubit systems. Eur. Phys. J. C **48**(3), 435–442 (2008)

5. W.K. Wootters, Entanglement of formation of an arbitrary state of two qubits. Phys. Rev. Lett. **80**(10), 2245 (1998)
6. Y. Guo, G. Gour, Monogamy of the entanglement of formation. Phys. Rev. A **99**(4), 042305 (2019)
7. M. Blasone, F. Dell'Anno, S. De Siena, M. Di Mauro, F. Illuminati, Multipartite entangled states in particle mixing. Phys. Rev. D **77**, 096002 (2008). https://doi.org/10.1103/PhysRevD.77.096002
8. A.K. Alok, S. Banerjee, S.U. Sankar, Quantum correlations in terms of neutrino oscillation probabilities. Nucl. Phys. B **909**, 65–72 (2016). https://doi.org/10.1016/j.nuclphysb.2016.05.001
9. S. Banerjee, A.K. Alok, R. Srikanth, B.C. Hiesmayr, A quantum information theoretic analysis of three flavor neutrino oscillations Eur. Phys. J. C **75**(10), 487 (2015). https://doi.org/10.1140/epjc/s10052-015-3717-x
10. J.A. Formaggio, D.I. Kaiser, M.M. Murskyj, T.E. Weiss, Violation of the Leggett-Garg inequality in neutrino oscillations. Phys. Rev. Lett. **117**(5), 050402 (2016). https://doi.org/10.1103/PhysRevLett.117.050402
11. J. Naikoo, A.K. Alok, S. Banerjee, S.U. Sankar, G. Guarnieri, C. Schultze, B.C. Hiesmayr, A quantum information theoretic quantity sensitive to the neutrino mass-hierarchy. Nucl. Phys. B **951**, 114872 (2020). https://doi.org/10.1016/j.nuclphysb.2019.114872
12. J. Naikoo, A.K. Alok, S. Banerjee, S.U. Sankar, Leggett-Garg inequality in the context of three flavour neutrino oscillation. Phys. Rev. D **99**(9), 095001 (2019). https://doi.org/10.1103/PhysRevD.99.095001
13. F. Ming, X.K. Song, J. Ling, L. Ye, D. Wang, Quantification of quantumness in neutrino oscillations. Eur. Phys. J. C **80**(3), 275 (2020). https://doi.org/10.1140/epjc/s10052-020-7840-y
14. D.S. Chattopadhyay, A. Dighe, Quantum mismatch: a powerful measure of "quantumness" in neutrino oscillations. Phys. Rev. D **108**(11), 112013 (2023). https://doi.org/10.1103/PhysRevD.108.112013
15. F.P. An et al., Observation of electron-antineutrino disappearance at Daya Bay. Phys. Rev. Lett. **108**, 171803 (2012). https://doi.org/10.1103/PhysRevLett.108.171803
16. F. An et al., Neutrino physics with JUNO. J. Phys. G **43**(3), 030401 (2016). https://doi.org/10.1088/0954-3899/43/3/030401
17. K. Eguchi et al., First results from KamLAND: evidence for reactor anti-neutrino disappearance. Phys. Rev. Lett. **90**, 021802 (2003). https://doi.org/10.1103/PhysRevLett.90.021802
18. K. Abe et al., The T2K experiment. Nucl. Instrum. Meth. A **659**, 106–135 (2011). https://doi.org/10.1016/j.nima.2011.06.067
19. D.G. Michael et al., Observation of muon neutrino disappearance with the MINOS detectors and the NuMI neutrino beam. Phys. Rev. Lett. **97**, 191801 (2006). https://doi.org/10.1103/PhysRevLett.97.191801
20. R. Acciarri et al., Long-baseline neutrino facility (LBNF) and deep underground neutrino experiment (DUNE): conceptual design report, volume 2: the physics program for DUNE at LBNF. [arXiv:1512.06148 [physics.ins-det]]
21. I. Esteban, M.C. Gonzalez-Garcia, M. Maltoni, I. Martinez-Soler, J. Salvado, Updated constraints on non-standard interactions from global analysis of oscillation data. J. High Energy Phys. **8**, 1–33 (2018)
22. I. Esteban, M.C. Gonzalez-Garcia, M. Maltoni, On the determination of leptonic CP violation and neutrino mass ordering in presence of non-standard interactions: present status. J. High Energy Phys. **6**, 1–24 (2019)
23. P. Coloma, I. Esteban, M.C. Gonzalez-Garcia, M. Maltoni, Improved global fit to Non-Standard neutrino Interactions using COHERENT energy and timing data. J. High Energy Phys. **2**, 1–30 (2020)
24. P. Coloma, M.C. Gonzalez-Garcia, M. Maltoni, J.P. Pinheiro, S. Urrea, Global constraints on non-standard neutrino interactions with quarks and electrons (2023). arXiv:2305.07698

25. T. Ohlsson, H. Snellman, Three flavor neutrino oscillations in matter. J. Math. Phys. **41**, 2768–2788 (2000). [erratum: J. Math. Phys. **42**, 2345 (2001)]. https://doi.org/10.1063/1.533270
26. P.F. de Salas et al., 2020 global reassessment of the neutrino oscillation picture. J. High Energy Phys. (2), 071 (2021)
27. L.J. Li, F. Ming, X.K. Song, L. Ye, D. Wang, Characterizing entanglement and measurement's uncertainty in neutrino oscillations. Eur. Phys. J. C **81**(8), 1–10 (2021)
28. L. Konwar, B. Yadav, NSI effects on tripartite entanglement in neutrino oscillations (2024). arXiv:2402.09952

Chapter 14
Exploring 3+1 Neutrino Models Under Modular S_3 symmetry

Mayengbam Kishan Singh and Ngangkham Nimai Singh

Abstract We study the minimal extended seesaw mechanism with one sterile neutrino in a 3+1 framework using modular S_3 symmetry. The active-sterile neutrino models are classified based on the assignments of S_3 representations and modular weights of the left-handed lepton doublets, right-handed neutrinos, and sterile neutrino. No scalar flavons are considered, and the flavour symmetry is broken by the vacuum expectation value (vev) of the modulus τ. For a particular set of representations of the Leptons and Higgs field, we develop eleven (11) different models based on different modular weights of charged lepton (k_L) and right-handed neutrinos (k_N). Out of these, we consider two models, which are discriminated by carrying out the numerical analysis so that the parameter space in each model can fit the latest neutrino oscillation data at 3σ. The Planck cosmological bound on the upper limit of the sum of the active neutrino masses $\sum m_i < 0.12$eV is also considered. Finally, the best-fit parameters of the neutrino observables and model predictions are evaluated using the minimum χ^2 analysis.

14.1 Introduction

Among various problems in the Standard Model (SM) of particle physics, the origin of neutrino masses and flavour mixing, mass hierarchy, Dirac CP-violating phase, and octant of θ_{23}, are some of the crucial areas that scientists are trying to explain both experimentally and theoretically. In addition, certain discrepancies in the results of LSND and MiniBooNE compelled the need to extend the three neutrino theory. These extra states of neutrinos do not have weak interaction; therefore, they are referred to as *sterile* neutrinos. However, sterile neutrinos can mix with the active neutrinos. Such neutrinos are incorporated with the three neutrino theory in various literatures

M. K. Singh (✉) · N. N. Singh
Department of Physics, Manipur University, Imphal 795003, India
e-mail: kishanmayengbam@gmail.com

N. N. Singh
Research Institute of Science and Technology (RIST), Manipur 795003, India

J. W. F. Valle et al. (eds.), *Proceedings of the International Conference on Future Prospects in Neutrino and Astroparticle Physics*, Springer Proceedings in Physics 361,
https://doi.org/10.1007/978-981-97-7441-8_14

in a 3+1 scheme, 3+1+1, 3+2. One of the most straightforward extensions is the 3+1 scheme, where a singlet sterile neutrino is added to the three active neutrinos, and the sterile neutrino gets mass through the minimal extended seesaw mechanism(MES) [1].

Neutrino mass models with non-Abelian discrete symmetries have been explored to address the problems of flavour mixing. However, such approaches require ad-hoc scalar fields called *flavons*, which are singlet under SM but transform non-trivially as irreducible representations of the symmetry. After the electroweak symmetry breaking, these flavons acquire vacuum expectation values along specific vacuum alignments, which shape the structure of fermion mass matrices. The main areas for improvement of such models are the presence of additional new flavons, difficulty in generating the right vacuum alignments of the scalars, and requirements of heavy fine-tuning of the parameters. Very recently, an interesting method was proposed in which modular flavour symmetry is used with the discrete symmetries as its subgroups [2]. In such frameworks, the Yukawa couplings transform non-trivially as modular forms. These modular forms depend on a single parameter τ called the modulus. Minimal or no scalar flavons are required to break the symmetry as the fermion masses are generated from the symmetry breaking by the vev of the modulus τ. The modular symmetry defined by a finite modular group Γ_N is isomorphic to the conventional discrete symmetry groups S_3, S_4, A_4, A_5, etc. For instance, the simplest group for $N = 2$, Γ_2 is isomorphic to S_3. Detailed analysis of modular groups and their application in neutrino model building are studied in Refs. [2, 3]. There are numerous studies based on modular groups such as S_3 [4], A_4 [5, 6] A_5 [7, 8], and S_4 [9, 10].

In modular S_3 symmetry, the Yukawa coupling transforms as an S_3 doublet represented by $Y_2^{(k)}$. Here, k denotes the weight of the modular form, and the subscript 2 denotes the doublet representation in S_3. The doublet modular form of lowest weight 2, $Y_2^{(2)} = (y_1, y_2)$ can be expressed in terms of the Dedekind-eta function $\eta(\tau)$ and its derivative $\eta'(\tau)$ as

$$\begin{aligned} y_1(\tau) &= \frac{i}{4\pi}\left(\frac{\eta'(\tau/2)}{\eta(\tau/2)} + \frac{\eta'((\tau+1)/2)}{\eta((\tau+1)/2)} - \frac{8\eta'(2\tau)}{\eta(2\tau)}\right) \\ y_2(\tau) &= \frac{\sqrt{3}i}{4\pi}\left(\frac{\eta'(\tau/2)}{\eta(\tau/2)} - \frac{\eta'((\tau+1)/2)}{\eta((\tau+1)/2)}\right) \end{aligned} \tag{14.1}$$

where $\eta(\tau)$ is defined as

$$\eta(\tau) = q^{1/24}\prod_{n=1}^{\infty}(1-q^n), \quad q \equiv e^{i2\pi\tau}. \tag{14.2}$$

Multiplication rules of S_3 can construct other higher-weighted modular forms. The modular forms of higher weights $k = 4, 6$ are given as

$$Y_1^{(4)} = y_1^2 + y_2^2, \quad Y_1^{(6)} = 3y_1^2 y_2 - y_2^3, \quad Y_{1'}^{(6)} = y_1^3 - 3y_1 y_2^2,$$
$$Y_2^{(4)} = \begin{pmatrix} 2y_1 y_2 \\ y_1^2 - y_2^2 \end{pmatrix}, \quad Y_2^{(6)} = \begin{pmatrix} y_1^3 + y_1 y_2^2 \\ y_2^3 + y_1^2 y_2 \end{pmatrix} \tag{14.3}$$

It is important to note that a doublet modular form must have weight $k \geq 2$ while the singlet modular form Y_1 cannot have weight 2. The lowest weight of an S_3 singlet modular form starts from 4. The number of independent modular forms of weight k increases with k/2+1. For minimality, we consider modular forms with a maximum weight of 6.

The present work will analyse active-sterile neutrino mass models based on modular S_3 symmetry in a minimal extended seesaw (MES) framework. The SM particle contents are extended by three right-handed neutrino fields $N_i (i = 1, 2)$, N_3, and a singlet sterile neutrino field S. We shall explore different models corresponding to different assignments of the fields' modular weight 'k'. We develop the neutrino mass models without using scalar flavons in the charged lepton and active neutrino sector, which is an advantage of using modular flavour symmetry.

14.2 Building Strategy of the Models

The S_3 group has a doublet (2) and two singlets $(1, 1')$ as its irreducible representations. We extend the SM particle contents by three right-handed neutrinos where two components N_i $(i = 1, 2)$ and a third component N_3 transform as a doublet and a singlet under S_3. A right-handed SM singlet sterile neutrino field S with modular weight k_S which transforms as a singlet under S_3 is also considered. The three generations of the SM lepton doublet L are also grouped into two irreps. of S_3 as doublet L_i $(i = 1, 2)$ and a singlet L_3. Similarly, the right-handed charged leptons are represented as singlets e_1, e_2, and e_3 of S_3, respectively. The Higgs represented as H_u and H_d having hypercharges $+1/2$ and $-1/2$ are taken as singlets of S_3. The Higgs develop VEVs after electroweak symmetry breaking along $H_u = \left(0, v_u/\sqrt{2}\right)^T$ and $H_d = \left(v_d/\sqrt{2}, 0\right)^T$ inducing SM lepton mass terms.

To simplify our analysis and to avoid trivial interaction terms in the superpotential, we consider that the weights of SM lepton doublets L_i, L_3 are denoted by k_L, charged lepton singlets e_1, e_2, e_3 are denoted by k_{e_i}, $(i = 1, 2, 3)$, respectively. The heavy right-handed neutrino fields N_i, N_3 have modular weight k_N. It should be noted that N_3 and S have the exact transformation charges under the S_3 symmetry. All the interactions allowed for N_3 will also be possible for S. To avoid weak interactions of the sterile neutrino field, we also assume that the weights of sterile neutrino and the heavy right-handed neutrino N_3 are different, i.e., $k_N \neq k_S$.We have fixed three

Table 14.1 Particle contents of the model and their group charges

Fields / Charges	L_i	L_3	e_1^c	e_2^c	e_3^c	$H_{u,d}$	N_i^c	N_3^c	S^c
$SU(2)_L$	2	1	1	1	1	1	1	1	1
$U(1)_Y$	−1/2	−1/2	1	1	1	± 1/2	0	0	0
S_3	2	1′	1	1	1′	1	2	1	1
$-k_i$	k_L	k_L	k_{e_1}	k_{e_2}	k_{e_3}	k_H	k_N	k_N	k_S

different weights $k_S = (1, 4, 5)$ for S field in our model.[1] The complete particle contents of the model and their group charges are given in Table 14.1.

The most general superpotential of the charged lepton sector invariant under S_3 symmetry without specific assignments of modular weights is given by

$$\begin{aligned} W_e = \ & \alpha(Y_2^{(k)} H_u \bar{L}_i)e_1 + \beta(Y_2^{(k)} H_u \bar{L}_i)e_2 + \gamma(Y_1^{(k)} H_u \bar{L}_3)e_3 \\ & + \alpha'(Y_1^{(k)} H_u \bar{L}_3)e_1 + \beta'(Y_{1'}^{(k)} H_u \bar{L}_3)e_2 + \gamma'(Y_2^{(k)} H_u \bar{L}_i)_{1'} e_3 \end{aligned} \tag{14.4}$$

Here, the superscript (k) in the Yukawa couplings denotes the weights of the modular forms while the subscripts 1, 1′, 2 denote the representation of the couplings under S_3 symmetry. The parameters $\alpha, \beta, \gamma, \alpha', \beta'$, and γ' are the dimensionless coefficients which are the free parameters of the charged lepton sector. Terms in the superpotential are invariant under a modular symmetry when the overall weight of the superfields in each term is zero, and the term is invariant under the discrete symmetry. Based on the assignments of modular weights k_i for each field, certain interaction terms are allowed or disallowed in the superpotential. For a minimal structure in the charged lepton sector, we adopt the following constraints regarding the weights of the right-handed charged lepton singlets

$$k_{e_1} = 4 - k_L \tag{14.5}$$

$$k_{e_2} = 2 - k_L \tag{14.6}$$

$$k_{e_3} = -k_L \tag{14.7}$$

We assume that $k_L \geq 0$ and the Higgs field have zero modular weight. As a consequence of the above constraints, the coefficients $\beta' = 0$ since there are no singlet modular forms of weight 2. Since doublet modular forms cannot have weight zero, the coefficient $\gamma' = 0$. Finally, the term $\alpha'(Y_{1'}^{(k)} H_u \bar{L}_3)e_1$ cannot exist because $Y_{1'}^{(4)} = 0$. Thus, only three terms remain in the charged lepton superpotential. After the electroweak symmetry breaking, the Higgs gets a vev, and the resulting charged lepton mass matrix is given by

[1] We have taken this choice to avoid Majorana mass terms of the sterile neutrinos of the form $\sim \Lambda YSS$.

$$m_L = \frac{v_u}{\sqrt{2}} \begin{pmatrix} \alpha 2y_1y_2 & \beta y_1 & 0 \\ \alpha(y_1^2 - y_2^2) & \beta y_2 & 0 \\ 0 & 0 & \gamma \end{pmatrix} \tag{14.8}$$

Similarly, in the case of the neutrino sector, the general superpotential is given by

$$\begin{aligned} W_\nu = \; & g_1 Y_1^{(k)} H_d \bar{L}_i N_i + g_2 Y_2^{(k)} H_d \bar{L}_i N_3 + g_3 Y_2^{(k)} H_d \bar{L}_3 N_i + g_4 Y_2^{(k)} H_d \bar{L}_i N_i \\ & + g_5 Y_{1'}^{(k)} H_d \bar{L}_3 N_3 + \lambda_1 \left(Y_1^{(k)} \bar{N}_i^c N_i + Y_2^{(k)} \bar{N}_i{}^c N_i \right) \\ & + \lambda_2 \left(Y_2^{(k)} \bar{N}_i{}^c N_3 + Y_2^{(k)} \bar{N}_3{}^c N_i \right) + \lambda_3 Y_1^{(k)} \bar{N}_3{}^c N_3 \\ & + \lambda_{s1} Y_1^{(k)} \bar{S}^c N_3 + \lambda_{s2} Y_2^{(k)} \bar{S}^c N_i, \end{aligned} \tag{14.9}$$

where the coefficients g_i $(i = 1, 2, 3, 4, 5)$ are the dimensionless coefficients. Other parameters λ_i $(i = 1, 2, 3)$, λ_{sj} $(j = 1, 2)$ are the constants having the dimension of mass, and they define the mass scales of the heavy right-handed Majorana neutrino and sterile neutrino, respectively. We have divided our models based on the choice of k_L as follows:

1. For $k_L = 0$, there are two possible models : $A1$ $(k_N = 0, k_S = 4)$ and $A2$ $(k_N = 2, k_S = 4)$. In model $A1$, the vanishing coefficients in (14.9) are $g_2 = g_3 = g_4 = 0$ and $\lambda_2 = 0$, whereas, in model $A2$, the coefficients $g_1 = g_5 = 0$.
2. For $k_L = 1$, the two models are $B1$ $(k_N = 1, k_S = 5)$ and $B2$ $(k_N = 3, k_S = 1)$. In model $B1$, the vanishing coefficients in (14.9) are $g_1 = g_5 = 0$ and $\lambda_3 = 0$ while in model $B2$, the vanishing coefficient is $g_5 = 0$.
3. For $k_L = 2$, two possible models are $C1$ $(k_N = 0, k_S = 4)$ and $C2$ $(k_N = 2, k_S = 4)$. In model $C4$, the coefficients $g_1 = g_5 = 0$ and $\lambda_2 = 0$. On the other hand, in model $C2$, all the coefficients in (14.9) are non-zero.
4. For $k_L = 3$, two models $D1$ $(k_N = 1, k_S = 5)$ and $D2$ $(k_N = 3, k_S = 1)$ are possible. Only the coefficient $g_5 = 0$ in (14.9) while all the terms are allowed for model $D2$.
5. For $k_L = 4$, two models $E1$ $(k_N = 0, k_S = 4)$ and $E2$ $(k_N = 2, k_S = 4)$ are possible. All the coefficients in (14.9) are allowed for these two models.
6. Finally, for $k_L = 5$, only one model is possible which is F $(k_N = 1, k_S = 5)$. All the terms in the superpotential W_ν are allowed in this model.

We have summarized these models in Table 14.2.

In the present work, we shall discuss the numerical results and consistency concerning neutrino oscillation data for only two models, A2 and B1. It is important to note that in the models A1, C1, and E1, the weight of heavy right-handed neutrino $k_N = 0$. This condition gives $\lambda_2 = 0$ in (14.9). The resulting heavy Majorana mass matrices M_R in each model are singular; hence, these models are impossible.

Table 14.2 Different models for different choices of k_L keeping the maximum weight of modular forms at 6

k_L	k_N	k_S	Models	Coefficients(g_i)
0	0	4	A1	$g_2 = g_3 = g_4 = 0$
	2	4	A2	$g_1 = g_5 = 0$
1	1	5	B1	$g_1 = g_5 = 0$
	3	1	B2	$g_5 = 0$
2	0	4	C1	$g_1 = g_5 = 0$
	2	4	C2	Nil
3	1	5	D1	$g_5 = 0$
	3	1	D2	Nil
4	0	4	E1	Nil
	2	4	E2	Nil
5	1	5	F	Nil

14.2.1 Model A2 ($k_L = 0, k_N = 2, k_S = 4$)

When we take $k_L = 0$, the possible choices of k_N are $0, 2, 4, 6$. However, $k_N = \{4, 6\}$ will be possible[2] only when the weight of the Yukawa, $k > 6$. For model A2, we consider $k_L = 0$, $k_N = 2$, and $k_S = 4$. The coefficients $g_1 = g_5 = 0$ and (14.9) becomes

$$\begin{aligned} W_\nu = {} & g_2 Y_2^{(2)} H_d \bar{L}_i N_3 + g_3 Y_2^{(2)} H_d \bar{L}_3 N_i + g_4 Y_2^{(2)} H_d \bar{L}_i N_i \\ & + \lambda_1 \left(Y_1^{(4)} \bar{N}_i^c N_i + Y_2^{(4)} \bar{N}_i{}^c N_i \right) + \lambda_2 \left(Y_2^{(4)} \bar{N}_i{}^c N_3 + Y_2^{(4)} \bar{N}_3{}^c N_i \right) \\ & + \lambda_3 Y_1^{(4)} \bar{N}_3{}^c N_3 + \lambda_{s1} Y_1^{(6)} \bar{S}^c N_3 + \lambda_{s2} Y_2^{(6)} \bar{S}^c N_i . \end{aligned} \tag{14.10}$$

After EWSB, the resulting Dirac, Majorana, and sterile neutrino mass matrices are

$$\begin{aligned} M_D &= \frac{v g_4}{\sqrt{2}} \begin{pmatrix} y_2 & y_1 & g_{o2} y_1 \\ y_1 & -y_2 & g_{o2}\, y_2 \\ -g_{o3}\, y_2 & g_{o3}\, y_1 & 0 \end{pmatrix}, \; M_S = \lambda_{s1} \left(y_1^3 + y_1 y_2^2 \;\; y_2^3 + y_1^2 y_2 \;\; \lambda(3 y_1^2 y_2 - y_2^3) \right) \\ M_R &= \lambda_1 \begin{pmatrix} 2y_1^2 & 2y_2 y_1 & \lambda_{2o} 2 y_1 y_2 \\ 2 y_2 y_1 & 2 y_2^2 & \lambda_{2o}(y_1^2 - y_2^2) \\ \lambda_{2o} 2 y_1 y_2 & \lambda_{2o}(y_1^2 - y_2^2) & \lambda_{3o}(y_1^2 + y_2^2) \end{pmatrix} \end{aligned} \tag{14.11}$$

---

[2] For instance, if $k_N = 6$, the Majorana mass terms $\lambda Y^{(k)} N_i^c N_i$ requires weight $k = 12$.

where the coefficients $g_{o2} = g_2/g_4$, $g_{o3} = g_3/g_4$, $\lambda = \lambda_{s1}/\lambda_{s2}$, $\lambda_{2o} = \lambda_2/\lambda_1$, and $\lambda_{3o} = \lambda_3/\lambda_1$.

The active and sterile neutrino masses in the MES mechanism are given by

$$m_\nu \simeq M_D M_R^{-1} M_S^T \left(M_S M_R^{-1} M_S^T\right)^{-1} M_S \left(M_R^{-1}\right)^T M_D^T - M_D M_R^{-1} M_D^T; \tag{14.12}$$

$$m_s \simeq -M_S M_R^{-1} M_S^T. \tag{14.13}$$

Using these, the resulting sterile neutrino mass is given by

$$m_s = -\frac{\lambda_{s1}^2}{2\lambda_1}(y_1^2 + y_2^2)^2. \tag{14.14}$$

14.2.2 Model B1 ($k_L = 1, k_N = 1, k_S = 5$)

In this case, the weights of the fermion fields will be $k_L = 1, k_N = 1$, and $k_S = 5$. These assignments of the modular weights imply that the coefficients $g_1 = g_5 = 0$ and $\lambda_3 = 0$ since all the singlet modular forms of weight 2 will vanish. The modular invariant terms of W_ν in (14.9) are

$$\begin{aligned} W_\nu &= g_2 Y_2^{(2)} H_d \bar{L}_i N_3 + g_3 Y_2^{(2)} H_d \bar{L}_3 N_i + g_4 Y_2^{(2)} H_d \bar{L}_i N_i \\ &+\lambda_1 Y_2^{(2)} \bar{N}_i^{\,c} N_i + \lambda_2 (Y_2^{(2)} \bar{N}_i^{\,c} N_3 + Y_2^{(2)} \bar{N}_3^{\,c} N_i) + \lambda_{s1} Y_1^{(6)} \bar{S}^c N_3 \\ &+\lambda_{s2} Y_2^{(6)} \bar{S}^c N_i. \end{aligned} \tag{14.15}$$

After EWSB, the Dirac, Majorana, and sterile neutrino mass matrices are obtained as

$$\begin{aligned} M_D &= \frac{v g_4}{\sqrt{2}} \begin{pmatrix} y_2 & y_1 & g_{o2} y_1 \\ y_1 & -y_2 & g_{o2}\, y_2 \\ -g_{o3}\, y_2 & g_{o3}\, y_1 & 0 \end{pmatrix}, \quad M_S = \lambda_{s1} \left(y_1^3 + y_1 y_2^2 \;\; y_2^3 + y_1^2 y_2 \;\; \lambda(3y_1^2 y_2 - y_2^3) \right) \\ M_R &= \lambda_1 \begin{pmatrix} y_2 & y_1 & \lambda_{2o} y_1 \\ y_1 & -y_2 & \lambda_{2o} y_2 \\ \lambda_{2o} y_1 & \lambda_{2o} y_2 & 0 \end{pmatrix}. \end{aligned} \tag{14.16}$$

Here, M_D and M_S in this case are same as model A1 but M_R is different. The resulting sterile neutrino mass matrix becomes

$$m_s = \frac{\lambda_{s1}^2}{\lambda_1} \frac{\lambda\, y_2}{\lambda_{2o}^2} (y_2^2 - 3y_1^2)(y_1^2 + y_2^2)(2\lambda_{2o} - \lambda). \tag{14.17}$$

Table 14.3 Ranges of input parameters used in numerical analysis

Free Parameters	Range
g_{o2}	1–10
g_{o3}	1–3
$\vert\lambda\vert$	0.01–10
ϕ_λ	$[-\pi, \pi]$
λ_{2o}	0.01–10
$\vert\lambda_{3o}\vert$	0.01–10
$\phi_{\lambda_{3o}}$	$[-\pi, \pi]$
m_s	(0.8–50) eV

14.3 Numerical Analysis

The steps of numerical analysis are conducted as follows. The modulus parameter τ is scanned in the fundamental domain as given below

$$Re[\tau] = [-0.5, 0.5], \ \ Im[\tau] = [0.6, 2]. \tag{14.18}$$

For the charged lepton sector, the unknown coefficients α, β, γ are solved using the identities

$$Tr[M_L] \quad = m_e^2 + m_\mu^2 + m_\tau^2, \tag{14.19}$$
$$Det[M_L] \quad = m_e^2 \times m_\mu^2 \times m_\tau^2, \tag{14.20}$$
$$Tr[M_L]^2/2 - Tr[M_L^2]/2 = m_e m_\mu + m_\mu m_\tau + m_\tau m_e \tag{14.21}$$

where m_e, m_μ, and m_τ are the charged lepton masses and $M_L = m_L m_L^\dagger$ is a hermitian matrix. The charged lepton mass matrix can be diagonalized using the relation $M_L^{diag} = U_L M_L U_L^\dagger$.

In the neutrino sector, the unknown free parameters of the model are randomly scanned in the ranges given in Table 14.3 along with the values of τ, which reproduce the charged lepton masses. Now, the active neutrino mass matrix can be numerically diagonalized using the relation $M_\nu^{diag} = U_\nu M_\nu U_\nu^\dagger$, where $M_\nu = m_\nu m_\nu^\dagger$ and $M_\nu^{diag} = diag(m_1^2, m_2^2, m_3^2)$. Then, the 3×3 active neutrino mixing matrix will be given by

$$\mathcal{U}_{PMNS} = U_L^\dagger U_\nu \tag{14.22}$$

Due to active-sterile mixing, the U_{PMNS} is no longer unitary. The full (4×4) mixing matrix in the 3+1 scheme is given by

$$U \simeq \begin{pmatrix} (1 - \frac{1}{2}RR^\dagger)\mathcal{U}_{PMNS} & R \\ -R^\dagger \mathcal{U}_{PMNS} & 1 - \frac{1}{2}R^\dagger R \end{pmatrix}, \tag{14.23}$$

where R gives the strength of active-sterile mixing given by

$$R = M_D M_R^{-1} M_S^T (M_S M_R^{-1} M_S^T)^{-1}. \tag{14.24}$$

The three active neutrino mixing angles are determined using the following relations:

$$\sin^2\theta_{12} = \frac{|U_{e2}|^2}{1-|U_{e4}|^2-|U_{e3}|^2}, \qquad \sin^2\theta_{13} = \frac{|U_{e3}|^2}{1-|U_{e4}|^2}, \tag{14.25}$$

$$\sin^2\theta_{23} = \frac{|U_{e3}|^2(1-|U_{e4}|^2)-|U_{e4}|^2|U_{\mu4}|^2}{1-|U_{e4}|^2-|U_{\mu4}|^2} + \frac{|U_{e1}U_{\mu1}+U_{e2}U_{\mu2}|^2(1-|U_{e4}|^2)}{1-|U_{e4}|^2-|U_{e3}|^2)(1-|U_{e4}|^2-|U_{\mu4}|^2}.$$

The free parameters of the model are filtered by constraining the three mixing angles and the ratio of two mass squared differences $r = \sqrt{\frac{\Delta m_{21}^2}{|\Delta m_{3l}^2|}} = m_2/m_3$ using the latest neutrino oscillation data from Nufit 5.20. The absolute scale of the active neutrino mass is fixed by the overall factor $g_4^2 v^2/2\lambda_1$. We have also imposed the Planck cosmological upper limit on the sum of active neutrino masses, $\sum m_i < 0.12$ eV.

14.4 Results

In this section, we discuss the results and predictions of neutrino observables of the model for the normal hierarchy(NH) of neutrino masses only. For model A2, the values of unknown parameters evaluated for the charged lepton sector are shown as variation plots between them in Fig. 14.1a, whereas the components of the Yukawa couplings are plotted in Fig. 14.1b. The variation of Jarlskog invariant J with the active neutrino mixing angles is shown in Fig. 14.2a. The absolute scale of the active neutrino masses is fixed by choosing the overall factor $g_4^2 v^2/2\lambda_1 = 0.01$ eV. Variation of active neutrino masses m_2, m_3 (where $m_1 = 0$ for NH in MES mechanism) with

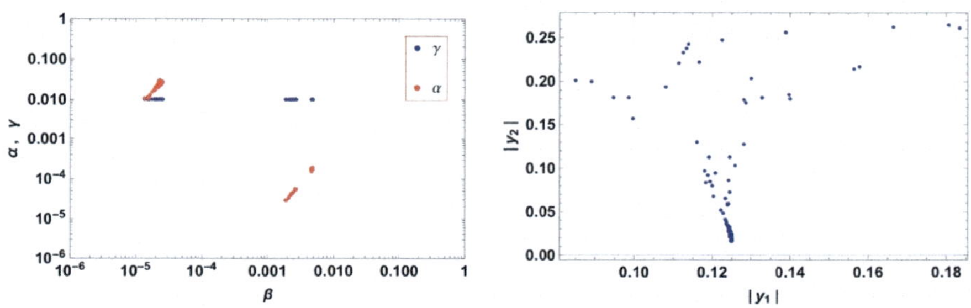

(a) (b)

Fig. 14.1 For model A2: **a**. Plot between coefficients α, β, and γ. **b**. Variation between the Yukawa components $|y_1|$ and $|y_2|$

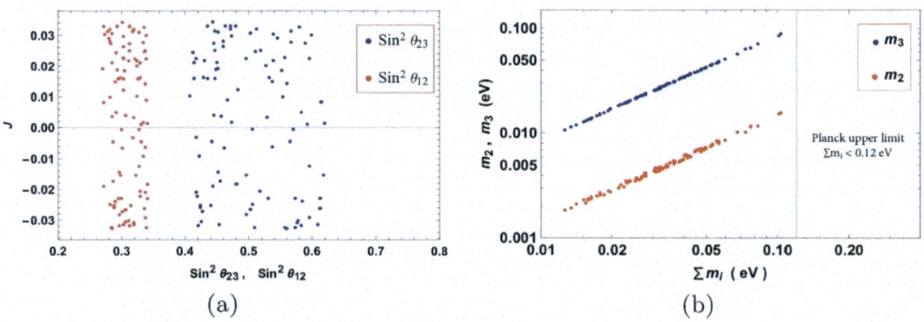

Fig. 14.2 **a**. Plot between Jarlskog invariant J with active neutrino mixing angles $\sin^2\theta_{23}$ and $\sin^2\theta_{12}$ predicted from the model A2. **b**. Variation between m_2, m_3 ($m_1 = 0$) with $\sum m_i$

sum of active neutrino mass $\sum m_i$ is shown in Fig. 14.2b. Similarly, the numerical results of model B1 are shown in Figs. 14.3 and 14.4. We have also calculated the best-fit values of the free parameters and predictions of neutrino observables based on the χ^2 function defined as

$$\chi^2(x_i) = \sum_j \left(\frac{y_j(x_i) - y_j^{bf}}{\sigma_j} \right)^2, \tag{14.26}$$

where x_i are the free parameters in the model and j is summed over the observables $\{\sin^2\theta_{12}, \sin^2\theta_{13}, \sin^2\theta_{23}, r\}$. Here, $y_j(x_i)$ denotes the model predictions for the observables, and y_j^{bf} are their best-fit values obtained from the global analysis. σ_j denotes the corresponding uncertainties obtained by symmetrizing 1σ range of the neutrino observables taken from Nufit 5.2 data. The best-fit values correspond to the condition when $\chi^2(x_i)$ is minimum. We have summarized these values for models A2 and B1 in Table 14.4.

14.5 Summary and Discussion

We have successfully developed a model-building strategy of leptons using modular S_3 symmetry without using any hypothetical flavons in the 3+1 framework of active-sterile neutrino mixing. We obtain a non-diagonal charged lepton mass matrix containing three unknown parameters. Based on assignments of modular weights of SM lepton doublet, we have classified eleven different models with the condition that the maximum allowed weight of the modular form of Yukawa coupling is 6. Out of these, only two models, A2 and B1, are discussed in the present work through numerical analysis constrained by the latest neutrino oscillation data and Planck upper limit on the sum of active neutrino masses $\sum m_i < 0.12$eV. In this work, only the normal

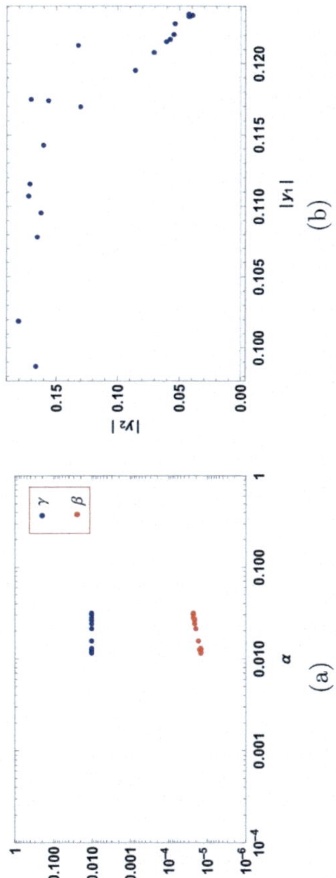

Fig. 14.3 Results for model B1: **a**. Plot between coefficients α, β, and γ. **b**. Variation between the Yukawa components $|y_1|$ and $|y_2|$

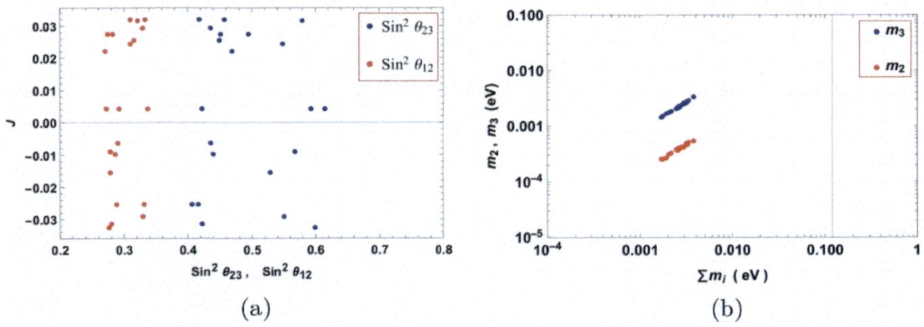

Fig. 14.4 **a**. Plot between Jarlskog invariant J with active neutrino mixing angles $\sin^2\theta_{23}$ and $\sin^2\theta_{12}$ predicted from the model B1. **b**. Variation between m_2, m_3 ($m_1 = 0$) with $\sum m_i$

Table 14.4 Best-fit values of the model parameters and the corresponding predictions of neutrino observables for $\chi^2_{min} = 0.74$ for model $A2$ and $\chi^2_{min} = 6.89$ for model $B1$. Here, the two Majorana phases are ϕ_1 and ϕ_2

Observables	Model A2	Model B1		Model parameters	Model A2	Model B1
$\sin^2\theta_{23}$	0.573	0.568		$\mathrm{Re}[\tau]$	0.438	0.364
$\sin^2\theta_{13}$	0.021	0.022		$\mathrm{Im}[\tau]$	1.14	0.739
$\sin^2\theta_{12}$	0.295	0.278		α	0.0289	0.15×10^{-4}
r	0.173	0.169		β	0.2×10^{-4}	0.0102
$\delta_{CP}/^o$	315.48	344.06		γ	0.0102	0.0123
$\phi_1/^o$	339.55	351.91		g_{o2}	5.996	3.32
$\phi_2/^o$	347.08	354.73		g_{o3}	1.608	1.98
m_2	0.0035 eV	0.00039 eV		λ	0.156	1.34
m_3	0.0202 eV	0.0023 eV		λ_{2o}	3.441	7.89
$\sum m_i$	0.0237 eV	0.00269 eV		λ_{3o}	1.232	–

hierarchy is discussed. The best-fit values of the model predictions of both models suggest $\sin^2\theta_{23} > 0.5$, implying the higher octant of atmospheric mixing angle.

In the numerical analysis, we performed a random scan of about 10^6 iterations of the free parameters and fitted the results of neutrino observables with experimental data. It is important to note that the number of allowed data points obtained from both the models are different: significantly less number of data points are obtained for model B1 compared to A2. Further analysis and phenomenological implications of all the models in normal and inverted hierarchies will be carried out in the future.

Acknowledgements One of the authors(MKS) would like to thank DST-INSPIRE, Govt. of India, for providing the fellowship for the research under INSPIRE fellowship(ID IF180349).

References

1. H. Zhang, Light sterile neutrino in the minimal extended seesaw. Phys. Lett. B **714**, 262–266 (2012)
2. F. Feruglio, Are neutrino masses modular forms? in *From My Vast RepertoireâŁ¦ Guido Altarelli's Legacy* (World Scientific, 2019), pp. 227–266
3. T. Kobayashi, N. Omoto, Y. Shimizu, K. Takagi, M. Tanimoto, T.H. Tatsuishi, Modular a4 invariance and neutrino mixing. J. High Energy Phys. **2018**(11), (2018)
4. D. Meloni, M. Parriciatu, A simplest modular s_3 model for leptons (2023). arXiv:2306.09028
5. D. Zhang, A modular a4 symmetry realization of two-zero textures of the Majorana neutrino mass matrix. Nucl. Phys. B **952**, 114935 (2020)
6. H. Okada, Y. Shoji, A radiative seesaw model with three Higgs doublets in modular a4 symmetry. Nucl. Phys. B **961**, 115216 (2020)
7. G.-J. Ding, S.F. King, X.-G. Liu, Neutrino mass and mixing with a 5 modular symmetry. Phys. Rev. D **100**(11), 115005 (2019)
8. X. Wang, B. Yu, S. Zhou, Double covering of the modular a 5 group and lepton flavor mixing in the minimal seesaw model. Phys. Rev. D **103**(7), 076005 (2021)
9. J.T. Penedo, S.T. Petcov, Lepton masses and mixing from modular s4 symmetry. Nucl. Phys. B **939**, 292–307 (2019)
10. T. Kobayashi, Y. Shimizu, K. Takagi, M. Tanimoto, T.H. Tatsuishi, A_4 lepton flavor model and modulus stabilization from S_4 modular symmetry. Phys. Rev. D **100**, 115045 (2019)

Chapter 15
A Modular A_4 Symmetric Minimal Model for One-Loop T4-2-*i* Topology

Monal Kashav and Surender Verma

Abstract The one-loop neutrino mass models based on the topological features of the dimension-5 Weinberg operator have been systematically categorized into three categories. Notably, within these topological categories, the extension of canonical seesaw scenarios at the one-loop level is of interest given the current LHC run. Besides the one-loop contribution, these extensions yield a prevalent tree-level contribution to neutrino masses. Achieving a dominant one-loop contribution necessitates the amalgamation of flavor symmetries and an expanded field content. Alternatively, we propose the realization of a specific topological Lorentz structure (T4-2-i) associated with the one-loop extension of Type-II seesaw employing modular A4 symmetry. In this realization, no additional fields are required beyond those allowed by the topology itself. The modular weights play a crucial role in suppressing tree-level terms and stabilizing the particles participating in the loop (such as N_i, ρ, and ϕ), rendering them potential dark matter candidates.

15.1 Introduction

The $d = 5$ Weinberg operator, as discussed in [1], allows for lepton number violation with $\Delta L = 2$ and provides insights into the generation of small neutrino masses. This operator gives rise to various UV completion paradigms, including Type-I/II/II seesaw mechanisms. Traditional seesaw mechanisms involve heavy and inaccessible new fields, demanding precise fine-tuned Yukawa couplings. However, a more promising alternative is the one-loop radiative generation framework, which simultaneously explains neutrino mass and dark matter and encompasses LHC-accessible alternative. The diverse avenues within this framework provide opportunities for

M. Kashav (✉) · S. Verma
Department of Physics and Astronomical Science, Central University of Himachal Pradesh, Dharamshala, India
e-mail: monalkashav@gmail.com

S. Verma
e-mail: s_7verma@hpcu.ac.in

J. W. F. Valle et al. (eds.), *Proceedings of the International Conference on Future Prospects in Neutrino and Astroparticle Physics*, Springer Proceedings in Physics 361,
https://doi.org/10.1007/978-981-97-7441-8_15

experimental exploration and hold the potential for significant understanding of neutrino mass generation.

The investigation of the Weinberg operator in both one-loop [2] and two-loop [3] analyses provide a systematic classification of neutrino mass models. These analyses categorize the models into six distinct topologies (T1 to T6) based on their underlying Lorentz structures. Categories i to iii address divergent topological diagrams and finite one-loop extensions of seesaw, respectively. The work proposes realizing the T4-2-*i* topology (contained in category iii) using modular symmetry (A_4), incorporating scalars fields : two inert doublets (ϕ, ρ), a triplet Higgs (Δ), and fermion field in the loop (ψ). A_4 modular symmetry efficiently suppresses tree-level contributions, enhancing the model's economic viability. Lepton number violation is introduced through the fermions as right-handed(RH) neutrinos in the loop and the work includes a detailed analysis of the model's viability based on current neutrino oscillation data. This proposed model grounded in modular symmetry presents a more economical solution compared to prior attempts [4] addressing neutrino mass and realizing the topological Lorentz structure.
The paper is structured with Sect. 15.2 discussing the proposed realization of T4-2-*i* topology using A_4 modular symmetry. This includes general features of the topology and a scenario for suppressing tree-level couplings. Section 15.3 presents numerical analysis based on the proposed model. Finally, we conclude in Sect. 15.4.

15.2 The Model

The T4-2-*i* topology, categorized within T4, serves as a one-loop extension within the Type-II seesaw framework. This configuration incorporates scalars fields, i.e., two inert doublets (ϕ, ρ), a triplet Higgs (Δ), and RH neutrinos (N_i) (refer to Fig. 15.1a). If ϕ, ρ acquires vev, the one-loop level structure simplifies to the standard Type-II seesaw's tree-level version, with both tree-level as well as one-loop level contributions to neutrino mass generation. To achieve the T4-2-*i* topology realization, inhibiting the tree-level mass term while allowing the $\mu_H H \Delta^\dagger H$ vertex under the new symmetry invariance is crucial for a dominant one-loop contribution.
Under the A4 modular symmetry, Yukawa couplings, typically treated as constants in symmetric models, are assigned even modular weights. In these models, matter fields possess integral modular weights to ensure the sum of weights is zero for each term in the invariant Lagrangian. In the context of the T4-2-*i* topology, characterized by a dominant one-loop contribution and the absence of tree-level mass term emanating from triplet Higgs, the presence of the $\mu_H H \Delta^\dagger H$ vertex is pivotal. To maintain the model's integrity, inhibiting Dirac neutrino mass terms originating from the Higgs field coupling with lepton doublets and RH Majorana neutrinos becomes necessary. Furthermore, vertices featuring couplings μ_Δ, y, and y' are indispensable for the one-loop generation of neutrino mass through the scalar triplet.

Within the framework of A4 modular symmetry, the model selectively incorporates specific fields in accordance with the T4-2-*i* topology namely, two inert doublets

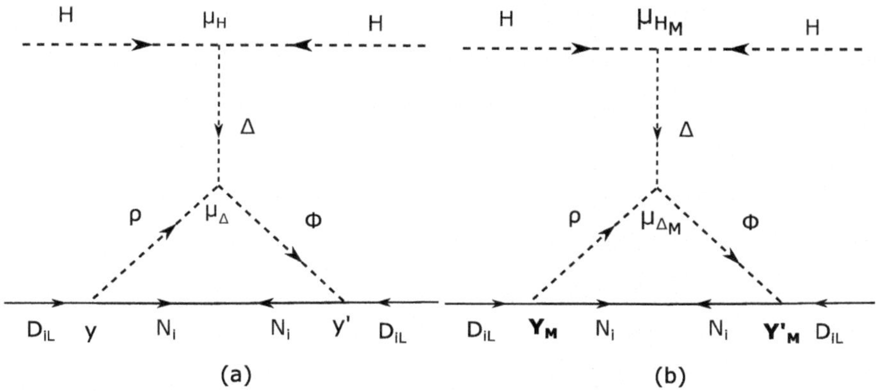

Fig. 15.1 **a** illustrates T4-2-*i* topological structure that exhibits constant Yukawa couplings, while **b** illustrates A4 modular symmetry that is employed to prevent tree-level contribution. The couplings $\mu_{\mathbf{H_M}}$, $\mu_{\Delta_\mathbf{M}}$, $\mathbf{Y_M}$, and $\mathbf{Y'_M}$ undergo transformations under A_4 modular symmetry

Table 15.1 The assignment of irreducible representations to fields employed in the model including their modular weights

Symmetry				
Particle content	$SU(2)_L$	$U(1)_Y$	A_4	$-k_I$
$(\bar{L}_{eL}, \bar{L}_{\mu L}, \bar{L}_{\tau L})$	(2,2,2)	(1,1,1)	$(1, 1'', 1')$	(2,2,2)
(e_R, μ_R, τ_R)	(1,1,1)	(–2,–2,–2)	$(1, 1', 1'')$	(–2,–2,–2)
(N_1, N_2, N_3)	(1,1,1)	(0,0,0)	$(1, 1', 1'')$	(–3,–3,–3)
$(H, \Delta, \tilde{\phi}, \rho)$	(2,3,2,2)	(1,2,–1,–1)	(1,1,1,1)	(0,0,–3,–3)

Table 15.2 The Yukawa couplings of different order having the A_4 irreducible representations and modular weight

Y_m^n	Y_1^4	$Y_{1'}^4$	Y_1^6	Y_1^8	$Y_{1'}^8$	$Y_{1''}^8$	Y_1^{10}	$Y_{1'}^{10}$
A_4	1	$1'$	1	1	$1'$	$1''$	1	$1'$
$-k_I$	4	4	6	8	8	8	10	10

(ϕ, ρ) and one triplet Higgs (Δ), RH Majorana neutrinos integrated into the loop structure [5]. The detailed field content and their respective modular weights can be found in Tables 15.1 and 15.2. Notably, to ensure stability and identify them as plausible dark matter candidates, odd weights are assigned to ϕ and ρ, as outlined in Table 15.1. Additionally, these assignments necessitate assigning a modular weight of 6 to the μ_{Δ_M} mass coupling, a requirement for enabling the $\mu_{\Delta_M}\Delta\phi^\dagger\rho$ vertex within the modular symmetry framework.

The Yukawa interactions concerning charged leptons within the framework of A_4 modular invariance result in a diagonal charged lepton mass matrix. By employing the modular transformations of fields outlined in Table 15.1 and Yukawa couplings in

Table 15.2, the Yukawa Lagrangian governing the one-loop contribution to neutrino mass is

$$\begin{aligned}\mathcal{L}_{II} &= p_1(\bar{L}_{eL}\rho N_1 Y_1^4) + p_2(\bar{L}_{\mu L}\rho N_2 Y_1^4) + p_3(\bar{L}_{\tau L}\rho N_3 Y_1^4)\\ &+ q_1(\bar{L}_{eL}\rho N_3 Y_{1'}^4) + q_2(\bar{L}_{\mu L}\rho N_1 Y_{1'}^4) + q_3(\bar{L}_{\tau L}\rho N_2 Y_{1'}^4)\\ &+ r_1(\bar{L}_{eL}\tilde{\phi} N_1 Y_1^4) + r_2(\bar{L}_{\mu L}\tilde{\phi} N_2 Y_1^4) + r_3(\bar{L}_{\tau L}\tilde{\phi} N_3 Y_1^4)\\ &+ s_1(\bar{L}_{eL}\tilde{\phi} N_3 Y_{1'}^4) + s_2(\bar{L}_{\mu L}\tilde{\phi} N_1 Y_{1'}^4) + s_3(\bar{L}_{\tau L}\tilde{\phi} N_2 Y_{1'}^4)\\ &+ M_1' \bar{N_1^c} N_1 Y_1^6 + M_2'(\bar{N_2^c} N_3 Y_1^6 + \bar{N_3^c} N_2 Y_1^6),\end{aligned} \tag{15.1}$$

such that p_i, q_i, r_i, and s_i, with $i = 1, 2, 3$, represent the coupling constants. The role of modular weights is analogous to that of Z_2 symmetry, playing a pivotal role in the stabilization of potential dark matter candidates and the inhibition at the tree-level neutrino mass term. Specifically, odd modular weights serve as Z_2 odd charges, ensuring the stability of fields dictated by the underlying topology. These fields encompass not only scalar doublets (ϕ, ρ) but also RH Majorana neutrinos (N_1, N_2, N_3) actively participating in the loop structure.

It's crucial to emphasize that the scalar doublets (ϕ, ρ) remain inert, devoid of any vacuum expectation value (vev), highlighting their unique characteristics within this theoretical framework. The formulation of the Lagrangian, expressed in detail in (15.1), yields the emergence of the subsequent Dirac Yukawa matrices, contributing to the intricate dynamics under consideration as

$$y_\rho = \begin{pmatrix} p_1 Y_1^4 & 0 & q_1 Y_{1'}^4 \\ q_2 Y_{1'}^4 & p_2 Y_1^4 & 0 \\ 0 & q_3 Y_{1'}^4 & p_3 Y_1^4 \end{pmatrix}, \quad y_\phi = \begin{pmatrix} r_1 Y_1^4 & 0 & s_1 Y_{1'}^4 \\ s_2 Y_{1'}^4 & r_2 Y_1^4 & 0 \\ 0 & s_3 Y_{1'}^4 & r_3 Y_1^4 \end{pmatrix}. \tag{15.2}$$

Furthermore, the matrix M_R representing the masses of RH Majorana neutrinos is

$$M_R = \begin{pmatrix} M_1' Y_1^6 & 0 & 0 \\ 0 & 0 & M_2' Y_1^6 \\ 0 & M_2' Y_1^6 & 0 \end{pmatrix}, \tag{15.3}$$

where M_k' (where $k = 1, 2$) represents mass scale of the bare mass terms. The masses of the RH neutrinos, M_k (where $k = 1, 2, 3$), are derived by diagonalizing M_R through the unitary matrix U_R, i.e., $diag(M_1, M_2, M_3) = U_R M_R U_R^T$. Following the diagonalization, the Dirac Yukawa matrices are $Y_\rho = y_\rho U_R$ and $Y_\phi = y_\phi U_R$, in the basis in which RH neutrino mass matrix is diagonal. The one-loop contribution to the neutrino mass devoid of tree-level contribution is expressed as [2]

$$M_\nu = -\sum_k \frac{\mu_{\Delta_M} v^2}{m_\Delta^2} \mu_{H_M} [Y_\rho (M_k) Y_\phi^T] \mathcal{I}(M_\rho^2, M_\phi^2, M_k^2), \tag{15.4}$$

where μ_{Δ_M} and μ_{H_M} are couplings at vertices $\mu_{H_M} H \Delta^{\dagger} H$ and $\mu_{\Delta_M} \Delta \phi^{\dagger} \rho$ (Fig. 15.1b), respectively. The loop factor in the (15.4) is given by

$$\mathcal{J}(M_{\rho}^2, M_{\phi}^2, M_k^2) = -\Big(\frac{1}{4\pi} \frac{M_{\rho}^2}{(M_{\rho}^2 - M_{\phi}^2)(M_{\rho}^2 - M_k^2)} \ln \frac{M_k^2}{M_{\rho}^2} + \frac{M_{\phi}^2}{(M_{\phi}^2 - M_{\rho}^2)(M_{\phi}^2 - M_k^2)} \ln \frac{M_k^2}{M_{\phi}^2}\Big). \quad (15.5)$$

The neutrino mass matrix derived from our analysis exhibits complex and asymmetric nature. Achieving a complex symmetric M_{ν} imposes specific conditions on the coupling constants, necessitating the $p_i = r_i$ and $q_i = s_i$ for $i = 1, 2, 3$. It is important to note that there exist no non-trivial relations among Yukawa couplings, as explored in [4], due to the distinctive transformation properties of these couplings under A_4 modular symmetry.

It is worth highlighting that the intricate dynamics of Yukawa couplings, discussed in [4], introduce non-trivial relations that remain unconsidered in our current analysis. This omission is attributed to the specific transformations that these relations undergo within the framework of A4 modular symmetry. An additional noteworthy aspect is the inherent odd modular weight assigned to loop particles, particularly RH Majorana neutrinos (Ni) and inert doublets (ρ, ϕ). This characteristic renders these particles potential candidates for dark matter, adding an intriguing dimension to our exploration of the theoretical landscape.

Hence, the expression governing neutrino masses, denoted as M_{ν}, is defined by the proportional relation $M_{\nu} = \mathcal{P}(Y_{\rho} Y_{\phi}^T)$, where the encompassing scale factor $\mathcal{P}$ is defined as $M_1 \mathcal{J}(M_{\rho}^2, M_{\phi}^2, M_1^2)$. This yields $M_{\nu} = \mathcal{P} \tilde{M}_{\nu}$, with $\tilde{M}_{\nu}$ representing the product $Y_{\rho} Y_{\phi}^T$. The resulting neutrino masses, denoted as $m_i = \mathcal{P} \tilde{m}_i$ for $i = 1, 2, 3$, are expressed in terms of the mass eigenvalues $\tilde{m}_i$ originating from $Y_{\rho} Y_{\phi}^T$.

In the scenario of normal hierarchy, the determination of the scale factor involves the atmospheric mass-squared difference Δm_{31}^2, specifically given by $\mathcal{P}^2 = \Delta m_{31}^2 / (\tilde{m}_3{}^2 - \tilde{m}_1{}^2)$. Subsequently, the solar mass-square splitting is related to $\mathcal{P}$ as $\Delta m_{21}^2 = \mathcal{P}^2 (\tilde{m}_2{}^2 - \tilde{m}_1{}^2)$. The mixing matrix ($U$) dictating neutrino flavor mixing involves the diagonalization of M_{ν} through the transformation $diag(m_1, m_2, m_3) = V^T M_{\nu} V$. Due to the diagonal charged lepton mass matrix M_l, the lepton mixing matrix corresponds directly to the neutrino mixing matrix and is represented as $V = U_{PMNS}$. Here, U_{PMNS} denotes the Pontecorvo-Maki-Nakagawa-Sakata matrix. The mixing angles are obtained as

$$\sin^2 \theta_{13} = |V_{13}|^2, \quad \sin^2 \theta_{12} = \frac{|V_{12}|^2}{1 - |V_{13}|^2}, \quad \sin^2 \theta_{23} = \frac{|V_{23}|^2}{1 - |V_{13}|^2}. \quad (15.6)$$

The Jarlskog rephasing CP invariant is expressed by $J_{CP} = \text{Im}\left[V_{11} V_{22} V_{12}^* V_{21}^*\right]$ whereas the CP invariants associated with (α, β) are $I_1 = \text{Im}\left[V_{11}^* V_{12}\right]$ and $I_2 = \text{Im}\left[V_{11}^* V_{13}\right]$. One crucial parameter worth investigating is the effective Majorana neutrino mass, represented by M_{ee}, offering valuable insights into the characterization of neutrinos as either Dirac or Majorana particles. The formulation for the effective Majorana mass is $M_{ee} = \left|\sum_{i=1}^{3} V_{1i} m_i\right|$.

15.3 Numerical Analysis

The theoretical framework discussed above governing neutrino masses involves several factors, encompassing a common scale factor, Yukawa couplings denoted as p_i and q_i for $i = 1, 2, 3$, and a modular weight of 4. This modular weight system, extending to higher orders such as 4, 6, 8..., intricately links with those of weight 2, all of which are intricately influenced by the complex modulus τ. The numerical analysis incorporates the random variation of coupling constants (p_i, q_i), real and imaginary parts of τ, and masses of beyond standard model (BSM) fields within predefined ranges. In particular, $p_i, q_i \in [0.01, 1]$ whereas real $(|Re(\tau)|)$ and imaginary part $(Im(\tau))$ of τ are varied in ranges $[0, 0.5]$ & $Im(\tau) \in [0, 1]$. Also, masses of RH neutrino masses (M_k) are varied in ranges $[1, 10^3]\,\text{GeV}(M_1 < M_2 < M_3)$ and masses of inert scalar fields are randomly varied as $M_{\rho,\phi} \in [1, 5]\,\text{TeV}$. To obtain the mass eigenvalues $\tilde{m}_i$ and diagonalize the neutrino mass matrix $\tilde{M}_\nu$, we determine the scale factor $\mathcal{K}$ by comparing it with the atmospheric mass-squared difference (Δm^2_{31}). The neutrino mass eigenvalues are then calculated as $m_i = \mathcal{K}\tilde{m}_i$, and the corresponding mixing angles are computed, ensuring conformity with the experimental constraints on neutrino mixing angles and the Δm^2_{21}. Moreover, the Yukawa couplings obey restrictions imposed by the perturbative threshold $Y \leq \sqrt{4\pi}$ [6]. The complex modulus parameter space (τ) is visualized in Fig. 15.2, highlighting that the imaginary part of τ surpassing one acts as a significant source of CP violation. The model predicts the mixing angles as represented in Fig. 15.3a and b, exhibiting consonance to the 3σ experimental ranges [7].

Moreover, in the investigation of M_{ee} depicted in the correlation plot of $\sum m_i$ and M_{ee} (Fig. 15.4), the model predicts typical characteristics observed in the $\sum m_i - M_{ee}$ correlation plot which suggests that M_{ee} might tend toward zero in proximity to the lower limit of $\sum m_i$. Additionally, as $\sum m_i$ approaches the cosmological upper limit [8, 9], there is an observable upward shift in the lower bound of M_{ee}.

Furthermore, CP violation in the model stems from the complex modulus τ, exerting influence on CP invariants such as the Jarlskog CP invariant J_{CP} and Majorana phases (α, β), illustrated by I_1, I_2 in Fig. 15.5. The graphical representations in Fig. 15.5a and b clearly indicate that the magnitudes of $|J_{CP}|$ and $|I_2|$ are constrained within the range of 0.15, while the magnitude of $|I_1|$ is limited to 0.45. The model anticipates the coexistence of both CP conserving as well as violating solutions. The proposed model does not satisfy the neutrino oscillation data for the inverted hierarchy.

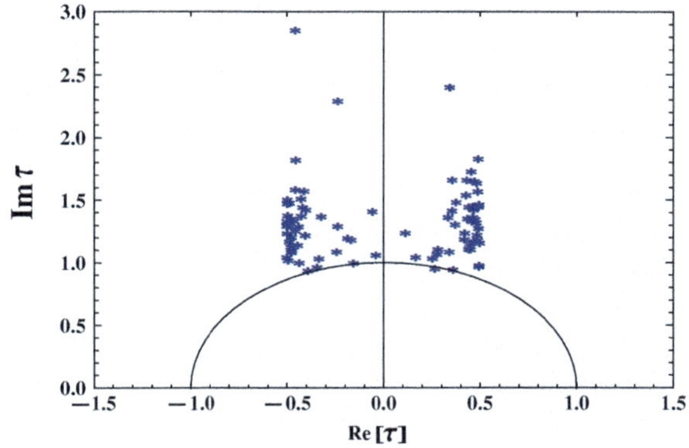

Fig. 15.2 The viable region for the complex modulus τ while respecting the fundamental domain of the complex plane

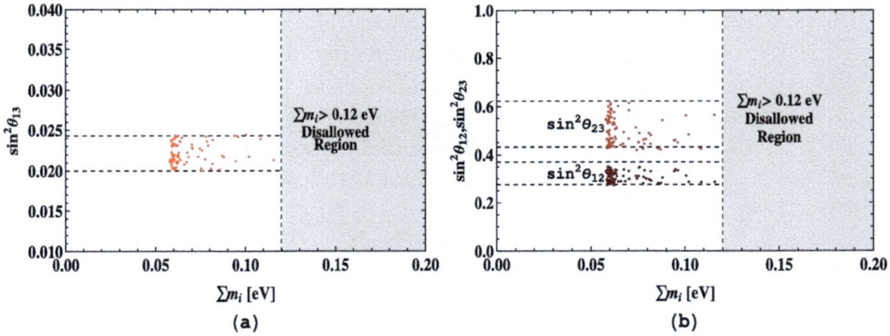

Fig. 15.3 The depiction shows the correlation between neutrino mixing angles and the sum of neutrino masses ($\sum m_i$) under the conditions of Normal hierarchy. The exclusion of the gray shaded area aligns with cosmological restrictions on the cumulative mass of neutrinos [8, 9]. Each horizontal line represents the 3σ intervals for respective mixing angles

15.4 Conclusions

Among the various one-loop Weinberg operator topologies in dimension-5, the T4 topology's finite Lorentz structures play a crucial role as a one-loop extension to conventional seesaw scenarios. Categorized under (iii), the T4 topology allows for a dominant tree-level contribution alongside the one-loop mechanism for neutrino mass, even when discrete or $U(1)$ symmetry is imposed [2]. Unlike previous strategies that required expanded field content or higher-dimensional Weinberg operators to suppress the tree-level contribution, our proposal utilizes modular symmetry to achieve the same goal without additional fields.

The proposed model demonstrates the symmetry realization of the T4-2-*i* Lorentz structure within the framework of A_4 modular symmetry, leading to the generation of

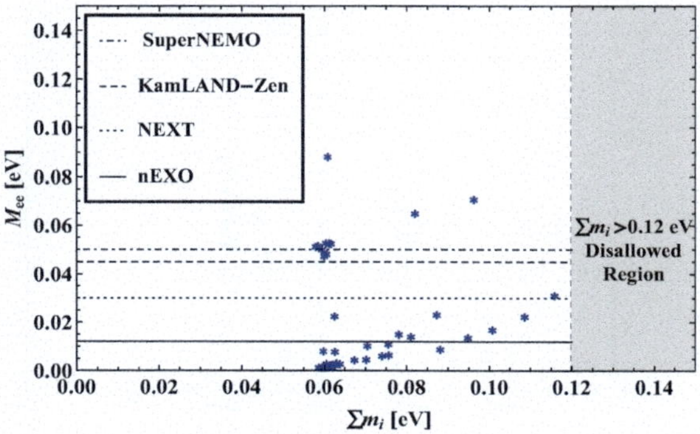

Fig. 15.4 The correlation plot illustrates the Effective Majorana mass (M_{ee}) with respect to the sum of neutrino masses ($\sum m_i$) in the context of normal hierarchical neutrino mass distribution. Expected sensitivities of $0\nu\beta\beta$ decay experiments such as SuperNEMO [10], KamLAND-Zen [11], NEXT [12, 13], and nEXO [14] are indicated by the horizontal lines

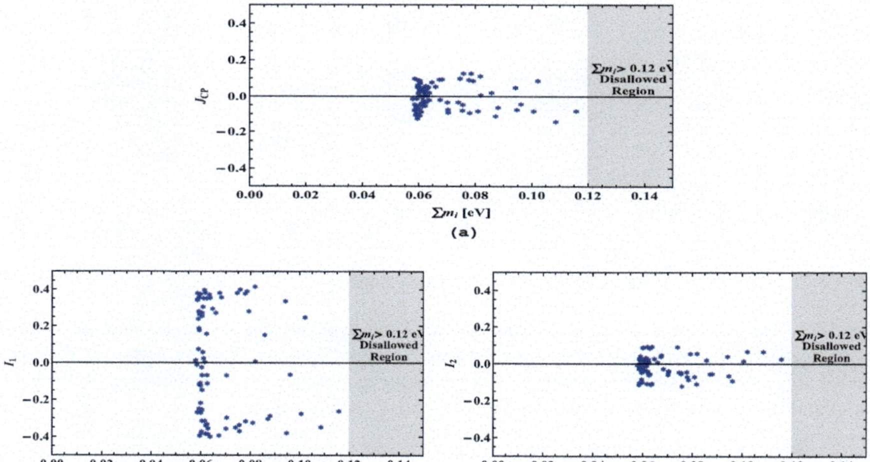

Fig. 15.5 The CP invariants, specifically J_{CP}, I_1, and I_2, in relation to the sum of neutrino masses, denoted as $\sum m_i$, are depicted

neutrino mass through a one-loop dimension-5 Weinberg operator. The intentional assignment of an odd modular weight to loop particles, including RH Majorana neutrinos (Ni) and inert doublets (ρ, ϕ), acts as a robust mechanism ensuring the stability of dark matter. Our model's predictions on neutrino mixing angles and mass-squared differences are in consonance with the latest neutrino oscillation data. In addition, we extensively explore the implications for the effective Majorana mass M_{ee} and CP invariants, as illustrated in Figs. 15.4 and 15.5, respectively. Notably, for a sum of neutrino masses ($\sum m_i$) totaling 0.10 eV, the effective Majorana mass M_{ee} exceeds 0.012 eV at 3σ, well within the sensitivity range of experiments probing $0\nu\beta\beta$ decay. This investigation into the intricate interplay among coupling constants, modular weights, and field stability is pivotal for advancing our comprehension of the proposed theoretical model.

Acknowledgements M. K. expresses gratitude for the financial backing received from the Department of Science and Technology (DST), Government of India, under Grant No. DST/INSPIRE Fellowship/2018/IF180327. The authors also extend their appreciation to the Department of Physics and Astronomical Science for facilitating the essential infrastructure for conducting this research.

References

1. S. Weinberg, Phys. Rev. Lett. **43**, 1566–1570 (1979)
2. F. Bonnet, M. Hirsch, T. Ota, W. Winter, JHEP **07**, 153 (2012)
3. D. Aristizabal Sierra, A. Degee, L. Dorame, M. Hirsch, JHEP **03**, 040 (2015)
4. M.A. Loualidi, M. Miskaoui, Nucl. Phys. B **961**, 115219 (2020)
5. M. Kashav, S. Verma, JCAP **03**, 010 (2023)
6. T. Nomura, H. Okada, O. Popov, Phys. Lett. B **803**, 135294 (2020)
7. I. Esteban, M.C. Gonzalez-Garcia, M. Maltoni, T. Schwetz, A. Zhou, JHEP **09**, 178 (2020)
8. E. Giusarma, M. Gerbino, O. Mena, S. Vagnozzi, S. Ho, K. Freese, Phys. Rev. D **94**, 083522 (2016)
9. N. Aghanim et al., Planck. Astron. Astrophys. **641**, A6 (2020)
10. A.S. Barabash, J. Phys. Conf. Ser. **375**, 042012 (2012)
11. A. Gando et al., KamLAND-Zen. Phys. Rev. Lett. **117**, 082503 (2016)
12. F. Granena et al., [NEXT], arXiv:0907.4054 [hep-ex]
13. J.J. Gomez-Cadenas et al., NEXT. Adv. High Energy Phys. **2014**, 907067 (2014)
14. C. Licciardi [nEXO], J. Phys. Conf. Ser. **888** 012237 (2017)

Chapter 16
Cosmological Signatures of Dirac Neutrino Portal Dark Matter

Nayan Das and Debasish Borah

Abstract We consider a scenario where light active neutrinos are of Dirac nature and their right chiral parts act as a portal between the standard model and dark matter (DM). Depending upon this portal coupling, DM can either be in thermal or non-thermal ballpark. A gauged $U(1)_{B-L}$ symmetry accommodates right chiral parts of Dirac neutrinos naturally and ensures DM stability. Due to portal and $U(1)_{B-L}$ interactions, it is possible to have additional relativistic degrees of freedom in the form of light Dirac neutrinos which can be probed by cosmic microwave background measurements of ΔN_{eff}. The parameter space also remains within reach of terrestrial experiments like collider.

16.1 Introduction

Within the standard model (SM) of particle physics, the particle nature of dark matter (DM) and the generation of non-zero neutrino mass stand out as two significant challenges [1, 2]. While the particle nature of DM and its interactions with the SM apart from gravity are not yet known, the nature of active neutrinos: Majorana or Dirac, also remains a mystery. While the origin of Majorana active neutrino mass requires the introduction of heavy fields for canonical seesaw mechanism, light Dirac neutrino scenario needs to accommodate the right chiral parts. The doubling of light neutrino degrees of freedom (dof) in Dirac neutrino scenario could increase the effective relativistic dof denoted by N_{eff} which can be probed at future cosmic microwave background (CMB) experiment.

In this paper, we consider the active neutrinos to be Dirac fermions and explore the connection of their Dirac nature to the origin of particle dark matter. In particular, the right chiral part of Dirac neutrino acts like a portal between the SM bath and DM, dubbed as the Dirac neutrino portal dark matter (DNPDM) [3, 4]. A gauged $B - L$ symmetry, one of the most popular beyond standard model (BSM) scenario,

N. Das (✉) · D. Borah
Department of Physics, Indian Institute of Technology Guwahati, Assam 781039, India
e-mail: nayan.das@iitg.ac.in

J. W. F. Valle et al. (eds.), *Proceedings of the International Conference on Future Prospects in Neutrino and Astroparticle Physics*, Springer Proceedings in Physics 361,
https://doi.org/10.1007/978-981-97-7441-8_16

is considered which automatically includes three copies of right-handed neutrinos (ν_R) from the requirement of anomaly cancellations [5]. The interaction between ν_R and DM, assumed to be a singlet Dirac fermion having vanishing $B-L$ charge, arises via a singlet scalar ϕ_1. Since ν_R, ϕ_1 couples to the SM bath via Yukawa, gauge as well as scalar portal interactions, the $\nu_R - \phi_1$-DM coupling provides the required DM-SM portal. The strength of this coupling can keep the DM in thermal or non-thermal ballpark giving rise to the popular DM scenarios namely the weakly interacting massive particle (WIMP) and feebly interacting massive particle (FIMP), respectively. While DM coupling can be small or large, we keep the $B-L$ portal interactions sizeable to keep the detection prospects at colliders promising. This also gives rise to a thermal contribution to $N_{\rm eff}$ due to thermal production of relativistic ν_R's. The ν_R coupling with DM can also provide another source of $N_{\rm eff}$. While direct detection cross-section of DM remains loop-suppressed, we can still put new constraints on the gauged $B-L$ parameter space, not obtained earlier.

This paper is structured as follows. In Sect. 16.2, we provide a concise overview of the model. Section 16.3 delves into the details of DM, Neff, relevant constraints, and phenomenology. The paper concludes in Sect. 16.4.

16.2 The Model

In the gauged $B-L$ extension of the SM considered here, we have two singlet scalars $\phi_{1,2}$, a singlet Dirac fermion ψ, and three species of right-handed neutrinos (ν_R), on top of the SM matter content. The three ν_R's with $B-L$ charges -1 each keep the model anomaly free and also lead to light Dirac neutrino mass via coupling with the Standard Model Higgs and ν_L. The singlet scalars ϕ_1, ϕ_2 and Dirac fermion ψ have $B-L$ charges 1, 3, and 0, respectively. The scalar part of the Lagrangian is

$$\mathcal{L}_{\rm s} = (D^{\mu}\Phi)^{\dagger}(D_{\mu}\Phi) + (D^{\mu}\phi_1)^{\dagger}D_{\mu}\phi_1 + (D^{\mu}\phi_2)^{\dagger}D_{\mu}\phi_2 - V(\Phi, \phi_1, \phi_2), \quad (16.1)$$

where

$$D_{\mu}\Phi = \left(\partial_{\mu} - i\frac{g}{2}\tau_a W_{\mu}^a - i\frac{g'}{2}B_{\mu}\right)\Phi,\ D_{\mu}\phi_1 = \left(\partial_{\mu} - ig_{BL}B'_{\mu}\right)\phi_1,$$
$$D_{\mu}\phi_2 = \left(\partial_{\mu} - i3g_{BL}B'_{\mu}\right)\phi_2, \quad (16.2)$$

denote the respective covariant derivatives. The scalar potential arising only from the new scalar fields and the SM Higgs Φ is

$$V(\Phi, \phi_1, \phi_2) = -\mu^2(\Phi^{\dagger}\Phi) + \mu_1^2(\phi_1^{\dagger}\phi_1) - \mu_2^2(\phi_2^{\dagger}\phi_2) + \lambda(\Phi^{\dagger}\Phi)^2 + \lambda_1(\phi_1^{\dagger}\phi_1)^2 + \lambda_2(\phi_2^{\dagger}\phi_2)^2$$
$$+ \lambda_{H\phi_1}(\Phi^{\dagger}\Phi)(\phi_1^{\dagger}\phi_1) + \lambda_{H\phi_2}(\Phi^{\dagger}\Phi)(\phi_2^{\dagger}\phi_2) + \lambda_{\phi_1\phi_2}(\phi_1^{\dagger}\phi_1)(\phi_2^{\dagger}\phi_2). \quad (16.3)$$

ϕ_2 gets non-zero vacuum expectation value (VEV), $\langle\phi_2\rangle = v_2$, leading to the spontaneous breaking of $U(1)_{B-L}$ gauge symmetry. ϕ_1 does not get any VEV and remains heavier than ψ. This makes ψ stable thereby making it a suitable dark matter candidate. Note that in principle ϕ_1 can be a dark matter candidate as well if $m_{\phi_1} < m_\psi$. However, through out the paper we consider ϕ_1 to be heavier than ψ, making ψ the only dark matter candidate. The Yukawa part of the Lagrangian is

$$-\mathcal{L}_{\rm Y} \supset Y_\nu \overline{L}\tilde{\Phi}\nu_R + y_{\phi_1}\overline{\psi}\phi_1\nu_R + m_\psi\overline{\psi}\psi. \tag{16.4}$$

The first term in the above equation gives mass to light Dirac neutrinos and it is not expected to play any phenomenological roles in DM as well as $N_{\rm eff}$ due to small Y_ν. The second term in (16.4) denotes the Yukawa coupling between new particles ϕ_1, ψ, ν_R and plays an important role for the phenomenology of the model. The relevant free parameters for our discussions are m_{ϕ_1}, $m_{Z'}$ (Z' denotes $U(1)_{B-L}$ gauge boson), m_ψ along with Higgs portal coupling $\lambda_{H\phi_1}$, $U(1)_{B-L}$ gauge coupling g_{BL}, and the Yukawa coupling y_{ϕ_1}.

16.3 Dark Matter with Observable $\Delta N_{\rm eff}$

The effective number of relativistic species is defined as

$$N_{\rm eff} \equiv \frac{8}{7}\left(\frac{11}{4}\right)^{4/3}\left(\frac{\rho_{\rm rad}-\rho_\gamma}{\rho_\gamma}\right), \tag{16.5}$$

where $\rho_{\rm rad}$ is the net radiation content of the universe and ρ_γ is the photon energy density. The change in the parameter ($\Delta N_{\rm eff} = N_{\rm eff} - N_{\rm eff}^{\rm SM}$) for 3 ν_R is $\Delta N_{\rm eff} = 3\left.\frac{\rho_{\nu_R}}{\rho_{\nu_L}}\right|_{10\rm MeV}$. For a light thermal BSM species, its energy density at 10 MeV can be written in terms of its decoupling temperature whereas for a non-thermal light BSM species, proper Boltzmann equation needs to be solved to find its energy density. Planck 2018 data gives current bound on $\Delta N_{\rm eff} < 0.28$ at 2σ. In future, CMB experiments like CMB Stage IV (CMB S4) can reach upto $\Delta N_{\rm eff} = 0.06$ [6]. Since the size of Yukawa coupling y_{ϕ_1} can significantly affect DM and $N_{\rm eff}$ phenomenology as well as relevant constraints, we discuss the two broad cases of FIMP and WIMP type DM separately.

16.3.1 FIMP Type DM

The production of dark matter ψ depends on the Yukawa coupling y_{ϕ_1}. For $y_{\phi_1} < 10^{-7}$, the dark matter is produced via freeze-in mechanism. Along with the dark matter, non-thermal ν_R are also produced from freeze-in. This gives rise to a non-thermal contribution to $\Delta N_{\rm eff}$ denoted by $\Delta N_{\rm eff}^{\rm non-th}$. However, due to the presence of $U(1)_{B-L}$ coupling, we can have thermalized ν_R also. The thermalization occurs via the resonant enhancement of the process $f\bar{f} \leftrightarrow \nu_R\bar{\nu}_R$, where f denotes SM fermions. So we have a thermal contribution (denoted by $N_{\rm eff}^{\rm th}$) along with the non-thermal one. Note that the non-thermal contribution arises only when the decoupling of thermalized ν_R precedes the production of non-thermal ν_R. Although thermal contribution depends on g_{BL} and $m_{Z'}$ only, the non-thermal contribution depends on other parameters of the model as well along with g_{BL} and $m_{Z'}$.

To study the non-thermal contribution, we consider a situation where ϕ_1 stays in bath initially and then it freezes-out. Due to Yukawa coupling y_{ϕ_1}, ϕ_1 decay and produce both the DM and dark radiation (or ν_R). The final freeze-out abundance of ϕ_1 depends on the parameters m_{ϕ_1}, $m_{Z'}$, $\lambda_{H\phi_1}$, and g_{BL}. To calculate the evolution of each species, we solve the respective Boltzmann equations as follows.

$$\frac{dY_{\phi_1}}{dx} = \frac{\beta s}{\mathcal{H}x}\left(-\langle\sigma v\rangle_{\phi_1\phi_1^\dagger\to X\bar{X}}\left((Y_{\phi_1})^2-(Y_{\phi_1}^{\rm eq})^2\right)-\frac{\Gamma_{\phi_1}}{s}\frac{K_1(x)}{K_2(x)}Y_{\phi_1}\right), \tag{16.6}$$

$$\frac{dY_\psi}{dx} = \frac{\beta}{x\mathcal{H}}\Gamma_{\phi_1}\frac{K_1(x)}{K_2(x)}Y_{\phi_1}, \tag{16.7}$$

$$\frac{d\widetilde{Y}}{dx} = \frac{\beta}{\mathcal{H}s^{1/3}x}\langle E\Gamma\rangle_{\phi_1}Y_{\phi_1}, \tag{16.8}$$

where Y_{ϕ_1} (Y_ψ) denotes comoving number density of ϕ_1 (ψ) and $\tilde{Y}$ denotes comoving energy density of ν_R. $\mathcal{H}$ represents the Hubble expansion rate. The variable x is defined as $x = m_{\phi_1}/T$ and $\beta = 1 + \frac{T}{3g_*^s}\frac{dg_*^s}{dT}$ with g_*^s denoting the effective relativistic entropy degrees of freedom and K_1 and K_2 are the modified Bessel functions of first and second kind, respectively. The decay rate and energy average decay rate are denoted by Γ_{ϕ_1} and $\langle E\Gamma\rangle_{\phi_1}$, respectively.

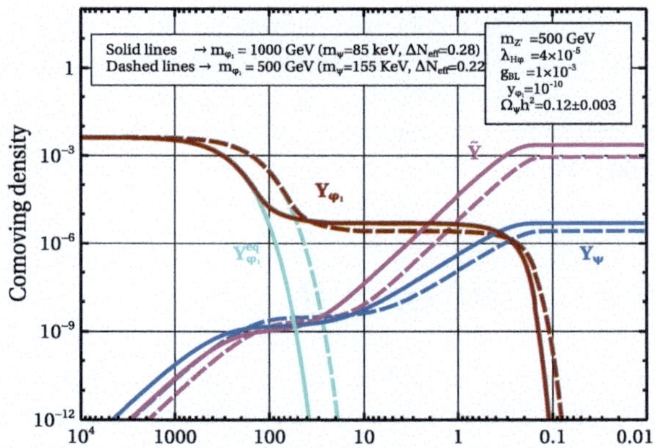

Fig. 16.1 Evolution of Y_{ϕ_1}, Y_ψ, and Y_{ν_R} for two different choices of m_{ϕ_1}

In Fig. 16.1, the comoving density evolution of ϕ_1, ψ, and ν_R is depicted for two different values of m_{ϕ_1}. The dark red and blue lines represent the comoving number density of ϕ_1 and ψ, respectively. The comoving energy density of ν_R is represented by the magenta lines. The cyan lines denote equilibrium comoving number densities of ϕ_1. The $\Delta N_{\rm eff}$ values in the legend bear both the thermal and non-thermal contribution. As the g_{BL} and $m_{Z'}$ are same in both solid and dashed lines, we have same thermal contribution for both the lines ($\Delta N_{\rm eff}^{\rm th} = 0.18$). Due to smaller comoving number abundance in dashed line, we have reduced non-thermal contribution ($\Delta N_{\rm eff}^{\rm non-th} = 0.04$) for the dashed line compared to that for the solid line ($\Delta N_{\rm eff}^{\rm th} = 0.10$).

Astrophysical structure formation can also provide strong constraint on the FIMP type DM scenario, as the DM production mechanism affects its free-streaming length (FSL). The FSL is defined as

$$\lambda_{\rm FSL} = \int_{t_{\rm prod}}^{t_{\rm eq}} \frac{\langle v \rangle}{a} dt = \int_{T_{\rm prod}}^{T_{\rm eq}} \frac{\langle v(T) \rangle}{a(T)} \frac{dt}{dT} dT. \tag{16.9}$$

In this context, $t_{\rm prod}$ ($T_{\rm prod}$) represents the time (temperature) at which the maximum production of dark matter takes place, while $t_{\rm eq}$ ($T_{\rm eq}$) signifies the time (temperature) of matter-radiation equality, marking the onset of structure formation. While hot dark matter with FSL $\gtrsim 0.1$ Mpc is ruled out from structure formation observations, warm dark matter with $0.1{\rm Mpc} \gtrsim \lambda_{\rm FSL} \gtrsim 0.01{\rm Mpc}$ is preferred over cold dark matter ($\lambda_{\rm FSL} \lesssim 0.01$ Mpc) due to small-scale issues associated with the latter. In Table 16.1, four benchmark points with their corresponding contribution to $\Delta N_{\rm eff}$ and FSL are shown. While the first two points are ruled out from FSL constraints, the last two points can be probed in future CMB S4 experiments.

Figure 16.2 shows a summary of parameter space in g_{BL}-$m_{Z'}$ plane. In the plot, m_{ϕ_1}, $\lambda_{H\phi_1}$, and y_{ϕ_1} are kept fixed. In the whole region, for the chosen parameters, non-thermal contribution to $\Delta N_{\rm eff}$ is small enough so that we can ignore it. The magenta coloured line separates the region of $\Delta N_{\rm eff} > 0.28$ and $\Delta N_{\rm eff} < 0.28$. Within the shaded grey area beneath the solid black line, g_{BL} does not determine the freeze-out abundance of ϕ_1 due to its small value. In that particular region, to get correct DM abundance, we need DM mass of 15 keV. Due to light mass, DM turns out to be

Table 16.1 Benchmark parameters for FIMP DM

	Parameters						$\Omega_{\rm DM}h^2$	$\Delta N_{\rm eff}$	FSL(Mpc)
	m_{ϕ_1}(GeV)	$\lambda_{H\phi_1}$	y_{ϕ_1}	$m_{Z'}$(GeV)	g_{BL}	m_ψ(keV)			
BP I	1000	5×10^{-5}	10^{-10}	500	0.001	126	0.12	0.251	3.09
BP II	500	5×10^{-5}	10^{-10}	500	0.001	233	0.12	0.210	1.32
BP III	1000	2×10^{-4}	10^{-9}	500	0.001	970	0.12	0.184	0.06
BP IV	500	2×10^{-4}	10^{-9}	500	0.001	938	0.12	0.184	0.05

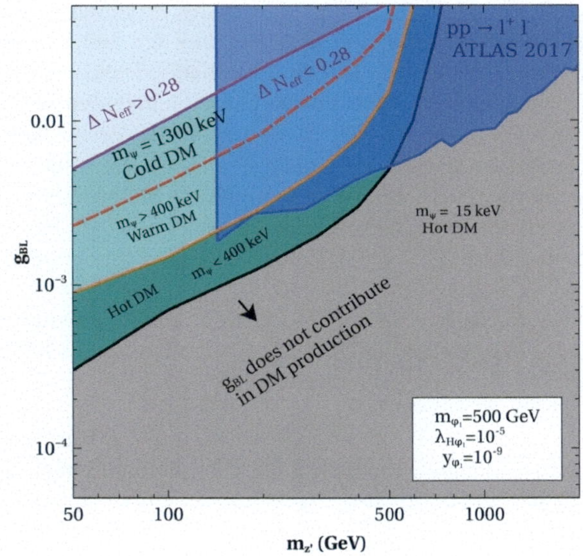

Fig. 16.2 Constraint on g_{BL} versus $m_{Z'}$ plane for fixed m_{ϕ_1}, $\lambda_{H\phi_1}$, and y_{ϕ_1}

hot there and hence is excluded from structure formation constraints. In the region between orange and black solid lines, DM has mass less than 400 keV, and in this region too, DM still remains in hot DM category. In the region between the dashed red and solid orange line, DM has a mass between 400 keV and 1300 keV giving warm dark matter. Cold dark matter with mass 1300 keV can be obtained in the region between magenta and dashed red line. In the whole parameter space below the magenta coloured line, $\Delta N_{\rm eff} > 0.14$ making the region probable in future CMB S4 experiment.

16.3.2 WIMP Type DM

Instead of having small Yukawa y_{ϕ_1}, it can be sufficiently large so that the BSM particles ϕ_1, ψ, and ν_R (with $m_{\phi_1} > m_\psi > m_{\nu_R}$, and ψ being the dark matter particle) can form a dark sector (DS) equilibrium among themselves after their decoupling from SM bath. Here, both the DM and ν_R can be produced thermally. From the interaction strength of dark sector particles with the SM bath, the dark sector decoupling temperature can be calculated. Hence, the ratio of the interaction rate between the dark sectors and the standard model bath to the Hubble expansion rate ($\mathcal{H}$) can be expressed as

$$\frac{\Gamma_{\rm total}}{\mathcal{H}} = \frac{1}{\mathcal{H}}\left[n_X^{\rm eq}\left(\langle\sigma v\rangle_{\phi_1 X\to\phi_1 X} + \langle\sigma v\rangle_{\phi_1 Z'\to\phi_1 Z'}\right) + n_{\nu_R}^{\rm eq}\langle\sigma v\rangle_{\nu_R\bar{\nu}_R\to X\bar{X}}\right]. \tag{16.10}$$

Here X denotes standard model particles. By comparing the total interaction rate with the Hubble expansion rate, the decoupling temperature can be calculated. After decoupling, the dark sector forms a separate bath among dark sector particles due to the large Yukawa coupling. So after decoupling, both the SM bath and dark sector bath evolve independently. To track the evolution of the DS bath, we define $\xi = \frac{T_{\rm DS}}{T}$, which denotes the deviation of dark sector temperature from SM bath temperature. Writing $Y = Y_{\phi_1} + Y_\psi$ and assuming $\frac{Y_i}{Y} \approx \frac{Y_i^{\rm eq}}{Y^{\rm eq}}$ $(i = \phi_1, \psi)$, the equations for Y and ξ become

$$\frac{dY}{dx} = \frac{1}{2}\frac{\beta s}{\mathcal{H}x}\langle\sigma v\rangle_{\rm eff}\left((Y^{\rm eq})^2 - Y^2\right), \tag{16.11}$$

$$\frac{d\xi}{dx} = \frac{1}{x}\left(-\frac{1}{2}\frac{\beta x^4 s^2}{4\alpha\xi^3 H m_{\phi_1}^4}\langle E\sigma v\rangle_{\rm eff}\left((Y^{\rm eq})^2 - Y^2\right) - (\beta - 1)\xi\right), \tag{16.12}$$

where

$$\langle\sigma v\rangle_{\rm eff} = \frac{(Y_{\phi_1}^{\rm eq})^2\langle\sigma v\rangle_{\phi_1\phi_1^\dagger\to X\bar{X},\nu_R\bar{\nu}_R,Z'Z'} + (Y_\psi^{\rm eq})^2\langle\sigma v\rangle_{\psi\bar{\psi}\to\nu_R\bar{\nu}_R}}{(Y_{\phi_1}^{\rm eq} + Y_\psi^{\rm eq})^2}, \tag{16.13}$$

and

$$\langle E\sigma v\rangle_{\rm eff} = \frac{(Y_{\nu_R}^{\rm eq})^2\langle E\sigma v\rangle_{\nu_R\bar{\nu}_R\to\psi\bar{\psi}} + (Y_{\nu_R}^{\rm eq})^2\langle E\sigma v\rangle_{\nu_R\bar{\nu}_R\to\phi_1\phi_1^\dagger}}{(Y_{\phi_1}^{\rm eq} + Y_\psi^{\rm eq})^2}. \tag{16.14}$$

Equations (16.11) and (16.12) are solved simultaneously from the decoupling temperature to 10 MeV and the corresponding DM abundance and contribution to the effective number of relativistic species are calculated. Figure 16.3 illustrates the

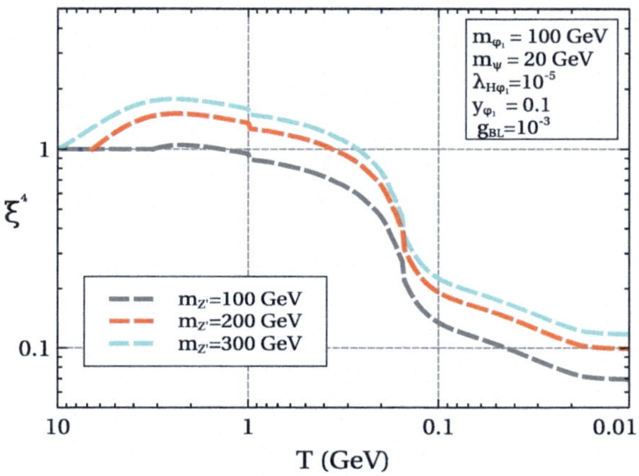

Fig. 16.3 Evolution of temperature ratio for three different $m_{Z'}$

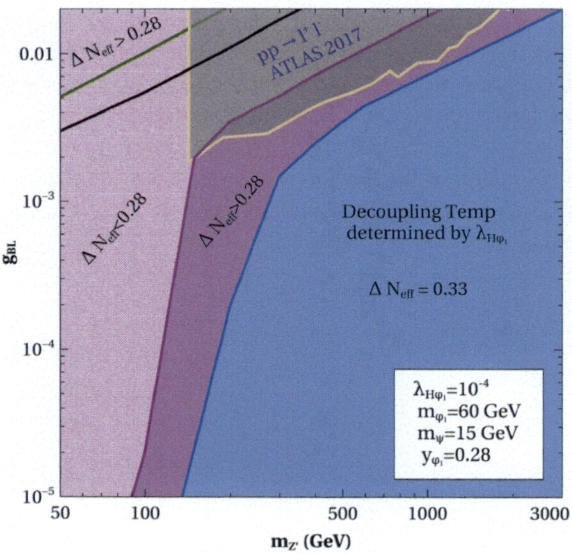

Fig. 16.4 Constraint on g_{BL} versus $m_{Z'}$ plane for fixed value of $\lambda_{H\phi_1}, m_{\phi_1}, m_\psi, y_{\phi_1}$. The grey shaded region shows the constraint from LHC

variation of the dark sector temperature in relation to the temperature of the standard model bath for three distinct values of $m_{Z'}$. Different $m_{Z'}$ gives different decoupling temperatures of the dark sector. As long as the decoupling temperature is less than the mass of ϕ_1, a higher decoupling temperature results in a higher ξ. This behaviour is visible from Fig. 16.3. The increase in dark sector temperature with respect to SM bath temperature has important consequences. Due to higher DS temperature, the contribution of ν_R to $\Delta N_{\rm eff}$ is enhanced. This gives a stronger cosmological constraint on the $U(1)_{B-L}$ parameter space.

The constraint that WIMP type DM puts on the parameter space for a particular value of benchmark points is shown in Fig. 16.4. In the portion above the black solid line, due to large $U(1)_{B-L}$ coupling, the chosen Yukawa coupling cannot create a separate dark sector bath after SM-DS decoupling. This implies that DM freezes-out when it is in the bath. Hence, in the region above the black solid line, we get the same constraint on g_{BL}-$m_{Z'}$ parameter space as for only thermalized ν_R via Z'. In the portion below the black solid line, a separate dark sector bath exists. For the portion below the solid blue line (light blue shaded region), the $U(1)_{B-L}$ coupling does not have any role in determining the decoupling temperature due to its small value. In that portion, the decoupling temperature is determined by the coupling $\lambda_{H\phi_1}$ and as the value of $\lambda_{H\phi_1}$ coupling is constant in that portion, we get same decoupling temperature there. In that portion, the contribution of ν_R to the effective number of relativistic species is 0.33. In the portion between the magenta and blue coloured lines, decoupling temperature is less than that in the blue portion, and we get $\Delta N_{\rm eff} > 0.28$ there. Finally, in the portion between the black and the magenta coloured lines, $\Delta N_{\rm eff} < 0.28$. This portion is allowed by the Planck 2018 data and can be probed by the future CMB S4 experiments.

16.4 Conclusion

We have discussed a light Dirac neutrino portal dark matter scenario focussing on cosmological signatures. A gauged $B - L$ symmetry at TeV scale not only accommodates right-handed neutrinos naturally from anomaly cancellation requirements but also keeps the model verifiable at different experiments. The right-handed neutrinos couple to the left-handed ones via the SM Higgs doublet realizing the light Dirac neutrino. The same right-handed neutrinos also act like a portal to DM, chosen to be a singlet Dirac fermion. Such a coupling is enabled by a singlet scalar ϕ_1 and the corresponding Yukawa coupling decides the thermal or non-thermal nature of DM. The same Yukawa coupling together with the gauged $B - L$ portal couplings can give rise to additional relativistic degrees of freedom by producing relativistic ν_R via thermal or non-thermal processes in ample amount. We discuss both WIMP and FIMP type DM scenarios and show the complementarity among detection prospects from cosmology, astrophysics, and collider experiments.

Acknowledgements The work of DB is supported by the Science and Engineering Research Board (SERB), Government of India, grant MTR/2022/000575. ND's work is sponsored by the Ministry of Education, Government of India, under the Prime Minister's Research Fellowship (PMRF) December 2021 scheme.

References

1. PARTICLE DATA GROUP collaboration, Review of Particle Physics. PTEP 2030, 083C01 (2020). https://doi.org/10.1093/ptep/ptaa104
2. PLANCK collaboration, Planck 2018 results VI: cosmological parameters. A&A 641, A6 (2020). https://doi.org/10.48550/arXiv.1807.06209
3. A. Biswas, D. Borah, D. Nanda, Light Dirac neutrino portal dark matter with observable $\Delta N_{\rm eff}$. JCAP 10, 002 (2021). https://doi.org/10.1088/1475-7516/2021/10/002
4. A. Biswas, D. Borah, N. Das, D. Nanda, Freeze-in dark matter via a light Dirac neutrino portal. Phys. Rev. D 107, 015015 (2023). https://doi.org/10.1103/PhysRevD.107.015015
5. D. Borah, N. Das, Light Dirac neutrino portal dark matter with gauged $U(1)_{B-L}$ symmetry. arXiv:2312.06777
6. K. Abazajian, others, CMB-S4 science case, reference design, and project plan. arXiv:1907.04473

Chapter 17
Sensitivities at Long-Baseline Experiments in the Light of Invisible Neutrino Decay

Zannatun Firdowzy Dey and Debajyoti Dutta

Abstract We study the impact of invisible neutrino decay in the measurements of octant and mass hierarchy sensitivities at long-baseline neutrino experiments. DUNE (Deep Underground Neutrino Experiment) and T2HKK (Tokai-to-Hyper-Kamiokande-to-Korea) are two upcoming long baseline experiments that can contribute a lot to precision measurements in the neutrino sector. Assuming ν_3 to be unstable, we show the effect of the invisible neutrino decay in the measurements of the sensitivities at the far detector of these experiments. We have also performed a combined analysis of the two experiments. The octant sensitivity increases, while mass hierarchy sensitivity changes very slightly in the presence of invisible neutrino decay. The value of the decay parameter τ_3/m_3 considered in this work is 2.2×10^{-11} s/eV.

17.1 Introduction

The observation of neutrino oscillations has led to the exploration of a new field of study that goes beyond the Standard Model (BSM) to account for the neutrino mass. Of the six oscillation parameters θ_{12}, θ_{13}, and Δm^2_{21} are measured accurately through data from various neutrino experiments, including solar, atmospheric, accelerator, and reactor experiments. However, certain parameters remain unknown, including: (i) the mass ordering of neutrinos, denoted as normal ordering (NO) for $\Delta m^2_{31} > 0$ or inverted ordering $\Delta m^2_{31} < 0$; (ii) the octant of the mixing angle θ_{23} categorized as the higher octant (HO) for $\theta_{23} > 45°$ or the lower octant (LO) for $\theta_{23} < 45°$, and (iii) the true value of δ_{cp} and its precision [1, 2]. Presently, data from ongoing

---

Z. F. Dey (✉)
Assam Don Bosco University, Tapesia Campus, Sonapur, Assam 782402, India
e-mail: zannatundey@gmail.com

D. Dutta
Bhattadev University, Pathsala, Bajali, Assam 781325, India
e-mail: phy.debajyoti@bhattadevuniversity.ac.in

J. W. F. Valle et al. (eds.), *Proceedings of the International Conference on Future Prospects in Neutrino and Astroparticle Physics*, Springer Proceedings in Physics 361,
https://doi.org/10.1007/978-981-97-7441-8_17

experiments suggest a tendency towards NO, HO, and $\delta_{cp} = -90°$ [3]. Numerous dedicated experiments aim to confirm and establish these indications definitively.

To effectively plan future experiments and optimize the running parameters of existing ones, it is crucial to anticipate the sensitivity of an experiment concerning its various running parameters. This includes a meticulous consideration of unknown parameters within the neutrino mixing matrix, which directly influence the predicted interaction rates of neutrinos in the experimental setup. Ongoing (T2K [4], NOνA [5, 6]) and upcoming long baseline neutrino oscillation experiments (DUNE [7], T2HK/T2HKK [8, 9], ESSnuSB [10], JUNO [11], PINGU [12], etc.) aim to investigate CP violation in the lepton sector, resolve the neutrino mass hierarchy, and achieve precise measurements of oscillation parameters.

Besides standard neutrino oscillation physics, numerous new physics scenarios can be investigated through long baseline neutrino oscillation experiments. One example of such a scenario is invisible neutrino decay [13]. This may occur if neutrinos are (i) Dirac particles and a light scalar boson are coupled to the neutrinos [14, 15]. This permits the decay $\nu_j \rightarrow \bar{\nu}_i R + \chi$, where χ is an iso-singlet scalar with a lepton number and $\bar{\nu}_i R$ is a right-handed singlet. (ii) Pseudo-scalar Majorana particles coupled to a Majoron [16, 17] and a sterile neutrino, resulting in the decay $\nu_j \rightarrow \nu_s + J$. In order to meet the restrictions derived from LEP data on the Z decay to invisible particles, the Majoron must be predominantly singlet. Models of this type have been covered in [18]. Every final state particle in both scenarios is invisible and sterile. The other scenario of decay, i.e., visible decay occurs when neutrinos turn into active neutrinos [19, 20]. The decay modes are either $\nu_j \rightarrow \nu_i + J$ or $\nu_j \rightarrow \bar{\nu}_i + J$.

In Ref. [21] authors have studied the invisible neutrino decay with T2K and NOνA data. Reference [22] studied how CP-sensitivity in the ESSnuSB experiment was affected by invisible neutrino decay. Their findings revealed an enhancement in CP-sensitivity when considering invisible neutrino decay in the context of the ESSnuSB experiment. Another investigation in Ref. [23] delved into the impact of invisible neutrino decay on the DUNE experiment. The results indicated improved octant sensitivity in the lower octant, accompanied by a reduction in the higher octant in the presence of decay. In Ref. [24] authors showed impact of invisible neutrino decay in T2HKK and ESSnuSB experiments. They separately showed the impact of invisible neutrino decay in both the detectors. In the context of invisible neutrino decay, octant sensitivity and CP violation sensitivity at DUNE and T2HKK experiments have been explored in Ref. [25]. In this work, we explore octant sensitivity and mass hierarchy within the context of invisible neutrino decay, focusing on the long baseline experiments DUNE and T2HKK. Additionally, we conduct a combined analysis of DUNE+T2HKK, considering the implications of invisible neutrino decay on the experimental outcomes.

This paper is organized in the following way. In the next section we discuss the theory of invisible neutrino decay. In Sect. 17.3, we discuss about the experiments and simulation details of our analysis. In Sect. 17.4 we give our results and we conclude our findings in Sect. 17.5.

17.2 Neutrino Propagation in Presence of Invisible Decay

We consider the scenario where the ν_3 state undergoes decay into a sterile neutrino and a singlet scalar J; $(\nu_3 \rightarrow \bar{\nu_4} + J)$. Additionally, we assume that the neutrino mass eigenstate experiences decay. As a result, it becomes possible to simultaneously diagonalize both the mass matrix and the decay matrix. With the inclusion of decay, the evolution equation in matter takes on the form:

$$i\frac{d}{dx}\begin{pmatrix}\nu_e\\ \nu_\mu\\ \nu_\tau\end{pmatrix} = \left[U\left[\frac{1}{2E}\begin{pmatrix}0 & 0 & 0\\ 0 & \Delta m^2_{21} & 0\\ 0 & 0 & \Delta m^2_{31}\end{pmatrix} - i\frac{m_3}{2E\tau_3}\begin{pmatrix}0&0&0\\0&0&0\\0&0&1\end{pmatrix}\right]U^\dagger + \begin{pmatrix}A_{cc}&0&0\\0&0&0\\0&0&0\end{pmatrix}\right]\begin{pmatrix}\nu_e\\ \nu_\mu\\ \nu_\tau\end{pmatrix} \tag{17.1}$$

Here, $A = 2\sqrt{2}G_F n_e E$ symbolizes the matter potential originating from neutrino-electron scattering within matter. In this expression, G_F corresponds to the Fermi coupling constant, E denotes the neutrino energy, and n_e represents the electron density. To obtain a numerical solution to Eq. 17.1, we employ the Runge-Kutta method with a constant matter density.

17.3 Experimental and Simulation Details

In this section, we give a brief description of the experiments used in this work. We have used GLoBES [26, 27] for simulating the two experiments. We estimate the statistical χ^2 function using

$$\chi^2_{stat} = 2\sum_i\left[N_i^{test} - N_i^{true} + N_i^{true} log\frac{N_i^{true}}{N_i^{test}}\right] \tag{17.2}$$

where N_i^{true} represents the number of true events and N_i^{test} represents the number of test events. We incorporate the systematics by the pull methods. In our study, we use 5 + 5 years runtime for DUNE and 2.5 + 7.5 years for T2HKK experiments. For DUNE experiment, we have included the neutral current along with charged current in the analysis. DUNE aims to explore fundamental questions about the nature of matter and the evolution of the universe through its investigation of neutrinos. DUNE [28–30] experiment involves two detectors strategically placed along the world's most intense neutrino beam. One detector close to the beam source at Fermi National Accelerator Laboratory in Batavia, Illinois. The second, significantly larger detector is situated over a kilometer underground at the Sanford Underground Research Laboratory in Lead, South Dakota, 1,300 km away from the source.

While T2HK exhibits impressive sensitivity when the mass hierarchy is known, its ability to detect CP violation diminishes when the mass hierarchy remains undetermined. This limitation arises from the degeneracy between the mass hierarchy and the δ_{cp} parameter[31]. To address this challenge, an alternative proposal known as T2HKK has been introduced. In the T2HKK plan, one of the water tanks will be relocated to Korea, approximately 1100 km from the source. This setup ensures that the flux at 295 km covers the first oscillation maximum, while the flux at 1100 km encompasses the second oscillation maximum.

17.4 Results

17.4.1 Octant Discovery in the Presence of Invisible Neutrino Decay

Figure 17.1 represents the octant discovery potential plots of DUNE (top panel), T2HKK (middle panel) and DUNE+T2HKK (bottom panel) experiments in the presence of invisible neutrino decay. Octant discovery potential of an experiment is the measure of excluding the wrong octant for the considered true octant, i.e., for a true higher octant, at what C. L. we can exclude the LO in the given experiment and vice-versa. Here in this work we show the effect of invisible neutrino decay in the octant discovery potential of DUNE, T2HKK, and their combinations. Here we consider three possible cases: (i) decay in data but not in fit or theory (ii) decay in fit or theory but not in data (iii) decay in both data and theory The value of decay parameter considered in this work is $\tau_3/m_3 = 2.2 \times 10^{-11}$ s/eV. The black plots represent the standard three flavor case, the green plots represent the case when we consider decay in data, the blue plots represent the case where we consider decay in fit or theory only. The red plot is for the case where we consider decay both in data and theory. The data is generated based on the oscillation parameters listed in Table 17.1. When analyzing octant discovery potential in the higher octant (HO = 48.44°), for any true value of θ_{23} in the higher octant (HO), we vary the test θ_{23} in LO within the range of [40°, 45°]. We show the effect of decay separately for appearance, disappearance and their total channels. From Fig. 17.1, we can make the following observations.

- In DUNE (top panel), the potential to discover the octant increases when including decay only in the fit but not in the data for all the cases. However, if decay is considered in both data and theory, the discovery potential decreases. Specifically, the potential for octant discovery reaches a significant 5σ level when accounting for both appearance and disappearance channels together. Notably, the discovery potential decreases in the appearance channel and overall, but increases in the disappearance channel when decay is included in the data but not in the theory.
- In the T2HKK (middle panel) experiment, the appearance channel is most sensitive when we consider decay in theory or fit. However, the discovery potential decreases in the appearance channel with decay in the data and in both data and theory. On the other hand, the disappearance channel in the T2HKK experiment doesn't show sensitivity without decay but becomes most sensitive when decay is considered in

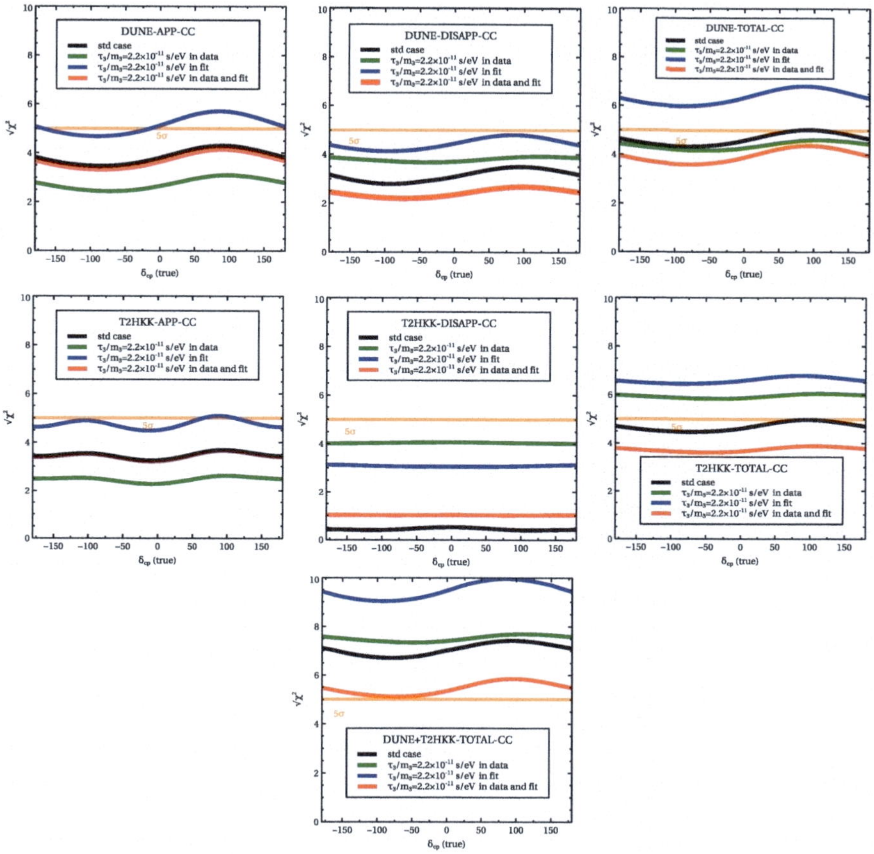

Fig. 17.1 Octant discovery potential ($\sigma = \sqrt{\chi^2}$) of DUNE (top panel), T2HKK (middle panel) and DUNE+T2HKK (bottom panel) experiments for various channels. The black plots represent the case for the standard case, the green plots represent the case for decay in data, the blue plots represent the case for decay in fit or theory, and the red plots represent the case for decay in both data and fit or theory

the data. The combined potential from both appearance and disappearance channel is highest with decay in theory or fit.

- Combining DUNE and T2HKK (bottom panel) experiments raises the octant discovery potential above 5σ level. The octant discovery potential increases when considering decay in data and decay in theory, but decreases when decay is considered in both data and theory. The maximum potential is achieved when decay is considered in theory.

Table 17.1 The oscillation parameters' values utilized in this analysis

Oscillation parameters	True Values	Marginalization Range
θ_{13}	8.6°	[8.1°, 8.9°]
θ_{12}	33.82°	Fixed
θ_{23}	41.95° (LO), 48.44° (HO)	[38°, 52°]
$\Delta m^2_{21}(eV^2)$	7.39×10^{-5}	Fixed
$\Delta m^2_{31}(eV^2)$	2.52×10^{-3}	$[2.3 \times 10^{-3}, 2.6 \times 10^{-3}]$
δ	−90°	[−180°, 180°]

17.4.2 *Mass-Hierarchy Sensitivity Analysis*

Figures 17.2 and 17.3 represent the impact of decay on expected mass hierarchy sensitivity of DUNE, T2HKK, and DUNE+T2HKK experiments respectively. To measure the mass hierarchy sensitivity of an experiment, for a fixed true normal (inverted) hierarchy, we consider the inverted (normal) hierarchy in the fit or theory. For a considered true NH, we vary Δm^2_{31} (in the 3 σ range) in the inverted hierarchy. We show the results for both normal and inverted hierarchy as the true hierarchy. Like the previous case, to see the effect of the invisible neutrino decay in the MH sensitivity, we consider three cases of decay: (i) decay in data but not in theory (ii) decay in fit or theory but not in data (iii) decay in both data and theory We consider the same value of the decay parameter, i.e., $\tau_3/m_3 = 2.2 \times 10^{-11}$ s/eV. For all the cases the black plots represent the MH sensitivity for the standard case, the green plots represent the case when we consider decay in data, the blue plots represent the case when we consider decay in fit or theory, and the red plots represent the case for decay in both data and fit. We can make the following observations from Figs. 17.2 and 17.3:

- In the DUNE experiment, when considering the normal hierarchy (NH) as true, the mass hierarchy sensitivity increases slightly in appearance, disappearance, and overall channels with decay in theory. However, when there is decay in the data, the sensitivity decreases in the appearance channel, and for some values of δ_{cp} it increases in the disappearance channel. If decay is considered in both data and theory, the mass hierarchy sensitivity decreases in all the cases. In the case of the inverted hierarchy (IH) being true, sensitivity is higher in the appearance channel and lower in the disappearance channel compared to the true NH. The change in sensitivity due to decay for the inverted hierarchy in this case is not much significant.
- In the T2HKK experiment, when treating NH as the true hierarchy and incorporating decay in the fit, there is an enhanced mass hierarchy sensitivity in the disappearance channel. Similarly, the MH sensitivity in the combined appearance and disappearance channel increases when decay is accounted for in the data, as well as in the case of decay in the fit. Conversely, assuming IH as true, the mass hier-

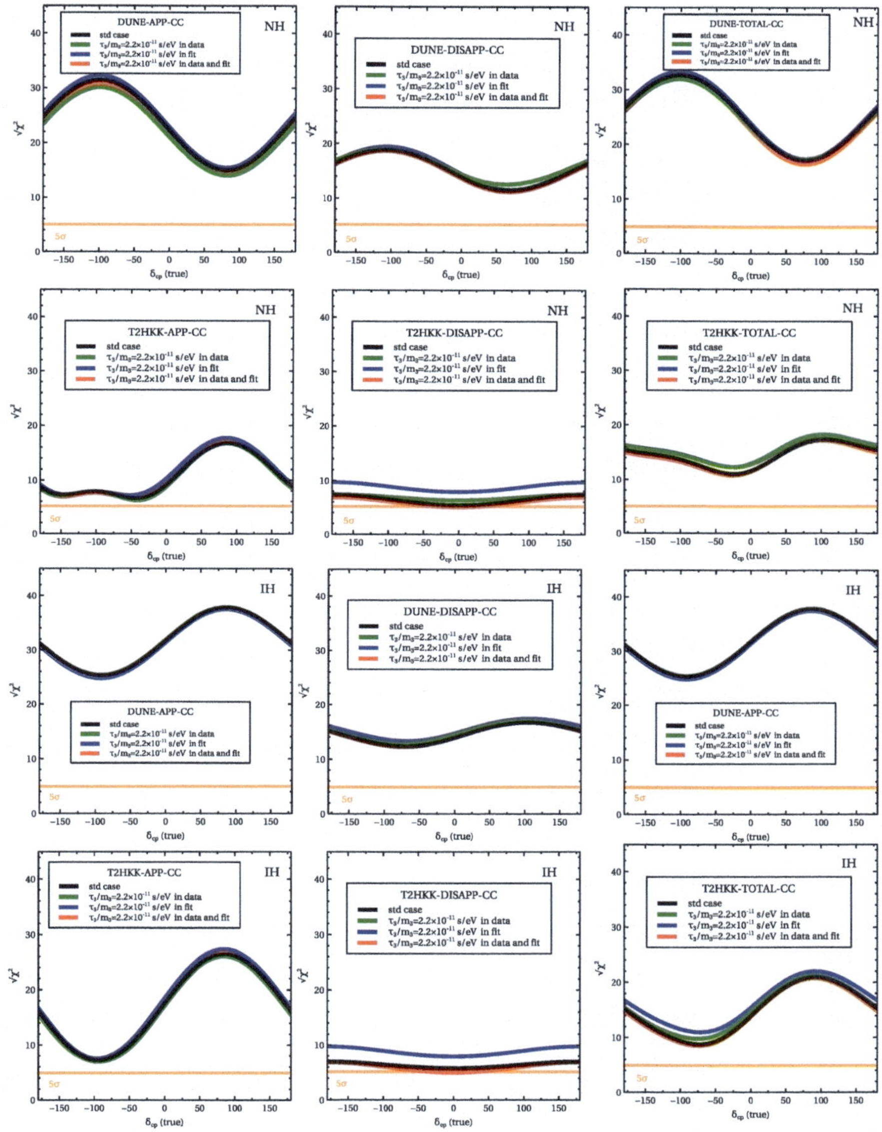

Fig. 17.2 Mass hierarchy sensitivity ($\sigma = \sqrt{\chi^2}$) as a function of δ_{cp} (true) for various channels of DUNE and T2HKK experiments. The black plots represent the MH sensitivity for the standard case, the green plots represent the case for decay in data, the blue plots represent the case for decay in fit, and the red plots represent the case for decay in both data and fit

archy sensitivity rises for all cases involving decay in the fit. Notably, the change in mass hierarchy for other decay parameter cases appears to be insignificant.

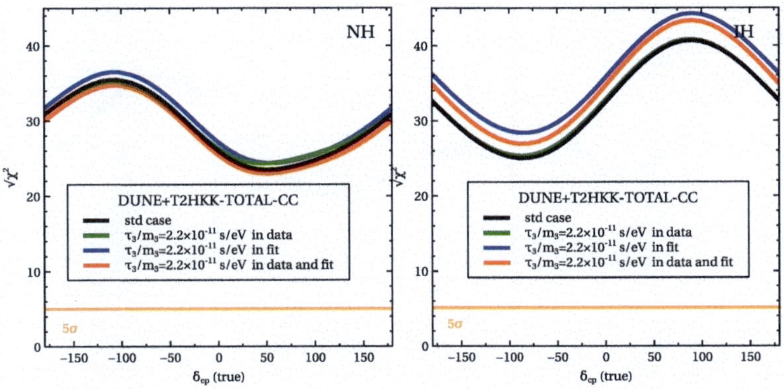

Fig. 17.3 Mass hierarchy sensitivity ($\sigma = \sqrt{\chi^2}$) plots for DUNE+T2HKK experiment. The black plots represent the MH sensitivity for the standard case, the green plots represent the case for decay in data, the blue plots represent the case for decay in fit, and the red plots represent the case for decay in both data and fit

- In the DUNE+T2HKK experiment (Fig. 17.3), considering NH as the true hierarchy, mass hierarchy (MH) sensitivity rises when decay is included in the fit or theory. MH sensitivity for decay in the data decreases for some δ_{CP} values and slightly increases for others. In the case of decay in both data and theory, MH sensitivity decreases for all true δ_{CP} values. Conversely, assuming IH as the true hierarchy, MH sensitivity increases with decay in the fit and decay in both data and theory. However, MH sensitivity remains unchanged with decay in the data. Also, Combining DUNE with T2HKK enhances the overall mass hierarchy sensitivity.

17.5 Conclusion

This study delved into the exploration of octant sensitivity and mass hierarchy in the DUNE and T2HKK experiments in the context of invisible neutrino decay. We examined the mass hierarchy and octant sensitivity for each experiment as a function of true δ_{cp} for the appearance and disappearance channels independently. The impact of invisible neutrino decay is studied for three scenarios of the decay parameter. It is observed that for all the cases the highest octant discovery potential achieved for the case of decay in theory, except for the disappearance channel of T2HKK. The disappearance channel of T2HKK does not show any sensitivity in the absence of invisible neutrino decay but exhibited maximum potential when decay was considered in the data. Additionally, it is noted that, for all the cases, octant discovery potential decreased when decay is considered in both data and theory.

Regarding mass hierarchy analysis, although we observed a slight improvement in the disappearance channel of T2HKK experiment with decay in test but we do not notice any significant improvement in the overall MH sensitivity of both the experiments. Also, we noticed that Combining DUNE with T2HKK enhances the overall mass hierarchy sensitivity. In this case for both NH and IH with decay in the fit the mass hierarchy sensitivity increases. Also, the mass hierarchy sensitivity increases for IH with decay in data and theory and decreases for NH for the same decay case.

References

1. F. Capozzi, E. Lisi, A. Marrone, D. Montanino, A. Palazzo, Nucl. Phys. B **908**, 218 (2016). https://doi.org/10.1016/j.nuclphysb.2016.02.016. (arXiv:1601.07777 [hep-ph])
2. I. Esteban, M.C. Gonzalez-Garcia, M. Maltoni, I. Martinez-Soler, T. Schwetz, JHEP **01**, 087 (2017). https://doi.org/10.1007/JHEP01(2017)087. (arXiv:1611.01514 [hep-ph])
3. I. Esteban, M. C. Gonzalez-Garcia, M. Maltoni, T. Schwetz, and A. Zhou, https://doi.org/10.1007/JHEP09(2020)178 JHEP **09**, 178 (2020), arXiv:2007.14792 [hep-ph]
4. K. Abe et al., collaboration T2K. Phys. Rev. Lett. **118**, 151801 (2017). https://doi.org/10.1103/PhysRevLett.118.151801. (arXiv:1701.00432 [hep-ex])
5. P. Adamson et al., collaboration NOvA. Phys. Rev. Lett. **116**, 151806 (2016). https://doi.org/10.1103/PhysRevLett.116.151806. (arXiv:1601.05022 [hep-ex])
6. P. Adamson et al., collaboration NOvA. Phys. Rev. Lett. **118**, 231801 (2017). https://doi.org/10.1103/PhysRevLett.118.231801. (arXiv:1703.03328 [hep-ex])
7. R. Acciarri et al. (collaboration DUNE), (2015). arXiv:1512.06148 [physics.ins-det]
8. K. Abe et al. (collaboration Hyper-Kamiokande Proto-), PTEP **2015**, 053C02 (2015). https://doi.org/10.1093/ptep/ptv061, arXiv:1502.05199 [hep-ex]
9. K. Abe et al. (collaboration Hyper-Kamiokande), PTEP **2018**, 063C01 (2018). https://doi.org/10.1093/ptep/pty044, arXiv:1611.06118 [hep-ex]
10. E. Baussan et al., collaboration ESSnuSB. Nucl. Phys. B **885**, 127 (2014). https://doi.org/10.1016/j.nuclphysb.2014.05.016. (arXiv:1309.7022 [hep-ex])
11. F. An et al. (collaboration JUNO), J. Phys. G **43**, 030401 (2016). https://doi.org/10.1088/0954-3899/43/3/030401 , arXiv:1507.05613 [physics.ins-det]
12. M.G. Aartsen et al. (collaboration IceCube-PINGU), (2014). arXiv:1401.2046 [physics.ins-det]
13. M. Lindner, T. Ohlsson, W. Winter, Nucl. Phys. B **607**, 326 (2001). https://doi.org/10.1016/S0550-3213(01)00237-1. (arXiv:hep-ph/0103170)
14. A. Acker, S. Pakvasa, J.T. Pantaleone, Phys. Rev. D **45**, 1 (1992). https://doi.org/10.1103/PhysRevD.45.R1
15. A. Acker, S. Pakvasa, Phys. Lett. B **320**, 320 (1994). https://doi.org/10.1016/0370-2693(94)90663-7. (arXiv:hep-ph/9310207)
16. G.B. Gelmini, M. Roncadelli, Phys. Lett. B **99**, 411 (1981). https://doi.org/10.1016/0370-2693(81)90559-1
17. Y. Chikashige, R.N. Mohapatra, R.D. Peccei, Phys. Lett. B **98**, 265 (1981). https://doi.org/10.1016/0370-2693(81)90011-3
18. S. Pakvasa, A.I.P. Conf, Proc. **542**, 99 (2000). https://doi.org/10.1063/1.1336244. (arXiv:hep-ph/0004077)
19. C.W. Kim, W.P. Lam, Mod. Phys. Lett. A **5**, 297 (1990). https://doi.org/10.1142/S0217732390000354
20. A. Acker, A. Joshipura, S. Pakvasa, Phys. Lett. B **285**, 371 (1992). https://doi.org/10.1016/0370-2693(92)91520-J
21. S. Choubey, D. Dutta, D. Pramanik, JHEP **08**, 141 (2018). https://doi.org/10.1007/JHEP08(2018)141. (arXiv:1805.01848 [hep-ph])

22. S. Choubey, M. Ghosh, D. Kempe, T. Ohlsson, JHEP **05**, 133 (2021). https://doi.org/10.1007/JHEP05(2021)133. (arXiv:2010.16334 [hep-ph])
23. S. Choubey, S. Goswami, D. Pramanik, JHEP **02**, 055 (2018). https://doi.org/10.1007/JHEP02(2018)055. (arXiv:1705.05820 [hep-ph])
24. K. Chakraborty, D. Dutta, S. Goswami, D. Pramanik, JHEP **08**, 136 (2021). https://doi.org/10.1007/JHEP08(2021)136. (arXiv:2012.04958 [hep-ph])
25. Z.F. Dey, D. Dutta, (2024). arXiv:2402.13235 [hep-ph]
26. P. Huber, M. Lindner, W. Winter, Comput. Phys. Commun. **167**, 195 (2005). https://doi.org/10.1016/j.cpc.2005.01.003. (arXiv:hep-ph/0407333)
27. P. Huber, J. Kopp, M. Lindner, M. Rolinec, W. Winter, Comput. Phys. Commun. **177**, 432 (2007). https://doi.org/10.1016/j.cpc.2007.05.004. (arXiv:hep-ph/0701187)
28. R. Acciarri et al. (collaboration DUNE), (2016). arXiv:1601.05471 [physics.ins-det]
29. J. Strait et al. (collaboration DUNE), (2016). arXiv:1601.05823 [physics.ins-det]
30. R. Acciarri et al. (collaboration DUNE), (2016). arXiv:1601.02984 [physics.ins-det]
31. T. Kajita, H. Minakata, S. Nakayama, H. Nunokawa, Phys. Rev. D **75**, 013006 (2007). https://doi.org/10.1103/PhysRevD.75.013006. (arXiv:hep-ph/0609286)

Chapter 18
Study of Traversable Wormhole in the Dark Matter Halo

Partha Pratim Nath and Debojit Sarma

Abstract In this work, we have carried out the study of a traversable wormhole in the dark matter halo. We have formulated the exact solution of the spherically symmetric traversable wormhole with isotropic pressure conditions by using the Jaffe dark matter density profile. The Jaffe profile produces a suitable shape function that meets all the requirements for presenting the wormhole geometries. We have investigated all the necessary energy conditions viz., null energy condition(NEC), weak energy condition(WEC), strong energy condition(SEC), and dominant energy conditions(DEC) for the considered wormhole. All the suitable stability conditions are analyzed and explained by related plots.

18.1 Introduction

Wormholes belong to a special class of solutions to Einstein's Field Equations which act as tube-like or bridge-like structures that connect two distinct points of the same space-time or two different universes. The tubular structure is considered to be asymptotically flat on both sides. They are very fascinating topic for physicists from the early 90s. Theoretically, significant progress has been made on this topic, yet no physical evidence has been found. One important property of wormholes is the wormhole throat, which can be defined as a 2-D hypersurface of minimal area or the point where the radius becomes minimum [1]. Traversability is also an important property of wormholes. If anything that enters through one side of the wormhole can exit through the other, then the wormhole is traversable. One of the most important features of a traversable wormhole is the absence of a horizon because the presence of the horizon would prevent two-way travel through the wormhole.

A considerable number of wormhole metrics have been constructed in the past. The very first wormhole solution was studied by Ludwig Flamm [2, 3] in Einstein's gravity, but his solution was not stable. In 1928, Hermann Weyl [4] proposed a

P. P. Nath (✉) · D. Sarma
Cotton University, Guwahati, India
e-mail: phy1891005_partha@cottonuniversity.ac.in

J. W. F. Valle et al. (eds.), *Proceedings of the International Conference on Future Prospects in Neutrino and Astroparticle Physics*, Springer Proceedings in Physics 361,
https://doi.org/10.1007/978-981-97-7441-8_18

wormhole hypothesis of matter in connection with mass analysis of electromagnetic field energy. However, Weyl uses the term 'one-dimensional tube' instead of the term 'wormhole'. Then Wheeler [5] named them 'Geons' and predicted the shape of the wormhole which offers a twofold space. Einstein and Rosen [6] constructed a solution known as the Einstein-Rosen bridge and they studied the solution in detail. These wormholes were unstable. In 1988, Thorne and his graduate student Morris [7] constructed a wormhole with a throat and two mouths and introduced the concept of static traversable wormholes, which require a kind of exotic matter [8, 9] to keep the wormhole's throat open. This leads to the violation of the energy conditions [7, 10, 11] near the throat region. Exotic matter is a form of dark energy(having an EoS with $\omega < -\frac{1}{3}$) that produces a repulsion. Wheeler and Misner [12] coined the term 'wormhole' and later their solutions were transformed into Euclidean wormholes by Hawking [13] and others. Later Ellis [14] introduced another term for wormhole known as 'Drainhole'. The possible existence of wormholes brings attention to many questions like causality violation, energy conditions, geometric structure, topology of space-time, etc.

To avoid energy violations, wormhole solutions have been developed using the background modified gravity theories such as Kaluza-Klein gravity [15], Born-Infeld theory [16], Brans-Dicke theory [17], $f(R)$ gravity [18], Einstein-Gauss-Bonnet theory [19], and Einstein-Cartan theory [20]. It has been shown in modified theories of gravity that the matter inside the wormhole may satisfy the necessary energy conditions but the effective stress-energy tensor [21] containing higher-order derivative terms is responsible for the violation of the energy conditions.

If we move to the galactic level, we observe some phenomena that cannot be explained by the standard model, for example, flat rotation curves in galaxies. These phenomena have led us to a hypothesis that galaxies and even the clusters of galaxies are filled by some non-luminous matter, now known as dark matter. It does not emit electromagnetic waves, nor does it interact with normal ordinary matter. Dark matter plays a vital role in the formation and evolution of galaxies. Many candidates for dark matter have been proposed over time, such as new particles predicted by supersymmetry, massive neutrinos known as WIMPs [22], global monopoles [23], geometric effects of $f(R)$ [24] gravity, etc. Given the ubiquity of galaxies and dark matter halos, it is important to think about the formation of wormholes in the dark matter halo and galaxy. Only a few works have been carried out in the literature on the possibilities of wormholes in halos using various density profiles [25, 26]. So this motivates us to study a spherically symmetric traversable wormhole metric in the dark matter halos using the Jaffe density profile under the isotropic pressure condition. We also analyze the energy as well as the stability conditions of the wormhole. The Jaffe profile is a spherically symmetric density profile. No one in the literature has ever studied wormholes using the Jaffe profile.

This paper is outlined in the following style. In Sect. 18.2, we introduce the dark matter density profile, and the Jaffe density profile is briefly presented. In Sect. 18.3, a traversable wormhole solution is obtained using isotropic pressure condition and its various properties such as shape function, energy condition, etc. will also be studied. In Sect. 18.4, we will study the stability condition for the obtained wormhole. Finally, a conclusion is presented in Sect. 18.5.

18.2 The Dark Matter Density Profile

The dark matter density profile refers to the distribution of dark matter within a galaxy, galaxy cluster, or other cosmic structure. The density profile describes how the density of dark matter varies as a function of the distance from the center of the cosmic structure. One of the very famous discoveries made in recent years in numerical simulations is the internal structure of dark matter haloes. Several heuristic fitting functions have been developed that capture the shape of individual dark matter halos, among others the Navarro-Frenk-White profile [27] and the Einasto profile [28], and the shape of individual luminous matter distributions, like the de-Vaucouleurs profile or the Jaffe profile [29]. Without a deeper understanding of how these profiles can be derived from more fundamental principles, it remains an open question why these formulae are good fits to the simulated and observed mass density distributions.

Many density profiles can be written as double power laws [30],

$$\rho_{DMprofile} = \rho_s(\frac{r}{R_s})^{-\gamma}(1+(\frac{r}{R_s})^{\alpha})^{\frac{\gamma-\beta}{\alpha}}, \tag{18.1}$$

where the parameters α, β, and γ can take various values.

Jaffe Density Profile

The values of α, β, and γ are 1, 4, and 2, respectively, in the Jaffe density profile, i.e.,

$$\rho_{Jaffe} = \frac{\rho_s}{(\frac{r}{R_s})^2(1+\frac{r}{R_s})^2}, \tag{18.2}$$

where ρ_s is the central density and R_s is the scale radius. Both ρ_s and R_s are positive constants. The values of these parameters are determined through a process of estimation and model fitting. The ranges of these parameters may be constrained by physical considerations, but they are not fixed and may change from one astrophysical system to another.

It is a spherically symmetric density distribution that goes to zero as $r \to \infty$. Jaffe density profile is used as a simple model for galaxy-scale dark matter halos in the absence of baryonic matter. Equation (18.2) has a well-defined core, given by the first bracket in the denominator, and only one scale radius R_s [31].

18.3 Traversable Wormhole

A spherically symmetric traversable wormhole metric can be expressed as [7],

$$ds^2 = -e^{2\Phi(r)}dt^2 + \frac{1}{1-\frac{b(r)}{r}}dr^2 + r^2(d\theta^2 + \sin^2\theta d\phi^2), \tag{18.3}$$

where $\Phi(r)$ and $b(r)$ are redshift and shape function, respectively. In order to be traversable, Eq. (18.3) should not contain an event horizon, i.e., $e^{2\Phi(r)} \neq 0$. Therefore, $\Phi(r)$ should be finite and tend to zero as $r \to \infty$. The shape function $b(r)$ determines the geometry of the wormhole. It should satisfy the boundary condition,

$$b(r_0) = r_0, \tag{18.4}$$

where b_0 is the radius of the wormhole throat. The radial coordinate r is non-monotonic such that it decreases from infinity to a minimum value r_0, representing the location of the throat of the wormhole, and then it increases from r_0 toward infinity. Apart from this, the shape function has to satisfy a set of conditions,

$$b(r) < r \quad as \quad r > r_0, \tag{18.5}$$

$$\frac{b(r)}{r} \to 0 \quad as \quad r \to \infty, \tag{18.6}$$

$$b'(r_0) < 1, \tag{18.7}$$

and

$$\frac{b(r) - rb'(r)}{b^2(r)} > 0 \quad at \quad r = r_0. \tag{18.8}$$

Here, Eqs. (18.5 and 18.6) give asymptotic flatness and Eqs. (18.7 and 18.8) represent flare-out condition.

Using the Einstein field equation, $G_{\mu\nu} = 8\pi T_{\mu\nu}$, where $T_{\mu\nu}$ is the energy-momentum tensor, which can be obtained by dark matter profile. The Einstein field equations can be simplified as

$$\rho(r) = \frac{1}{8\pi}\frac{b'(r)}{r^2}, \tag{18.9}$$

$$P_r(r) = \frac{1}{8\pi}\left[-\frac{b(r)}{r^3} + 2\left(1 - \frac{b(r)}{r}\right)\frac{\Phi'(r)}{r}\right], \tag{18.10}$$

$$P_\theta = P_\phi = \frac{1}{8\pi}\left(1 - \frac{b(r)}{r}\right)\left[\Phi''(r) + \Phi'^2(r) - \frac{rb'(r) - b(r)}{2r(r - b(r))}\Phi'(r) - \frac{rb'(r) - b(r)}{2r^2(r - b(r))} + \frac{\Phi'(r)}{r}\right] \quad (18.11)$$

where $\rho(r) \rightarrow$ dark matter energy density, $P_r(r) \rightarrow$ radial pressure, and $P_\theta(r)$, $P_\phi(r) \rightarrow$ tangential pressures. Now, from the energy-momentum conservation law, $T^{\mu\nu}_{,\nu} = 0$, the hydrostatic equation for dark matter comes out as

$$P_r' = -(P_r + \rho)\Phi' + \frac{2(P_\theta - P_r)}{r} \quad (18.12)$$

Isotropic Pressure

We assume that the dark matter pressure is isotropic, i.e., $P_r = P_\theta = P_\phi = P$, Eq. (18.12) becomes

$$P' = -(P + \rho)\Phi'. \quad (18.13)$$

From the above equation, we can obtain the redshift function $\Phi(r)$ with the state equation $P = \omega\rho$. Equation(18.13) reduces to

$$P' = -(\frac{1+\omega}{\omega})P\Phi'. \quad (18.14)$$

Now integrating the above Eq. (18.14),

$$P(r) = \omega\rho(r) = -\frac{1}{8\pi r^2}e^{-\frac{(1+\omega)}{\omega}[\Phi(r) - \Phi(r_0)]}. \quad (18.15)$$

Here we will use $\Phi(r_0) = 0$ without a significant loss of generality. Now using Eq. (18.15) in Eq. (18.9), we get

$$b'(r) = -\frac{1}{\omega}(\frac{r}{r_0})^2 e^{-(\frac{1+\omega}{\omega})\Phi(r)}, \quad (18.16)$$

which can be simplified as

$$\Phi(r) = -\left(\frac{\omega}{1+\omega}\right)\ln\left[-\omega b'(r)(\frac{r_0}{r})^2\right]. \quad (18.17)$$

Shape Function and Redshift Function

Now, we will derive the traversable wormhole's metric based on the Jaffe dark matter density profile. Now using Eqs. (18.2) and (18.9) and the boundary condition $b(r_0) = r_0$, we can obtain the shape function as

$$b(r) = r_0 + 8\pi\rho_s R_s^4 (\frac{1}{r_0 + R_s} - \frac{1}{r + R_s}). \tag{18.18}$$

Now using Eq. (18.18) in condition Eq. (18.7),

$$b'(r_0) = \frac{8\pi\rho_s R_s^4}{(r + R_s)^2} < 1, \tag{18.19}$$

we can get the range of b_0 as

$$r_0 > 2R_s^2\sqrt{2\pi\rho_s} - R_s. \tag{18.20}$$

We can find the expression for the redshift function by using Eq. (18.18) in Eq. (18.17),

$$\Phi(r) = -\left(\frac{\omega}{1+\omega}\right)\ln\left[-\omega\frac{8\pi\rho_s R_s^4}{(r + R_s)^2}\left(\frac{r_0}{r}\right)^2\right]. \tag{18.21}$$

Since we have considered that $\Phi(r_0) = 0$, from Eq. (18.21), we get

$$-\frac{1}{\omega} = \frac{8\pi\rho_s R_s^4}{(r_0 + R_s)^2}. \tag{18.22}$$

Now using the above relation, the final expression for the redshift function comes out as

$$\Phi(r) = \ln\left[\left(\frac{r(r + R_s)}{r_0(r_0 + R_s)}\right)^{\frac{2\omega}{1+\omega}}\right] \tag{18.23}$$

Using Eqs. (18.18 and 18.23) in Eq. (18.3), the space-time metric can be written as

$$ds^2 = -\left(\frac{r(r + R_s)}{r_0(r_0 + R_s)}\right)^{\frac{4\omega}{1+\omega}} dt^2 + \frac{1}{1 - \frac{r_0 + 8\pi\rho_s R_s^4(\frac{1}{r_0+R_s} - \frac{1}{r+R_s})}{r}} dr^2 \tag{18.24}$$
$$+ r^2(d\theta^2 + \sin^2\theta d\phi^2).$$

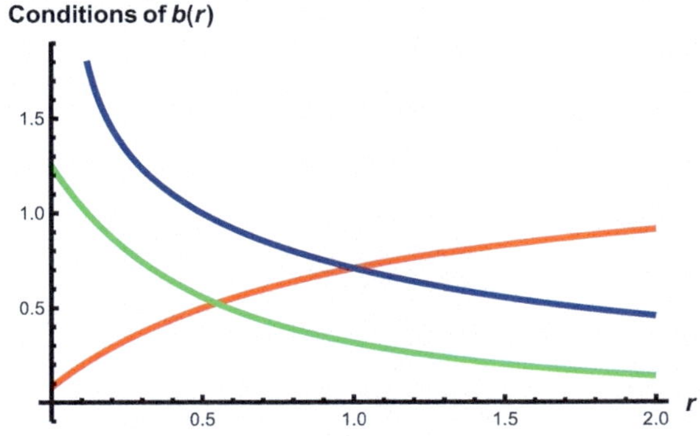

Fig. 18.1 Plots for conditions of shape function $b(r)$. Here ■ → $b(r)$; ■ → $\frac{b(r)}{r}$; ■ → $b'(r)$

The stress-energy scenario is presented as

$$\rho(r) = \frac{\rho_s R_s^4}{r^2(r + R_s)^2}, \tag{18.25}$$

$$P(r) = \omega\rho(r) = -\frac{1}{8\pi r^2}\left(\frac{r_0 + R_s}{r + R_s}\right)^2 \tag{18.26}$$

All the required properties of the shape function $b(r)$ are shown in Fig. 18.1. For all the plots shown in this paper, we have considered $R_s = 1\,\text{kpc}$, $\rho_s = 0.05 M_\odot\,\text{kpc}^{-3}$, and $r_0 = 0.5kpc$.

Energy Conditions

We consider the matter content of the wormhole to check whether it violates the Null Energy Condition(NEC). According to the general theory of relativity, the violation of NEC is necessary to sustain the wormhole structure and to keep the wormhole's throat open. For any null vector k^a, $T_{\mu\nu}k^\mu k^\nu \geq 0$ or $\rho + P \geq 0$. That means the stress energy experienced by a light ray must not be negative. In our case,

$$\rho(r) + P(r) = \frac{\rho_s R_s^4}{r^2(r + R_s)^2} - \frac{(r_0 + R_s)^2}{8\pi r^2(r + R_s)^2} \tag{18.27}$$

From Fig. 18.2, it is seen that the wormhole violates the NEC in the whole space-time. The wormhole has a positive energy density and a negative pressure (shown in Fig. 18.3), causing a repulsive effect, and preventing the wormhole from collapsing. Since the wormhole violates the NEC, which is the weakest of the four main pointwise energy conditions, it will automatically violate all the other two energy conditions,

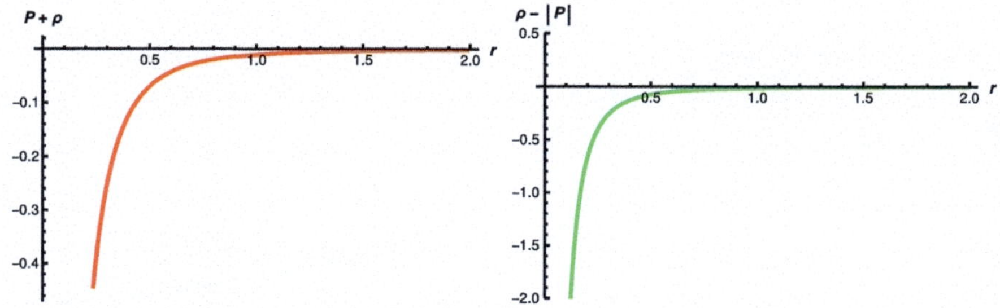

Fig. 18.2 Plots for $\rho + P$ and $\rho - |P|$ for examining NEC and DEC, respectively

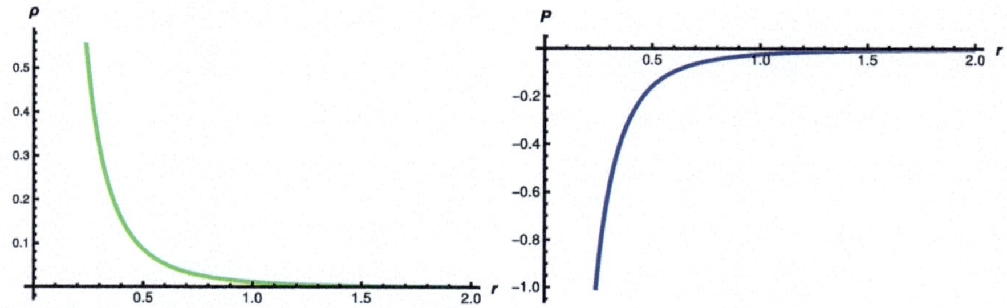

Fig. 18.3 Plots for energy density ρ and pressure P

viz., Weak Energy Condition (WEC) and Strong Energy Condition (SEC). The violation of SEC promises a repulsive nature. From Fig. 18.2, it is also clear that the wormhole also violates the Dominant Energy Condition (DEC).

18.4 Stability

Flare-Out Condition

The flare-out condition can be easily understood through the embedding geometry. The embedded space-time of the metric Eq. (18.3) at t = constant and $\theta = \frac{\pi}{2}$ is

$$\begin{aligned} ds_2^2 &= \left(1 - \frac{b(r)}{r}\right)^{-1} dr^2 + r^2 d\phi^2 = dz^2 + dr^2 + r^2 d\phi^2 \\ &= \left[1 + \left(\frac{dz}{dr}\right)^2\right] dr^2 + r^2 d\phi^2, \end{aligned} \tag{18.28}$$

where $\frac{dz}{dr} = \pm \frac{1}{\sqrt{\frac{r}{b(r)} - 1}}$ or $\frac{dr}{dz} = \pm \sqrt{\frac{r}{b(r)} - 1}$. Now, the flare-out condition is given by the minimality of the wormhole throat as [7, 32],

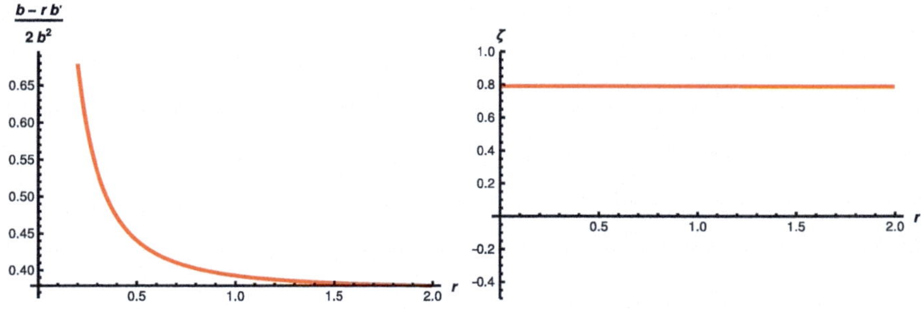

Fig. 18.4 Plots for flare-out condition and exoticity function ζ

$$\frac{d}{dz}\left(\frac{dr}{dz}\right) = \frac{b(r) - rb'(r)}{2(b(r))^2} > 0. \tag{18.29}$$

It is independent of $\Phi(r)$, solely determined by $b(r)$. In our case, using Eq. (18.18), we observe that $\frac{dz}{dr} \to 0$ as $r \to \infty$. This implies that the space is asymptotically flat [32].

From Fig. 18.4, we can see that Eq. (18.29) is valid, i.e., $\frac{d}{dz}(\frac{dr}{dz}) > 0$ for our wormhole.

Again, the exoticity function ζ is used for the flare-out condition, which is given as

$$\zeta = \frac{\tau - \rho}{|\rho|} > 0, \tag{18.30}$$

where τ is the surface tension. In our case, the value of surface tension is $\tau = \frac{1}{8\pi r2}\left(\frac{r_0+R_s}{r+R_s}\right)^2$. So the value of ζ comes out as

$$\zeta = \frac{(r_0 + R_s)^2}{8\pi\rho_s R_s^4} - 1 = constant. \tag{18.31}$$

From condition Eq. (18.7), we get $(r_0 + R_s)^2 > 8\pi\rho_s R_s^4$, so the exoticity function $\zeta > 0$, i.e., the metric obeys flare-out conditions everywhere. The behavior of ζ is shown in Fig. 18.4. It was assumed that the wormhole should have a large surface tension compared to the energy density to continue the wormhole geometry. This violates the Weak Energy Condition(WEC) or averaged Weak Energy Condition to minimize the use of exotic matter.

Tolman-Oppenheimer-Volkov (TOV) Equation

Here we will consider the generalized Tolman-Oppenheimer-Volkov (TOV) equation [33, 34] to realize the stability of our obtained wormhole solutions. The generalized

TOV equation is given as

$$\Phi'(\rho + P_r) + \frac{dP_r}{dr} + \frac{2}{r}(P_r - P_\theta) = 0. \quad (18.32)$$

The TOV equation is an important method and can be used to investigate the stability of any kind of astrophysical solution including wormhole/compact objects. It provides information about the equilibrium condition for the wormhole subjected to gravitational (F_g), hydrostatic (F_h), and another force due to the anisotropic nature (F_a) of the matter constituting the wormhole. For equilibrium, Eq. (18.32) takes the form,

$$F_g + F_h + F_a = 0, \quad (18.33)$$

where

$$F_g = -\Phi'(\rho + P_r), \quad (18.34)$$

$$F_h = -\frac{dP_r}{dr}, \quad (18.35)$$

$$F_a = -\frac{2}{r}(P_r - P_\theta). \quad (18.36)$$

Since, we have considered the isotropic pressure condition, i.e., $P_r = P_\theta = P_\phi = P$, therefore, the anisotropic force becomes zero. So, the TOV equation simplifies to

$$F_g + F_h = 0. \quad (18.37)$$

The profiles of F_g and F_h for the matter distribution in the galactic halo region are shown in Fig. 18.5. From the figure, it is clear that the equilibrium stage can be achieved due to the combined effect of gravitational and hydrostatic forces. These two forces act nearly opposite to each other to make the system balanced.

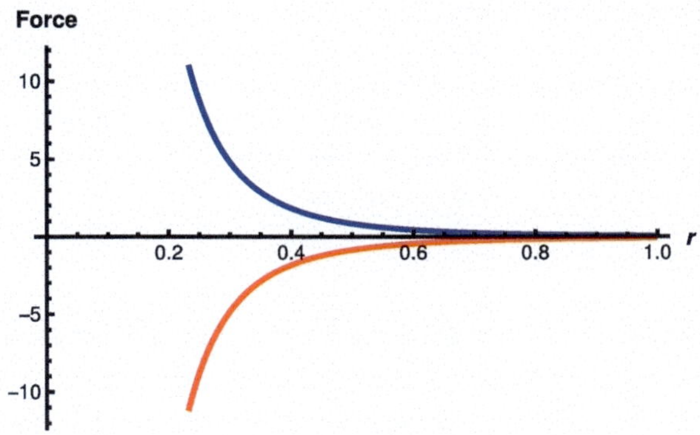

Fig. 18.5 Plots for F_g and F_h. Here ■ → F_h; ■ → F_g

18.5 Conclusion

In this work, we have studied the spherically symmetric traversable wormhole metric in dark matter halo region with Jaffe density profile. We obtained the exact solution of the traversable wormhole for the dark matter halo under the isotropic pressure condition. The obtained shape and redshift functions are found to be consistent with desired properties.

It has been observed that the matter content of the wormhole violates all the physical energy conditions. The violation of NEC, WEC, and SEC represents the repulsive nature of the matter present inside the throat, which is essential to keep the wormhole throat open and also to prevent it from collapsing.

The geometry of the obtained wormhole obeys the flare-out condition and the exoticity function is found to be positive everywhere in the wormhole spacetime. It also shows the asymptotical flatness. Finally, the Tolman-Oppenheimer-Volkov(TOV) equation showed the stability of our wormhole solution.

Considering all these factors we can conclude that the galactic halo possesses some of the characteristics needed to support a stable traversable wormhole.

References

1. D. Hochberg, M. Visser, Geometric structure of the generic static traversable wormhole throat. Phys. Rev. D **56**(8), 4745 (1997)
2. L. Flamm, Beiträge zur Einsteinschen gravitationstheorie. Hirzel (1916)
3. L. Flamm, Republication of: contributions to Einstein's theory of gravitation. Gen. Relativ. Gravit. **47**(6), 72 (2015)
4. H. Weyl, Der Zusammenhang zwischen der symmetrischen und der linearen Gruppe. Ann. Math. 499–516 (1928)
5. J.A. Wheeler, Geons. Phys. Rev. **97**(2), 511 (1955)
6. A. Einstein, N. Rosen, The particle problem in the general theory of relativity. Phys. Rev. **48**(1), 73 (1935)
7. M.S. Morris, K.S. Thorne, Wormholes in spacetime and their use for interstellar travel: a tool for teaching general relativity. Am. J. Phys. **56**(5), 395–412 (1988)
8. M.S. Morris, K.S. Thorne, U. Yurtsever, Wormholes, time machines, and the weak energy condition. Phys. Rev. Lett. **61**(13), 1446 (1988)
9. M. Visser, Lorentzian Wormholes. From Einstein to Hawking. Woodbury (1995)
10. D. Hochberg, A. Popov, S.V. Sushkov, Self-consistent wormhole solutions of semiclassical gravity. Phys. Rev. Lett. **78**(11), 2050 (1997)
11. D. Hochberg, M. Visser, Dynamic wormholes, antitrapped surfaces, and energy conditions. Phys. Rev. D **58**(4), 044021 (1998)
12. C.W. Misner, J.A. Wheeler, Classical physics as geometry. Ann. Phys. **2**(6), 525–603 (1957)
13. S.W. Hawking, Wormholes in spacetime. Phys. Rev. D **37**(4), 904 (1988)
14. H.G. Ellis, Ether flow through a drainhole: a particle model in general relativity. J. Math. Phys. **14**(1), 104–118 (1973)
15. V. Dzhunushaliev, D. Singleton, Wormholes and flux tubes in 5D Kaluza-Klein theory. Phys. Rev. D **59**(6), 064018 (1999)
16. M.G. Richarte, C. Simeone, Wormholes in Einstein-Born-Infield theory. Phys. Rev. D **80**(10), 104033 (2009)

17. A.G. Agnese, M. La Camera, Wormholes in the Brans-Dicke theory of gravitation. Phys. Rev. D **51**(4), 2011 (1995)
18. N.M. Garcia, Francisco SN Lobo, Nonminimal curvature-matter coupled wormholes with matter satisfying the null energy condition. Class. Quantum Gravity **28**(8), 085018 (2011)
19. M.R. Mehdizadeh, M.K. Zangeneh, F.S.N. Lobo, Einstein-Gauss-Bonnet traversable wormholes satisfying the weak energy condition. Phys. Rev. D **91**(8), 084004 (2015)
20. K.A. Bronnikov, A.M. Galiakhmetov, Wormholes without exotic matter in Einstein-Cartan theory. Gravitation Cosmol. **21**(4), 283–288 (2015)
21. E. Kocuper, J. Matyjasek, K. Zwierzchowska, Stress-energy tensor of quantized massive fields in static wormhole spacetimes. Phys. Rev. D **96**(10), 104057 (2017)
22. E.W. Kolb, M.S. Turner, J. Silk, The early universe 106–107 (1991)
23. U. Nucamendi, M. Salgado, D. Sudarsky, Nonminimal global monopoles and bound orbits. Phys. Rev. Lett. **84**(14), 3037 (2000)
24. C.G. Böhmer, T. Harko, F.S.N. Lobo, Dark matter as a geometric effect in f (R) gravity. Astroparticle Phys. **29**(6), 386–392 (2008)
25. Z. Xu et al. Possibility of traversable wormhole formation in the dark matter halo with isotropic pressure. Eur. Phys. J. C **80**, 1–7 (2020)
26. F. Rahaman et al., Possible existence of wormholes in the galactic halo region. Eur. Phys. J. C **74**(2), 2750 (2014)
27. J.F. Navarro, The structure of cold dark matter halos. *Symposium-international Astronomical Union*, vol. 171. (Cambridge University Press, 1996)
28. J. Einasto, On the construction of a composite model for the galaxy and on the determination of the system of galactic parameters. Trudy Astrofizicheskogo Instituta Alma-Ata **5**, 87–100 (1965)
29. W. Jaffe, A simple model for the distribution of light in spherical galaxies. Mon. Not. R. Astron. Soc. **202**(4), 995–999 (1983)
30. H. Mo, F. Van den Bosch, S. White, *Galaxy Formation and Evolution* (Cambridge University Press, 2010)
31. J. Wagner, Cosmic structures from a mathematical perspective 1: dark matter halo mass density profiles. Gen. Relativ. Gravit. **52**(6), 61 (2020)
32. S.-W. Kim, Flare-out condition of a Morris-Thorne wormhole and finiteness of pressure. J. Korean Phys. Soc. **63**, 1887–1891 (2013)
33. J.R. Oppenheimer, G.M. Volkoff, On massive neutron cores. Phys. Rev. **55**(4), 374 (1939)
34. R.C. Tolman, Static solutions of Einstein's field equations for spheres of fluid. Phys. Rev. **55**(4), 364 (1939)

Chapter 19
Neutrino Mass Model Based on $\Delta(27)$ Symmetry with Additional Higgs Doublets

Ph. Wilina and Ngangkham Nimai Singh

Abstract We develop a neutrino mass model based on $\Delta(27)$ discrete symmetry by using three Higgs doublets, three right-handed neutrinos, and two $\Delta(27)$ triplets to explain neutrino oscillation parameters consistent with the Planck cosmological bound on the sum of the three absolute neutrino mass eigenvalues, $\sum_i |m_i| < 0.12$ eV. The detailed numerical analysis is carried out by applying input model parameters and the predicted neutrino oscillation parameters are in agreement with the global neutrino oscillation data for both normal (NH) and inverted (IH) hierarchical model. We also calculate the effective neutrino masses for neutrinoless double beta decay (m_{ee}), tritium beta decay (m_β), and Jarlskog invariant (J_{CP}) for CP violation as $m_{ee} \leq 30.6$ meV, $m_\beta \leq 30.5$ meV, and $J_{CP} \leq 0.0217$ for NH, and $m_{ee} \leq 51.7$ meV, $m_\beta \leq 51.4$ meV, and $J_{CP} \leq 0.0222$ for IH.

19.1 Introduction

The origin of non-zero tiny neutrino masses being an unsolved mystery of the so-called Standard Model (SM) has been an important evidence which supports beyond standard model (BSM). Neutrino oscillation experiments such as Daya Bay, T2K, MINOS, Double CHOOZ, and RENO. [1–3] have provided certain constraints on the neutrino oscillation parameters and leptonic mixing parameters. Various neutrino mass models based on different discrete symmetries such as S_3 [4], S_4 [5], A_4 [6], $\Delta(27)$ [7, 8], $\Delta(54)$, $\Delta(96)$ [9], and A_5 [10] have been proposed to study tiny neutrino masses. Motivations to extend SM with non-abelian discrete flavor symmetries over continuous symmetries are discussed in [11].

Ph. Wilina (✉)
Department of Physics, Manipur University, Imphal 795003, India
e-mail: wilina.phd.phy@manipuruniv.ac.in

N. N. Singh
Research Institute of Science and Technology, Imphal 795003, India
e-mail: nimai03@yahoo.com

J. W. F. Valle et al. (eds.), *Proceedings of the International Conference on Future Prospects in Neutrino and Astroparticle Physics*, Springer Proceedings in Physics 361,
https://doi.org/10.1007/978-981-97-7441-8_19

Table 19.1 NuFIT data on neutrino oscillation parameters

NH	IH
$sin^2\theta_{12}$ = [0.270, 0.341]	$sin^2\theta_{12}$ = [0.270, 0.341]
$sin^2\theta_{13}$ = [0.02029, 0.2391]	$sin^2\theta_{13}$ = [0.02047, 0.02396]
$sin^2\theta_{23}$ = [0.406, 0.620]	$sin^2\theta_{23}$ = [0.412, 0.623]
$\delta = [108°, 404°]$	$\delta = [192°, 360°]$
Δm^2_{21}=[6.82, 8.03]$\times 10^{-5} eV^2$	Δm^2_{21}=[6.82, 8.03]$\times 10^{-5} eV^2$
Δm^2_{31}=[+2.428, +2.597]$\times 10^{-3} eV^2$	Δm^2_{32}=[-2.581, -2.408]$\times 10^{-3} eV^2$

In addition, the famous seesaw mechanism with type-I, type-II, and type-III has also been employed in many of the neutrino mass model building structure to obtain experimentally allowed neutrino oscillation parameters, thereby predicting effective neutrino masses for neutrinoless double beta decay and tritium beta decay and evaluating the CP violating parameter. Experiments like GERDA, KamLAND-Zen, CUORE, CUPID-0, CUPID-Mo, etc. [12, 13] are vigorously attempting to observe neutrinoless double beta decay. The current neutrino oscillation experimental data provides the parameters given in Table 19.1.

In the paper, we adopt $\Delta(27)$ discrete symmetry as it is the smallest group of the type $\Delta(3n^2)$, supplemented by Z_2 symmetry to obtain neutrino oscillation parameters in agreement with the experimental data and Planck cosmological bound on the sum of the three absolute neutrino mass eigenvalues, $\sum_i |m_i| < 0.12$ eV [14]. Due to the presence of two triplets (3 and $\bar{3}$), it is easier to construct neutrino mass matrix which will help in analyzing several neutrino phenomenology in the future. In addition, we consider additional Higgs doublets so as to get a diagonal charged-lepton mass matrix so that contribution to U_{PMNS} comes only from the neutrino sector.

The paper is organized as follows. Section 19.2 gives the model description along with necessary Lagrangian. Numerical analysis and results are presented in Sect. 19.3.

19.2 Model Description

We extend the SM field content by introducing two additional Higgs doublets to correct charged-lepton masses, three right-handed neutrinos plus two $\Delta(27)$ triplets (ϕ, χ) for implementation of type-I seesaw mechanism to obtain light neutrino masses. The field contents under $SU(2)_L \times \Delta(27) \times Z_2$ symmetry have the following transformation shown in Table 19.2.

With the field content given in Table 19.2 and the tensor product of $\Delta(27)$ symmetry [15], the Yukawa interactions for lepton sector can be written as

Table 19.2 Field contents of the model under $SU(2)_L \times \Delta(27) \times Z_2$ symmetry. Here, j=1,2,3

Field	$(\bar{L}_{eL}, \bar{L}_{\mu L}, \bar{L}_{\tau L})$	(e_R, μ_R, τ_R)	H_1	H_2	H_3	ν_{Rj}	ϕ	χ
$SU(2)_L$	2	1	2	2	2	1	2	1
$\Delta(27)$	3	$\bar{3}$	1_1	1_2	1_3	3	3	3
Z_2	1	1	1	1	1	−1	−1	1

$$\begin{aligned}-\mathcal{L}^l_{yuk} &= h_1\,(\overline{L}_{iL}l_R)_{\mathbf{1_1}}H_1 + h_2(\overline{L}_{iL}l_R)_{\mathbf{1_3}}H_2 + h_3(\overline{L}_{iL}l_R)_{\mathbf{1_2}}H_3\\ &+ \tfrac{y_1}{2}(\overline{L}_{iL}\nu_{Rj})_{\mathbf{3_{s_1}}}\phi + \tfrac{y_2}{2}(\overline{L}_{iL}\nu_{Rj})_{\mathbf{3_{s_2}}}\phi + \tfrac{y_3}{2}(\overline{L}_{iL}\nu_{Rj})_{\mathbf{3_a}}\phi\\ &+ \tfrac{\lambda_1}{2}(\overline{\nu}^c_{Rj}\nu_{Rj})_{\mathbf{3_{s_1}}}\chi + \tfrac{\lambda_2}{2}(\overline{\nu}^c_{Rj}\nu_{Rj})_{\mathbf{3_{s_2}}}\chi + \tfrac{\lambda_3}{2}(\overline{\nu}^c_{Rj}\nu_{Rj})_{\mathbf{3_a}}\chi + H.c.\end{aligned} \tag{19.1}$$

Choosing the vev alignments $\langle\phi\rangle = (\langle\phi_1\rangle, \langle\phi_2\rangle, \langle\phi_1\rangle)$ with $\langle\phi_1\rangle = v_{\phi_1}$, $\langle\phi_2\rangle = v_\phi$, and $\langle\chi\rangle = v_\chi(1,1,0)$, the charged-lepton mass matrix can be written as

$$M_l = \begin{pmatrix} h_1v_1 + h_2v_2 + h_3v_3 & 0 & 0\\ 0 & h_1v_1 + \omega^2h_2v_2 + \omega h_3v_3 & 0\\ 0 & 0 & h_1v_1 + \omega h_2v_2 + \omega^2h_3v_3\end{pmatrix} \tag{19.2}$$

where $v_{1,2,3}$ are the vevs of $H_{1,2,3}$. For the neutrino sector, the Dirac and Majorana neutrino mass matrices can be written as

$$m_D = \begin{pmatrix} x & y & z\\ y & t & y\\ z & y & x\end{pmatrix}; \qquad M_R = \begin{pmatrix} a & 0 & b\\ 0 & a & b\\ b & b & 0\end{pmatrix} \tag{19.3}$$

where $x = \frac{y_1 v_{\phi_1}}{2}$, $y = \frac{y_2 v_{\phi_1}}{2}$, $z = \frac{y_2 v_\phi}{2}$, $t = \frac{y_1 v_\phi}{2}$ and $a = \frac{\lambda_1 v_\chi}{2}$, $b = \frac{\lambda_2 v_\chi}{2}$.

Then, the neutrino mass matrix generated by type-I seesaw mechanism, $M_\nu = -m_D^T M_R^{-1} m_D$ is

$$\begin{aligned}M_\nu &= \begin{pmatrix} A & B & B\\ B & C & D\\ B & D & C\end{pmatrix} + \begin{pmatrix} 0 & a_1 & a_2\\ a_1 & b_1 & 0\\ a_2 & 0 & b_2\end{pmatrix}\\ &\equiv M' + \delta M'\end{aligned} \tag{19.4}$$

$A = \frac{-4byz+az^2}{2b^2}$; $B = \frac{ayz-b(2y^2+yz)}{2b^2}$; $C = \frac{a^2y^2-2aby^2}{2ab^2}$;
$D = \frac{a^2y^2-b^2(t-y)(y-z)-aby(t+2y+z)}{2ab^2}$; $a_1 = \frac{-btz}{2b^2}$; $a_2 = \frac{-bz^2}{2b^2}$; $b_1 \frac{-b^2(t-y)^2-2abty}{2ab^2}$; $b_2 = \frac{-b^2(y-z)^2-2abyz}{2ab^2}$. The eigenvalues and eigenvectors of the first matrix of Eq. (19.4) are given by

$$\begin{aligned} m'_{1,2} &= \tfrac{1}{2}\left[A + C + D \pm \sqrt{(A-C-D)^2 + 8B^2}\right], \\ m'_3 &= C - D \end{aligned} \tag{19.5}$$

$$U' = \begin{pmatrix} cos\theta & sin\theta & 0 \\ -\frac{sin\theta}{\sqrt{2}} & \frac{cos\theta}{\sqrt{2}} & \frac{-1}{\sqrt{2}} \\ -\frac{sin\theta}{\sqrt{2}} & \frac{cos\theta}{\sqrt{2}} & \frac{1}{\sqrt{2}} \end{pmatrix} \tag{19.6}$$

where $\theta = \arcsin\left(\frac{\mathcal{K}}{\sqrt{\mathcal{K}^2+2}}\right)$; $\mathcal{K} = \frac{A-C-D-\sqrt{(A-C-D)^2+8B^2}}{2B}$. Considering the second matrix of Eq. (19.4) as perturbed matrix, the eigenvalues and eigenvectors of Eq. (19.4) are given by

$$m_1 = m'_1 + \frac{1}{2}sin\theta\left[(b_1+b_2)sin\theta - 2\sqrt{2}(a_1+a_2)cos\theta\right],$$

$$m_2 = m'_2 + \frac{1}{2}cos\theta\left[(b_1+b_2)cos\theta + 2\sqrt{2}(a_1+a_2)sin\theta\right],$$

$$m_3 = m'_3 + \frac{1}{2}(b_1+b_2). \tag{19.7}$$

$$\begin{aligned} U_\nu = U' + \delta U &= \begin{pmatrix} cos\theta & sin\theta & 0 \\ -\frac{sin\theta}{\sqrt{2}} & \frac{cos\theta}{\sqrt{2}} & \frac{-1}{\sqrt{2}} \\ -\frac{sin\theta}{\sqrt{2}} & \frac{cos\theta}{\sqrt{2}} & \frac{1}{\sqrt{2}} \end{pmatrix} + \begin{pmatrix} \delta U_{11} & \delta U_{12} & \delta U_{13} \\ \delta U_{21} & \delta U_{22} & \delta U_{23} \\ \delta U_{31} & \delta U_{32} & \delta U_{33} \end{pmatrix} \\ &= \begin{pmatrix} U_{11} & U_{12} & U_{13} \\ U_{21} & U_{22} & U_{23} \\ U_{31} & U_{32} & U_{33} \end{pmatrix} \end{aligned} \tag{19.8}$$

where the elements of δU are given in Appendix 19.4. As neutrino mixing angles (θ_{12}, θ_{13}, θ_{23}) and Dirac CP-phase (δ) are related to the neutrino mixing matrix elements [16], we can write

$$U_{11} = \frac{1}{cos\theta + t_{12}sin\theta}; \qquad U_{12} = \frac{t_{12}}{cos\theta + t_{12}sin\theta} \tag{19.9}$$

with t_{12}= tan θ_{12}. As $U_{11} \in (0.803, 0.845)$, we have $t_{12} \in (0.6429, 0.7426)$ for sin θ = $\frac{1}{\sqrt{3}}$ and cos θ = $\sqrt{\frac{2}{3}}$, leading to $U_{12} \in (0.5413, 0.5964)$. Also using the relations

$$U_{11} = U'_{11} + \delta U_{11}; \qquad U_{13} = U'_{13} + \delta U_{13}, \tag{19.10}$$

the parameters a_1 and a_2 can be expressed as

$$\begin{aligned} a_2 = & \frac{(sin\theta - t_{12}cos\theta)(m_1' - m_2')}{\sqrt{2}(cos\theta + t_{12}sin\theta)(cos^2\theta - sin^2\theta)} + \frac{(b_1 + b_2)sin\theta cos\theta}{2\sqrt{2}(cos^2\theta - sin^2\theta)} \\ & + \frac{2\sqrt{2}(m_3' - m_1')(m_3' - m_2')sin\theta_{13}e^{-i\delta} - (b_1 - b_2)(m_1' - m_2')sin\theta cos\theta}{4(-m_1' sin^2\theta - m_2' cos^2\theta + m_3')}, \end{aligned} \tag{19.11}$$

$$a_1 = a_2 - \frac{2\sqrt{2}(m_3' - m_1')(m_3' - m_2')sin\theta_{13}e^{-i\delta} - (b_1 - b_2)(m_1' - m_2')sin\theta cos\theta}{2(-m_1' sin^2\theta - m_2' cos^2\theta + m_3')}. \tag{19.12}$$

The effective neutrino masses for neutrinoless double beta decay [17] and tritium beta decay [18] are given by

$$m_{ee} = |\sum_{i=1}^{3} U_{ei}^2 m_i|,$$

$$m_\beta = \left(\sum_{i=1}^{3} |U_{ei}^2| m_i^2\right)^{1/2}.$$

The Jarlskog invariant [19] is given by

$$\begin{aligned} J_{CP} &= Im(U_{e1}U_{\mu 2}U_{e2}^* U_{\mu 1}^*) \\ &= s_{12}c_{12}s_{23}c_{23}c_{13}^2 s_{13} sin\delta. \end{aligned}$$

19.3 Numerical Analysis and Results

In our model, numerical analysis is carried out by applying random range on the parameters b_1 and b_2 for both normal and inverted hierarchical case. These random range of b_1 and b_2 are applied in such a way that they generate neutrino oscillation parameters consistent with 3σ bound of current neutrino oscillation data and latest Planck cosmological data on the sum of three absolute neutrino masses.

Normal Hierarchy

For normal hierarchical case, we choose $b_1 \in [-1, -9] \times 10^{-4}$ eV, $b_2 \in [1, 9] \times 10^{-4}$ eV, $m_1' \in [0.0001, 0.029]$ eV with $\sin\theta = \frac{1}{\sqrt{3}}$ and $\cos\theta = \sqrt{\frac{2}{3}}$ as U' of Eq. (19.6) is very near to TBM mixing pattern. The correlations among the model parameters are plotted in Figs. 19.1, 19.2, and 19.3. Table 19.3 represents the model parameter values for b_1 = -8.2 $\times 10^{-4}$ eV, b_2 = 7.8 $\times 10^{-4}$ eV, m_1' = 0.02 eV, δ = 200°.

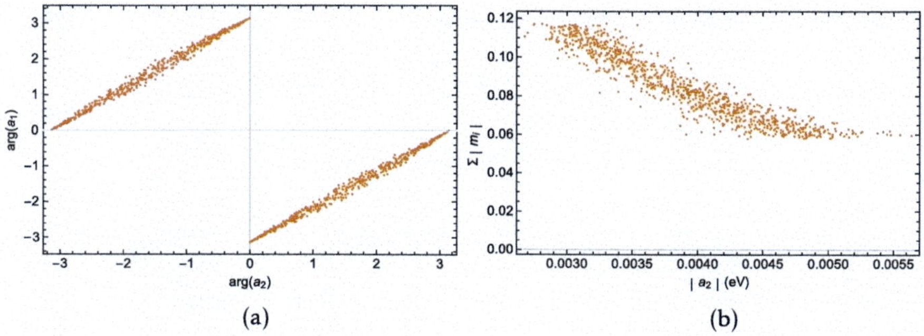

Fig. 19.1 Correlation plot between the model parameters **a** $\arg(a_1)$ and $\arg(a_2)$, **b** $\sum_i |m_i|$ and $|a_2|$ for normal hierarchy

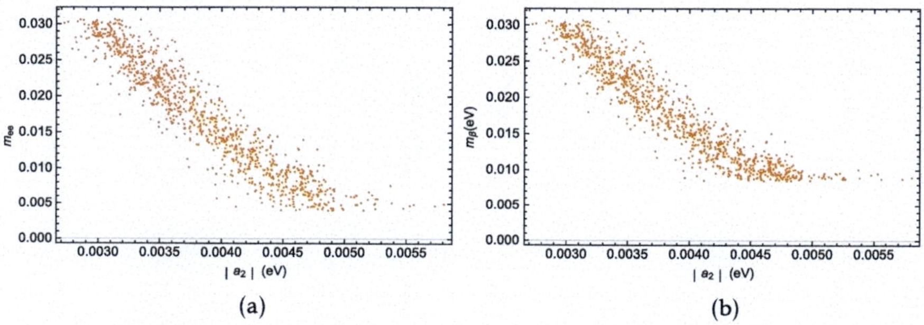

Fig. 19.2 Correlation plot between the model parameters **a** m_{ee} and $|a_2|$ **b** m_β and $|a_2|$ for normal hierarchy

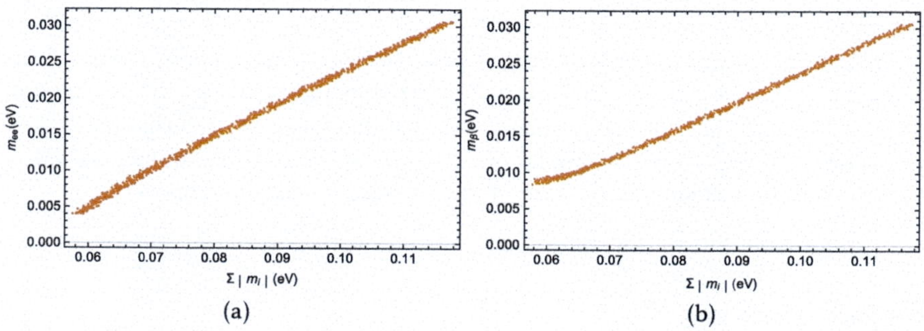

Fig. 19.3 Correlation plot between the model parameters **a** m_{ee} and $\sum_i |m_i|$ **b** m_β and $\sum_i |m_i|$ for normal hierarchy

Table 19.3 Model parameter values for $b_1 = -8.2\times10^{-4}$ eV, $b_2 = 7.8\times10^{-4}$ eV, $m'_1 = 0.02$ eV, δ=200° for normal hierachy

Parameters	Output
a_1	$(-1.72\text{–}3.1i)\times10^{-3}$ eV
a_2	$(1.74 + 3.1i)\times10^{-3}$ eV
m_1	0.01998 eV
m_2	0.02179 eV
m_3	0.05461 eV
Δm^2_{21}	$7.58\times10^{-5} eV^2$
Δm^2_{31}	$2.51\times10^{-3} eV^2$
$\sum_i \lvert m_i\rvert$	0.096 eV
m_{ee}	21.8 meV
m_β	22.1 meV

The leptonic mixing matrix in normal hierarchical case takes the form

$$|U| = \begin{pmatrix} 0.803 \rightarrow 0.842 & 0.5413 \rightarrow 0.5963 & 0.142 \rightarrow 0.155 \\ 0.3047 \rightarrow 0.5114 & 0.5077 \rightarrow 0.6691 & 0.6836 \rightarrow 0.7116 \\ 0.2922 \rightarrow 0.5018 & 0.4930 \rightarrow 0.6569 & 0.7027 \rightarrow 0.7301 \end{pmatrix}, \quad (19.13)$$

leading to $sin^2\theta_{12} \in (0.299, 0.364)$, $sin^2\theta_{23} \in (0.467, 0.506)$. With this, the effective neutrino masses for neutrinoless double beta decay and tritium beta decay are found to be $m_{ee} \leq 30.6$ meV and $m_\beta \leq 30.5$ meV. The Jarlskog invariant for CP violation is found to be $J_{cp} \leq 0.0217$.

Inverted Hierarchy

In inverted hierarchical case, we choose the random range of b_1, b_2 as in normal hierarchical case but with $m'_2 \in (0.05, 0.0518)$ eV. The correlations among the model parameters in this case are shown in Figs. 19.4 and 19.5. Table 19.4 represents the model parameter values for b_1 = -8.2×10^{-4} eV, $b_2 = 7.8\times10^{-4}$ eV, $m'_2 = 0.051$ eV, $\delta = 200°$.

The leptonic mixing matrix in inverted hierarchical case takes the form

$$|U| = \begin{pmatrix} 0.803 \rightarrow 0.842 & 0.5414 \rightarrow 0.5964 & 0.142 \rightarrow 0.155 \\ 0.2884 \rightarrow 0.5037 & 0.4970 \rightarrow 0.6493 & 0.7087 \rightarrow 0.7269 \\ 0.3000 \rightarrow 0.5126 & 0.5127 \rightarrow 0.6671 & 0.6873 \rightarrow 0.7055 \end{pmatrix}, \quad (19.14)$$

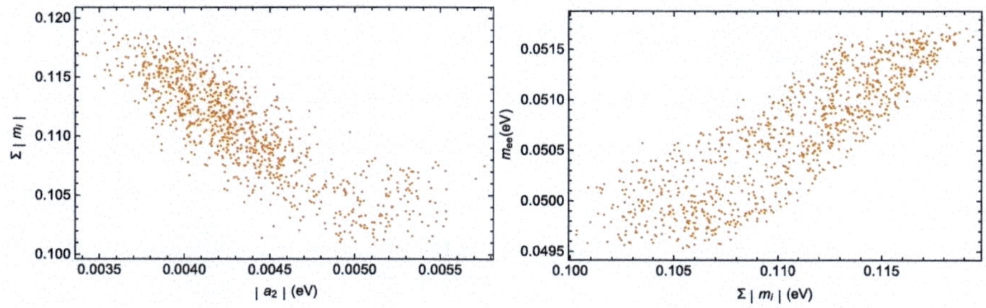

(a) (b)

Fig. 19.4 Correlation plot between the model parameters **a** $\sum_i |m_i|$ and $|a_2|$, **b** m_{ee} and $\sum_i |m_i|$ for inverted hierarchy

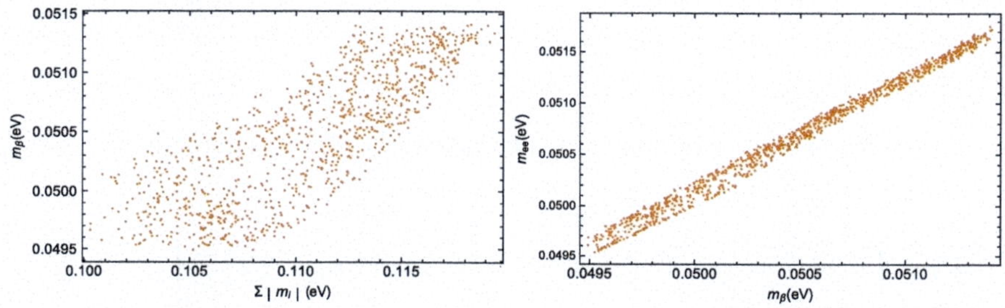

(a) (b)

Fig. 19.5 Correlation plot between the model parameters **a** m_β and $\sum_i |m_i|$, **b** m_{ee} and m_β for inverted hierarchy

Table 19.4 Model parameter values for b_1 = -8.2×10^{-4} eV, b_2 = 7.8×10^{-4} eV, m_2' = 0.051 eV, δ=200° for inverted hierarchy

Parameters	Output
a_1	$(1.94 + 3.5i)\times10^{-3}$ eV
a_2	$(-1.96–3.5i)\times10^{-3}$ eV
m_1	0.05022 eV
m_2	0.05098 eV
m_3	0.01227 eV
Δm_{21}^2	$7.64\times10^{-5} eV^2$
Δm_{31}^2	$-2.45\times10^{-3} eV^2$
$\sum_i \vert m_i \vert$	0.096 eV
m_{ee}	50.7 meV
m_β	50.5 meV

leading to $sin^2\theta_{12} \in (0.299, 0.364)$, $sin^2\theta_{23} \in (0.5129, 0.5413)$. Also, the effective neutrino masses for neutrinoless double beta decay and tritium beta decay are found to be $m_{ee} \leq 51.7$ meV and $m_\beta \leq 51.4$ meV. The Jarlskog invariant for CP violation is found to be $J_{cp} \leq 0.0224$.

19.4 Elements of δU

The elements of δU are given by

$$\begin{aligned}
\delta U_{11} &= \tfrac{1}{m_1'-m_2'}\Big[\tfrac{1}{2}\Big(-(b_1+b_2)sin^2\theta cos\theta + \sqrt{2}a_1(cos^2\theta - sin^2\theta)sin\theta \\
&\quad + \sqrt{2}a_2(cos^2\theta - sin^2\theta)sin\theta\Big)\Big],
\end{aligned}$$

$$\begin{aligned}
\delta U_{21} &= \tfrac{1}{m_1'-m_2'}\Big[\tfrac{1}{2\sqrt{2}}\Big(-(b_1+b_2)cos^2\theta sin\theta + \sqrt{2}b_2(cos^2\theta - sin^2\theta)cos\theta \\
&\quad + \sqrt{2}b_1(cos^2\theta - sin^2\theta)cos\theta\Big)\Big] + \tfrac{1}{m_1'-m_3'}\Big[\tfrac{1}{2\sqrt{2}}\Big(\sqrt{2}a_1cos\theta - \sqrt{2}b_1cos\theta - (b_1-b_2)sin\theta\Big)\Big],
\end{aligned}$$

$$\begin{aligned}
\delta U_{31} &= \tfrac{1}{m_1'-m_2'}\Big[\tfrac{1}{2\sqrt{2}}\Big(-(b_1+b_2)cos^2\theta sin\theta + \sqrt{2}a_1(cos^2\theta - sin^2\theta)cos\theta \\
&\quad + \sqrt{2}a_2(cos^2\theta - sin^2\theta)cos\theta\Big)\Big] + \tfrac{1}{m_1'-m_3'}\Big[\tfrac{1}{2\sqrt{2}}\Big(-\sqrt{2}a_1cos\theta \\
&\quad + \sqrt{2}a_2cos\theta + (b_1-b_2)sin\theta\Big)\Big],
\end{aligned}$$

$$\delta U_{12} = \tfrac{1}{m_2'-m_1'}\Big[\tfrac{1}{2}\Big(-(b_1+b_2)cos^2\theta sin\theta + \sqrt{2}cos\theta(cos^2\theta - sin^2\theta)(a_1+a_2)\Big)\Big],$$

$$\begin{aligned}
\delta U_{22} &= \tfrac{1}{m_2'-m_1'}\Big[\tfrac{1}{2\sqrt{2}}(b_1+b_2)sin^2\theta cos\theta - \sqrt{2}sin\theta(cos^2\theta - sin^2\theta)(a_1+a_2)\Big] \\
&\quad + \tfrac{1}{m_2'-m_3'}\Big[\tfrac{1}{2\sqrt{2}}\Big((b_1-b_2)cos\theta - \sqrt{2}(-a_1+a_2)sin\theta\Big)\Big],
\end{aligned}$$

$$\begin{aligned}
\delta U_{32} &= \tfrac{1}{m_2'-m_1'}\Big[\tfrac{1}{2\sqrt{2}}(b_1+b_2)sin^2\theta cos\theta - \sqrt{2}sin\theta(cos^2\theta - sin^2\theta)(a_1+a_2)\Big] \\
&\quad + \tfrac{1}{m_2'-m_3'}\Big[\tfrac{1}{2\sqrt{2}}\Big((-b_1+b_2)cos\theta + \sqrt{2}(-a_1+a_2)sin\theta\Big)\Big],
\end{aligned}$$

$$\begin{aligned}
\delta U_{13} &= \tfrac{1}{m_3'-m_1'}\Big[\tfrac{1}{2}\Big(\sqrt{2}cos^2\theta(-a_1+a_2) + (b_1-b_2)sin\theta cos\theta\Big)\Big] \\
&\quad + \tfrac{1}{m_3'-m_2'}\Big[\tfrac{1}{2}\Big((-b_1+b_2)sin\theta cos\theta + \sqrt{2}(-a_1+a_2)sin^2\theta\Big)\Big],
\end{aligned}$$

$$\begin{aligned}
\delta U_{23} &= \tfrac{1}{m_3'-m_1'}\Big[\tfrac{-1}{2\sqrt{2}}\Big(\sqrt{2}sin\theta cos\theta(-a_1+a_2) + (b_1-b_2)sin^2\theta\Big)\Big] \\
&\quad + \tfrac{1}{m_3'-m_2'}\Big[\tfrac{1}{2\sqrt{2}}\Big((-b_1+b_2)cos^2\theta + \sqrt{2}(-a_1+a_2)sin\theta cos\theta\Big)\Big],
\end{aligned}$$

$$\delta U_{33} = \delta U_{23} \tag{19.15}$$

References

1. I. Girardi, D. Meloni, S.T. Petcov, The Daya bay and T2K results on $sin^2 2\theta_{13}$ and non-standard neutrino interactions. Nucl. Phys. B **886**, 31–42 (2014). https://doi.org/10.1016/j.nuclphysb.2014.06.014
2. T. Nakaya, R.K. Plunkett, New J. Phys. **18**, 015009 (2016). https://doi.org/10.1088/1367-2630/18/1/015009
3. Y. Abe, C. Aberle, et al., First measurement of θ_{13} from delayed neutron capture on hydrogen in the Double Chooz experiment. Phys. Lett. B, **723**(1-3), 66–70
4. R.N. Mohapatra, S. Nasri, H.B. Yu, S_3 symmetry and tri-bimaximal mixing. Phys. Lett. B **639**(3–4), 318–321 (2006)
5. X. Wang, S. Zhou, The minimal seesaw model with a modular S_4 symmetry. J. High Energy Phys. **5**, 1–38 (2020)
6. G.-J. Ding, S.F. King, A.J. Stuart, Generalised CP and A_4 family symmetry. J. High Energy Phys. **2013**, 6 (2013)
7. E. Ma, Neutrino mass matrix from Δ (27) symmetry. Mod. Phys. Lett. A **21**(25), 1917–1921 (2006)
8. E. Ma, Neutrino mass matrix from S_4 symmetry. Phys. Lett. B **632**(2–3), 352–356 (2006)
9. N. Gautam, R. Krishnan, R. M.K. Das, Effect of sterile neutrino on low-energy processes in minimal extended seesaw with $\Delta(96)$ symmetry and TM_1 mixing. Front. Phys. **9**, 703266
10. G.J. Ding, L.L. Everett, A.J. Stuart, Golden ratio neutrino mixing and A_5 flavor symmetry. Nucl. Phys. B **857**, 219–253 (2012)
11. Z.Z. Xing, Z.H. Zhao, A review of μ-τ flavor symmetry in neutrino physics. Rep. Prog. Phys. **79**(7), 076201 (2016)
12. M. Agostini, M. Allardt et al., Results on neutrinoless double-β Decay of Ge 76 from phase I of the GERDA experiment. Phys. Rev. Lett. **111**(12), 122503
13. E. Armengaud, C. Augier et al., New limit for neutrinoless double-beta decay of Mo 100 from CUPID-Mo experiment. Phys. Rev. Lett. **126**(18), 181802 (2021)
14. Planck Collaboration, N. Aghanim et al., Planck 2018 results. I. Overview and the cosmological legacy of Planck. Astronomy Astrophys. **641**, 1 (2020)
15. V.V. Vien, A.C. Hernández, H.N. Long, The Δ (27) flavor 3-3-1 model with neutral leptons. Nucl. Phys. B **913**, 792–814 (2016)
16. C. Giganti, S. Lavignac, M. Zito, Neutrino oscillations: the rise of the PMNS paradigm. Prog. Part. Nucl. Phys. **98**, 1–54 (2018)
17. J.D. Vergados, H. Ejiri, F. Šimkovic, Theory of neutrinoless double-beta decay. Rep. Prog. Phys. **75**(10), 106301 (2012)
18. E.W. Otten, C. Weinheimer, Neutrino mass limit from tritium β decay. Rep. Prog. Phys. **71**(8), 086201 (2008)
19. P.B. Denton, S.J. Parke, Simple and precise factorization of the Jarlskog invariant for neutrino oscillations in matter. Phys. Rev. D **100**(5), 053004 (2019)

Chapter 20
A Thought Experiment on Neutrino Oscillation

Pralay Chakraborty and Subhankar Roy

Abstract We present a thought experiment and study the neutrino oscillation probability inside an infinite square well potential. Following Pontecorvo's approach, neutrino oscillation is visualized with respect to time for both relativistic and non-relativistic scenarios. In this regard, the dependence of the oscillation probability on the principle quantum number is seen. In addition, a contrasting result of non-zero oscillation probability is observed even if the neutrino mass eigenstates were degenerate.

20.1 Introduction

In the realm of particle physics, neutrinos emerge as mysterious entities that can transform from one flavour to another during their journey through space. This phenomenon is known as neutrino oscillation [1]. This concept was first introduced by Bruno Pontecorvo in 1957 which is consistent with the experiments [2–6]. The neutrinos have very low interaction cross-section with matter and the world essentially appears transparent to them. Looking at this fact, it is needless to mention that the likelihood of capturing a neutrino is exceedingly minimal. However, in the present work, we aim to diverge a little from the reality and delve into the implications of neutrino oscillation if, hypothetically, the neutrino could be trapped within an infinite square well potential. It is crucial to emphasize that the present work is solely based on a thought experiment and its realization is subjected to the future unfolding of the hidden secrets of Nature.

The neutrino oscillation arises from the phenomenon known as neutrino mixing, the inherent nature of three neutrino flavour states ($\nu_{l=e,\mu,\tau}$), which are the admixture of three mass eigenstates ($\nu_{i=1,2,3}$) with mass eigenvalues ($m_{i=1,2,3}$). Bridging the gap

P. Chakraborty (✉) · S. Roy
Department of Physics, Gauhati University, Guwahati, India
e-mail: pralay@gauhati.ac.in

S. Roy
e-mail: subhankar@gauhati.ac.in

J. W. F. Valle et al. (eds.), *Proceedings of the International Conference on Future Prospects in Neutrino and Astroparticle Physics*, Springer Proceedings in Physics 361,
https://doi.org/10.1007/978-981-97-7441-8_20

between flavour and mass states is the Pontecorvo-Maki-Nakagawa-Sakata (PMNS) matrix [7], a mathematical representation that provides the framework to understand how the neutrinos oscillate as they traverse space. A parametrization known as standard parametrization for U is adopted by the Particle Data Group (PDG) [8], where U is parametrized using three mixing angles $(\theta_{12}, \theta_{13}, \theta_{23})$, three CP phases (δ, α, β), and three unphysical phases (ϕ_1, ϕ_2, ϕ_3) as shown below,

$$V = P_\phi.\tilde{U}.P_M \tag{20.1}$$

where $P_\phi = (e^{i\phi_1}, e^{i\phi_2}, e^{i\phi_3})$, $P_M = (e^{i\alpha}, e^{i\beta}, 1)$, and

$$\tilde{U} = \begin{bmatrix} c_{12}c_{13} & c_{13}s_{12} & s_{13}e^{-i\delta} \\ -c_{23}s_{12} - c_{12}s_{13}s_{23}e^{i\delta} & c_{12}c_{23} - s_{12}s_{13}s_{23}e^{i\delta} & c_{13}s_{23} \\ s_{12}s_{23} - c_{12}c_{23}s_{13}e^{i\delta} & -c_{12}s_{23} - s_{12}s_{13}c_{23}e^{i\delta} & c_{13}c_{23} \end{bmatrix}, \tag{20.2}$$

where $c_{ij} = \cos\theta_{ij}$ and $s_{ij} = \sin\theta_{ij}$.

For simplicity, we shall stick to two-flavour mixing which can be visualized as shown in the following:

$$\begin{bmatrix} \nu_e \\ \nu_\mu \end{bmatrix} = \begin{bmatrix} \cos\theta & \sin\theta \\ -\sin\theta & \cos\theta \end{bmatrix} \begin{bmatrix} \nu_1 \\ \nu_2 \end{bmatrix}, \tag{20.3}$$

where θ is the mixing angle.

We know that the standard neutrino oscillation formula is derived for free ultra-relativistic neutrinos. In the present work, we shall explore both relativistic and non-relativistic scenarios based on speculation. Unlike in free space, the energy of neutrino mass eigenstates will be quantized within the well depicted by the principal quantum number n.

20.2 The Potential Well

The relativistic or non-relativistic behaviour of the particle inside the potential well can be linked to the length of the potential well and it is governed by the parameter $L_c = \frac{\lambda_c}{2\pi}$ [9–11], where λ_c is the Compton wavelength of the particle. The Compton wavelength λ_c is expressed as shown in the following:

$$\lambda_c = \frac{h}{mc}, \tag{20.4}$$

where m represents the mass of the particle. In general, we define the infinite square well potential in the following way:

$$V(x) = \begin{cases} 0 & 0 < x < L \\ \infty & \text{otherwise} \end{cases}, \tag{20.5}$$

where L represents the well size.

The particle inside the well is identified as non-relativistic if $L >> L_c$ while it is considered as relativistic if $L \sim L_c$. Looking at the fact that the spin of neutrino is not directly involved in the oscillation, we employ Schrödinger and Klein-Gordon equation. Taking into account the neutrino's spin 1/2 nature, the Dirac equation is employed. In addition, there is also a possibility that the neutrinos might resemble their own anti-particle state. In this regard, we adhere to the Majorana equation as well.

To calculate the parameter L_c, we consider the normal hierarchy of neutrino masses ($m_3 > m_2 > m_1$). Accordingly, we set $m_3 = 0.0527\,eV$. Using the mass squared differences (Δm_{21}^2 and Δm_{31}^2) [12], we calculate $m_1 = 0.0321eV$ and $m_2 = 0.0352eV$. It is important to mention that the sum of three neutrino masses Σm_i is consistent with the cosmological data [13]. In SI unit, the mass eigenvalues comes out of the order of $10^{-38}\,kg$. From Eq. (20.4), we calculate the Compton wavelength which is approximately $10^{-5}, m$. In this regard, we set $L = 1\,m$ for non-relativistic case ($>> L_c$) and $L = 10^{-5}\,m$ for the relativistic case ($\sim L_c$). We highlight that the angle θ is fixed at 45°.

20.3 Neutrino Oscillation Inside the Potential Well

The neutrino oscillation is expressed in terms of the probability of conversion from one flavour to another. First, we deal with the non-relativistic scenario. We solve the Schrödinger equation and obtain the neutrino mass eigenstates as shown in the following:

$$\nu_1 = \sqrt{\frac{2}{L}} \sin\left(\frac{n_1 \pi x}{L}\right) e^{-\frac{i}{\hbar} E_{n_1} t}, \tag{20.6}$$

$$\nu_2 = \sqrt{\frac{2}{L}} \sin\left(\frac{n_2 \pi x}{L}\right) e^{-\frac{i}{\hbar} E_{n_2} t}. \tag{20.7}$$

It is to be noted that the quantum numbers n_1 and n_2 correspond to mass eigenstates ν_1 and ν_2. Inside the well, the neutrino mass eigenstates can have quantized energies of the form as shown below,

$$E_{n_i} = \frac{n_i^2 \pi^2 \hbar^2}{2 m_i L^2}. \tag{20.8}$$

Following Pontecorvo's approach, the probability that an electron type neutrino is converted to muon type ($P_{\nu_e \to \nu_\mu}$) is given by

$$P_{\nu_e \to \nu_\mu} = \sin^2(2\theta)\, \sin^2\left[\frac{t}{2\hbar}(E_{n_1} - E_{n_2})\right]. \tag{20.9}$$

Using Eq. (20.8), the expression of $P_{\nu_e \to \nu_\mu}$ is obtained as shown in the following:

$$P_{\nu_e \to \nu_\mu} = \sin^2(2\theta)\, \sin^2\left[\frac{\pi^2 \hbar t}{4L^2}\left(\frac{n_1^2}{m_1} - \frac{n_2^2}{m_2}\right)\right]. \tag{20.10}$$

It is evident from Eq. (20.10) that the oscillation frequency depends on the principle quantum number n.

In principle, the mass eigenstate ν_1 corresponds to various states such as ν_1^1, ν_1^2, ν_1^3, ... with corresponding values of $n_1 = 1, 2, 3, \ldots$. This principle similarly applies to ν_2. In this context, the flavour states can be obtained as follows:

$$\nu_e = c_1^1 \nu_1^1 + c_1^2 \nu_1^2 + c_1^3 \nu_1^3 + \cdots + c_2^1 \nu_2^1 + c_2^2 \nu_2^2 + c_2^3 \nu_2^3 + \cdots \tag{20.11}$$

$$\nu_\mu = d_1^1 \nu_1^1 + d_1^2 \nu_1^2 + d_1^3 \nu_1^3 + \cdots + d_2^1 \nu_2^1 + d_2^2 \nu_2^2 + d_2^3 \nu_2^3 + \cdots \tag{20.12}$$

In this regard, the neutrino mixing can be visualized as shown below,

$$\begin{pmatrix} \nu_e \\ \nu_\mu \end{pmatrix} = \tilde{U} \begin{pmatrix} \nu_1^1 \\ \nu_1^2 \\ \nu_1^3 \\ \vdots \\ \nu_2^1 \\ \nu_2^2 \\ \nu_2^3 \\ \vdots \end{pmatrix}. \tag{20.13}$$

From Eq. (20.13), it is evident that $\tilde{U}$ is a rectangular matrix and the transformation is not unitary. For the mixing matrix to become unitary, it is necessary that at any given moment, only two states can contribute towards the mixing process. Thus, the mixing can occur in three possible ways,

1. **case-I :** The mixing between two states having identical quantum numbers but distinct masses ($n_1 = n_2,\ m_1 \neq m_2$).
2. **case-II :** The mixing between the states having different quantum numbers as well as the masses ($m_1 \neq m_2,\ n_1 \neq n_2$).
3. **case-III :** The mixing between two states having degenerate mass eigenvalues but distinct quantum numbers ($m_1 = m_2,\ n_1 \neq n_2$).

In the relativistic regime, it is not that straightforward to deal with problems like trapping potential because of the conjecture known as the Klein Paradox [14]. To be more precise, due to the emergence of the negative energy solutions, the reflected coefficient becomes greater than one, transforming a single-particle system into a multi-particle system. However, we can avoid this paradox by replacing the potential with a concept of position-dependent mass [9–11] as shown below,

$$m(x) = \begin{cases} m & 0 < x < L \\ \infty & \text{otherwise} \end{cases}. \tag{20.14}$$

It is observed that the mass eigenstates obtained by solving the Klein-Gordon equation closely resemble those of the non-relativistic scenario, provided the quantized energy is of the form,

$$E_{n_i} = \sqrt{\frac{n_i^2 \pi^2 \hbar^2 c^2}{L^2} + m_i^2 c^4}, \tag{20.15}$$

where $i = 1, 2$.

The probability $P_{\nu_e \to \nu_\mu}$ is obtained as shown below,

$$P_{\nu_e \to \nu_\mu} = \sin^2(2\theta) \sin^2 \Big[\frac{t}{2\hbar} \Big\{ \Big(\frac{n_1^2 \pi^2 \hbar^2 c^2}{L^2} + m_1^2 c^4 \Big)^{\frac{1}{2}} - \Big(\frac{n_2^2 \pi^2 \hbar^2 c^2}{L^2} + m_2^2 c^4 \Big)^{\frac{1}{2}} \Big\} \Big]. \tag{20.16}$$

Under non-relativistic limit, Eq. (20.16) reduces to the Schrödinger case.

Again, under the ultra-relativistic limit, Eq. (20.16) takes the form as shown below,

$$P_{\nu_e \to \nu_\mu}^{URL(KG)} = \sin^2(2\theta) \sin^2 \Big[\frac{t}{2\hbar} \Big\{ \frac{h\pi c}{L}(n_1 - n_2) + \frac{c^3 L}{2\pi \hbar} \Big(\frac{m_1^2}{n_1} - \frac{m_2^2}{n_2} \Big) \Big\} \Big]. \tag{20.17}$$

Taking into account the neutrino's inherent spin-half nature, we adhere to the Dirac equation. By solving the Dirac equation, we obtain the mass eigenstates of the following form [9–11]:

$$\nu_i = B e^{i \frac{k_i L}{2}} \begin{pmatrix} 2\cos\big(k_i x - \frac{k_i L}{2}\big)\chi \\ 2i\, r \sin\big(k_i x - \frac{k_i L}{2}\big)\sigma_x \chi \end{pmatrix} e^{-\frac{i}{\hbar} E_i t}, \qquad i = 1, 2. \tag{20.18}$$

where χ is either $(1, 0)^T$ or $(0, 1)^T$ and $r = \frac{c k_i \hbar}{E + mc^2}$. The wave number k_i associated with the quantum number ν_i takes discrete values governed by the transcendental equation,

$$\tan(k_i L) = -\frac{\hbar k_i}{m_i c}. \tag{20.19}$$

Table 20.1 Numerical values of k_1 and k_2 with respect to m_1 and m_2, respectively

Mass Eigenstates	$k_1(m^{-1})$	$k_2(m^{-1})$	$k_3(m^{-1})$
m_1	220445	502424	805247
m_2	224098	505054	807060

We calculate the k_i values with respect to two neutrino mass eigenstates (see Table 20.1).

The discrete energy eigenvalues E_i can be obtained from the following relation:

$$E_i = \sqrt{\hbar^2 c^2 k_i^2 + m_i^2 c^4}, \quad k_i \neq 0. \tag{20.20}$$

The expression for $P_{\nu_e \to \nu_\mu}$ is derived by using Eq. (20.20) as shown below,

$$P_{\nu_e \to \nu_\mu} = \sin^2(2\theta)\,\sin^2\left[\frac{t}{2\hbar}\left\{\left(\hbar^2 c^2 k_1^2 + m_1^2 c^4\right)^{\frac{1}{2}} - \left(\hbar^2 c^2 k_2^2 + m_2^2 c^4\right)^{\frac{1}{2}}\right\}\right]. \tag{20.21}$$

In addition, considering the electromagnetic neutrality of neutrinos, there arises the possibility that neutrino might be a Majorana Particle. However, we obtain the same transcendental equation like the Dirac scenario for the allowed wave number k_i [15]. Also the expression of $P_{\nu_e \to \nu_\mu}$ obtained is identical to the Dirac case.

20.4 Important Observations

In the present work, it is seen that the energy of the neutrino mass eigenstates depends on the quantum number n. In this regard, it is expected that the oscillation frequency will vary with respect to the quantum number n. For this, we consider the **case-I** and **case-II** as mentioned above. From the parameter $f' = \frac{df}{dn}$, we can visualize the variation of the oscillation frequency with respect to n.[1]

For the non-relativistic regime, we obtain the expression of oscillation frequency f_{NR} as shown below,

$$f_{NR} = \frac{\pi\,\hbar}{8\,L^2}\left(\frac{n_1^2}{m_1} - \frac{n_2^2}{m_2}\right). \tag{20.22}$$

[1] Here, f' is obtained to visualize how the oscillation frequency varies with respect to the quantum number n though n is a discrete variable.

In this connection, the f' is found to be

$$f' = \frac{n\pi\hbar}{4L^2}\Big(\frac{1}{m_1} - \frac{1}{m_2}\Big). \tag{20.23}$$

It is evident from Eq. (20.23), the f' is directly proportional to the n. Hence, for **case-I**, it is seen that the oscillation frequency increases as quantum number n increases (see Fig. 20.1).

For the **case-II**, we notice a shift in the oscillation frequency (see Fig. 20.2) as altering the energy states shifts the frequency due to the distinct masses.

For the relativistic regime, we first discuss the Klein-Gordon scenario. The oscillation frequency for the Klein-Gordon scenario is given by

$$f_{KG} = \frac{1}{4\pi\hbar}\Big[\Big(\frac{n_1^2\pi^2\hbar^2c^2}{L^2} + m_1^2c^4\Big)^{\frac{1}{2}} - \Big(\frac{n_2^2\pi^2\hbar^2c^2}{L^2} + m_2^2c^4\Big)^{\frac{1}{2}}\Big]. \tag{20.24}$$

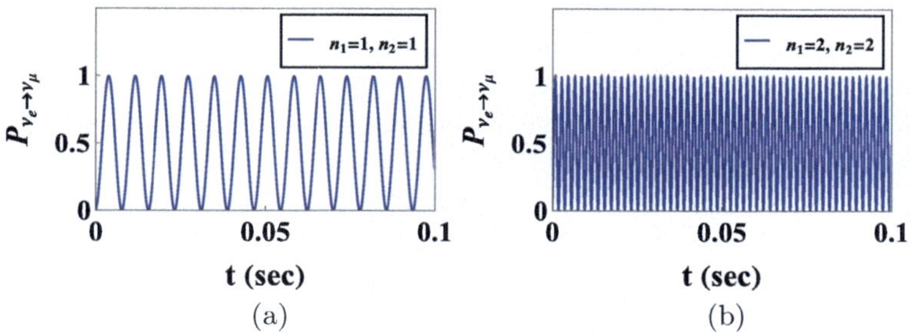

Fig. 20.1 Presents the plot of $P_{\nu_e\to\nu_\mu}$ against t (non-relativistic scenario) for **a** $n_1 = 1$ and $n_2 = 1$. **b** $n_1 = 2$ and $n_2 = 2$. From the plots, it is seen that that the oscillation frequency increases with increase in the energy of the mass eigenstates

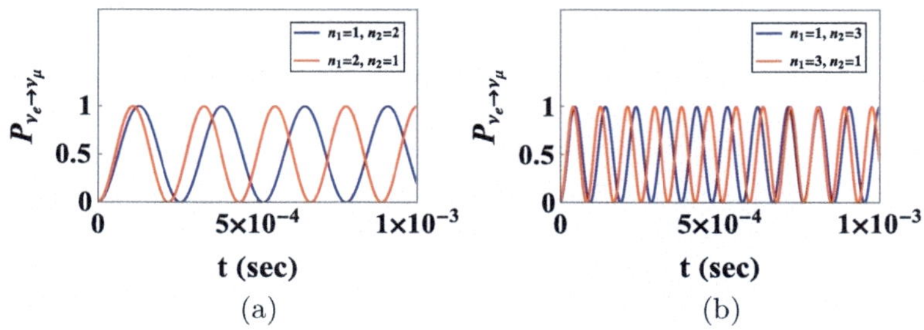

Fig. 20.2 Shows the plot of $P_{\nu_e\to\nu_\mu}$ against t (non-relativistic scenario) for **a** $n_1 = 1$ and $n_2 = 2$ and vice-versa. **b** $n_1 = 1$ and $n_2 = 3$ and vice-versa

Under the relativistic limit, the parameter f' is found to be

$$f' = \frac{Lc^3}{8\pi^2\hbar^2 n^2}(m_2^2 - m_1^2). \tag{20.25}$$

For the relativistic regime, we first discuss the Klein-Gordon scenario. The oscillation frequency for the Klein-Gordon scenario is given by

$$f_{KG} = \frac{1}{4\pi\hbar}\left[\left(\frac{n_1^2\pi^2\hbar^2c^2}{L^2} + m_1^2c^4\right)^{\frac{1}{2}} - \left(\frac{n_2^2\pi^2\hbar^2c^2}{L^2} + m_2^2c^4\right)^{\frac{1}{2}}\right]. \tag{20.26}$$

Under the relativistic limit, the parameter f' is found to be

$$f' = \frac{Lc^3}{8\pi^2\hbar^2 n^2}(m_2^2 - m_1^2). \tag{20.27}$$

From Eq. (20.27), it is seen that the f' is inversely proportional to n^2. Thus, for the **case-I**, we see that the oscillation frequency decreases with increase in the energy of the mass eigenstates (see Fig. 20.3).

For the **case-II**, we observe a comparable situation to the non-relativistic scenario, where the oscillation frequency changes as the energy states are altered (see Fig. 20.4).

In the Dirac framework, we observe a situation akin to the Klein-Gordon scenario. For the **case-I**, we see that oscillation frequency decreases as the energy of the mass eigenstates increases(see Fig. 20.5). However, for the Dirac case, the parameter f' cannot be calculated as the two mass eigenstates cannot possess the same wave number due to different masses.

For the **case-II**, we obtain the similar results where a shift in the oscillation frequency is observed (see Fig. 20.6).

In the standard theory of neutrino oscillation, it is clear that the neutrino mass eigenstates cannot be degenerate; otherwise, there will be no mixing as well as oscillation. However, within the potential well, mass is not the sole parameter distinguishing the two mass eigenstates. Another quantum number n serves to discriminate

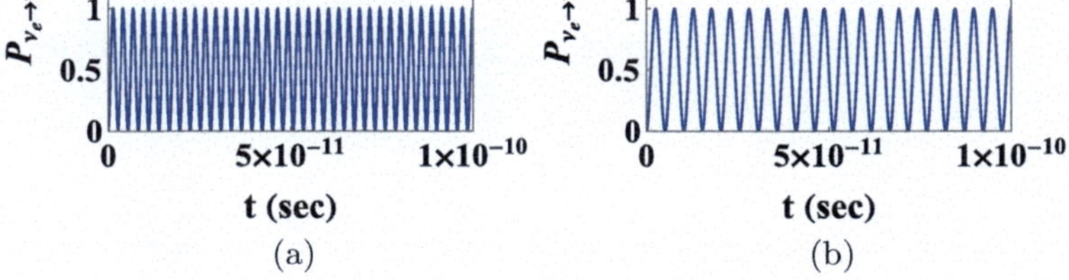

Fig. 20.3 Manifests the plot of $P_{\nu_e\to\nu_\mu}$ against t (Klein-Gordon scenario) for **a** $n_1 = 1$ and $n_2 = 1$. **b** $n_1 = 2$ and $n_2 = 2$

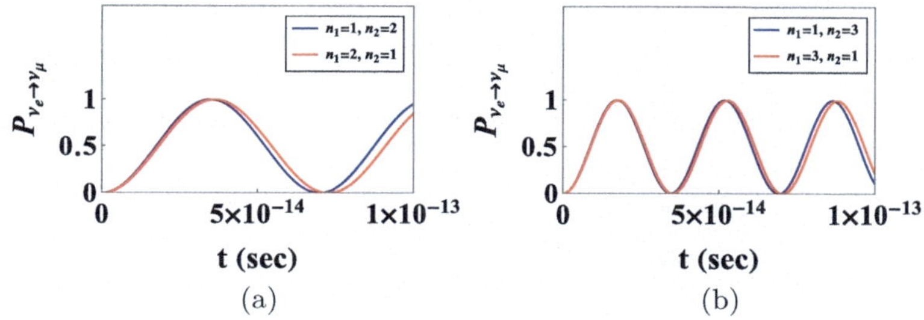

Fig. 20.4 Shows the plot of $P_{\nu_e \to \nu_\mu}$ against t (Klein-Gordon scenario) for **a** $n_1 = 1$ and $n_2 = 2$ and vice-versa. **b** $n_1 = 1$ and $n_2 = 3$ and vice-versa

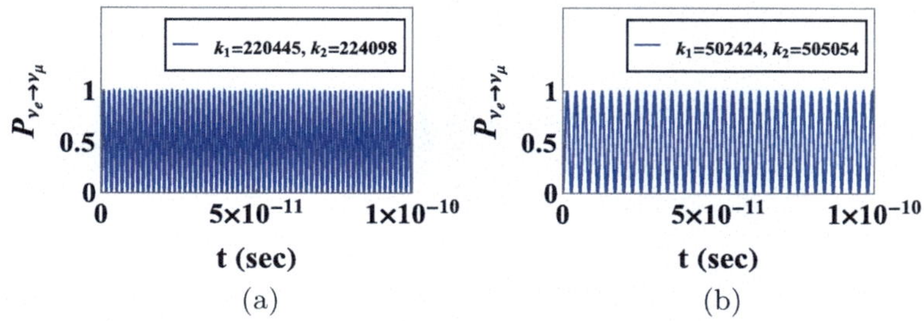

Fig. 20.5 Presents the plot of $P_{\nu_e \to \nu_\mu}$ against t (Dirac scenario) for **a** $k_1 = 220445\,m^{-1}$ and $k_2 = 224098\,m^{-1}$. **b** $k_1 = 502424\,m^{-1}$ and $k_2 = 505054\,m^{-1}$

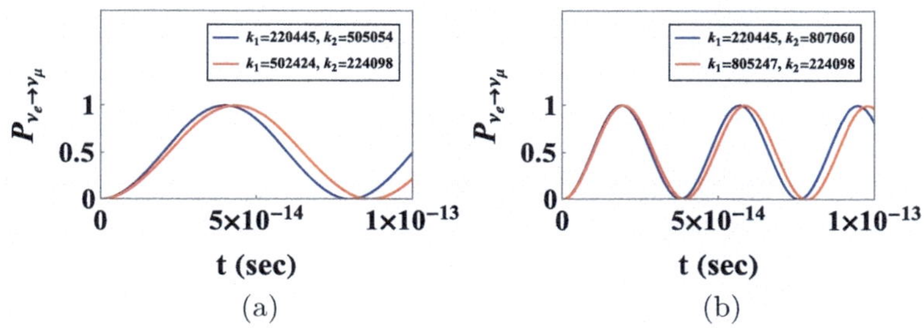

Fig. 20.6 Gives the plot of $P_{\nu_e \to \nu_\mu}$ against t (Dirac scenario) for **a** $k_1 = 220445\,m^{-1}$, $k_2 = 505054\,m^{-1}$ and $k_1 = 502424\,m^{-1}$, $k_2 = 224098\,m^{-1}$. **b** $k_1 = 220445\,m^{-1}$, $k_2 = 807060\,m^{-1}$ and $k_1 = 805247\,m^{-1}$, $k_2 = 224098\,m^{-1}$

between the two. Thus, the **case-III** tells us that within the well, neutrino mixing and oscillation persist even in the scenario of degenerate mass eigenstates (see Fig. 20.7).

However, there are some conditions for which the oscillation will not occur. From Eq. (20.10), we get the following relation for which the oscillation stops for non-relativistic case:

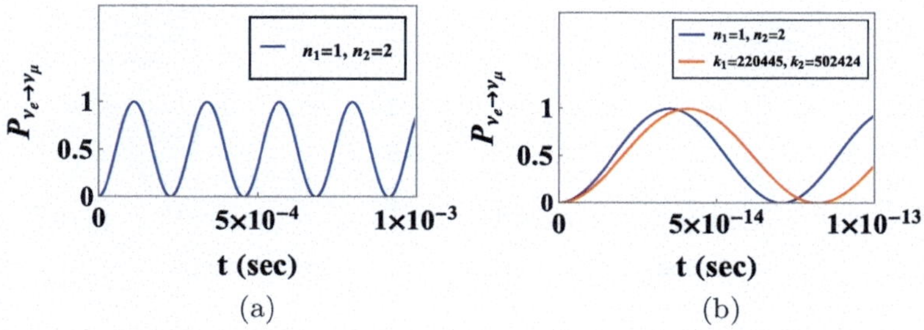

Fig. 20.7 Shows the plot of $P_{\nu_e \to \nu_\mu}$ against t in case of $m_1 = m_2$ for **a** non-relativistic scenario. **b** Klein-Gordon and Dirac scenario where the blue curve represents the Klein-Gordon scenario while the red curve signifies the Dirac one

$$m_1 = \left(\frac{n_1^2}{n_2^2}\right) m_2. \tag{20.28}$$

In the Klein-Gordon scenario, the oscillation will stop if

$$\Delta m_{21}^2 = \frac{\pi^2 \hbar^2}{L^2 c^2} (n_1^2 - n_2^2). \tag{20.29}$$

Similarly, for the Dirac scenario, the conditions are as shown below,

$$\Delta m_{21}^2 = \frac{\hbar^2}{c^2} (k_1^2 - k_2^2). \tag{20.30}$$

It is to be emphasized that the present work only deals with two-flavour mixing and can be extended to the three-flavour case. We highlight that similar results are observed for the three-flavour case, akin to those obtained in the two-flavour scenario. A comprehensive study on neutrino oscillation inside an infinite square well potential can be found in Ref [16].

20.5 Summary and Discussion

In the present work, by following Pontecorvo's approach we present a thought experiment and study neutrino oscillation inside an infinite square well potential. In this regard, we explore both relativistic and non-relativistic scenarios. We investigate the variation of the oscillation frequency with respect to the quantum number n. In addition, we obtain a contrasting result of non-zero oscillation probability even if the mass eigenstates are degenerate. Furthermore, we highlight some conditions for which the neutrino oscillation stops.

The neutrinos interact with matter very infrequently, making it challenging to trap them. However, extremely compact objects, such as the core of a neutron star and the core collapsing supernova, may trap neutrinos for very small amount of time due to high density and immense gravitational pull [17, 18]. Though the present work is based on a speculation, this study could become relevant for understanding these systems.

Acknowledgements Pralay Chakraborty would like to thank Innovation in Science Pursuit for Inspired Research (INSPIRE), Department of Science and Technology, Government of India, New Delhi, for providing the financial support vide grant No. IF190651.

References

1. B. Pontecorvo, Sov. Phys. JETP **6**, 429 (1957)
2. Y. Fukuda, et al., Phys. Rev. Lett. **81**, 1158 (1998). doi:https://doi.org/10.1103/PhysRevLett.81.1158. [Erratum: Phys.Rev.Lett. 81, 4279 (1998)]
3. Y. Fukuda et al., Phys. Rev. Lett. **81**, 1562 (1998). https://doi.org/10.1103/PhysRevLett.81.1562
4. J. Boger et al., Nucl. Instrum. Meth. A **449**, 172 (2000). https://doi.org/10.1016/S0168-9002(99)01469-2
5. Q.R. Ahmad et al., Phys. Rev. Lett. **89**, 011301 (2002). https://doi.org/10.1103/PhysRevLett.89.011301
6. Q.R. Ahmad et al., Phys. Rev. Lett. **89**, 011302 (2002). https://doi.org/10.1103/PhysRevLett.89.011302
7. Z. Maki, M. Nakagawa, S. Sakata, Prog. Theor. Phys. **28**, 870 (1962). https://doi.org/10.1143/PTP.28.870
8. P.A. Zyla, et al., PTEP **2020**(8), 083C01 (2020). doi:https://doi.org/10.1093/ptep/ptaa104
9. P. Alberto, C. Fiolhais, V.M.S. Gil, European Journal of Physics **17**(1), 19 (1996). doi:https://doi.org/10.1088/0143-0807/17/1/004. URL https://doi.org/10.1088/0143-0807/17/1/004
10. P. Alberto, S. Das, E.C. Vagenas, Phys. Lett. A **375**, 1436 (2011). https://doi.org/10.1016/j.physleta.2011.02.030
11. P. Alberto, S. Das, E.C. Vagenas, Eur. J. Phys. **39**(2), 025401 (2018). https://doi.org/10.1088/1361-6404/aa9b43
12. I. Esteban, M.C. Gonzalez-Garcia, M. Maltoni, T. Schwetz, A. Zhou, JHEP **09**, 178 (2020). https://doi.org/10.1007/JHEP09(2020)178
13. N. Aghanim, et al., Astron. Astrophys. **641**, A6 (2020). doi:https://doi.org/10.1051/0004-6361/201833910. [Erratum: Astron.Astrophys. 652, C4 (2021)]
14. O. Klein, Z. Phys. **53**, 157 (1929). https://doi.org/10.1007/BF01339716
15. S. De Vincenzo, Results Phys. **15**, 102598 (2019). https://doi.org/10.1016/j.rinp.2019.102598
16. P. Chakraborty, S. Roy, Phys. Lett. A **484**, 129070 (2023). https://doi.org/10.1016/j.physleta.2023.129070
17. S. Reddy, M. Prakash, Astrophys. J. **478**, 689 (1997). https://doi.org/10.1086/303804
18. Z. Stuchlik, J. Hladik, M. Urbanec, Gen. Rel. Grav. **43**, 3163 (2011). https://doi.org/10.1007/s10714-011-1229-z

Chapter 21
The Possibility of Other Forces and Particles in Lepton Sector

Sagar Tirtha Goswami and Subhankar Roy

Abstract With the assumption that neutrinos are not Majorana fermions and the right-handed neutrinos do not exist, a novel model is presented where the second and third generations of leptons are treated as composites, while the first generation remains fundamental. These composite states are formed through the interaction of fundamental leptons with two new hidden scalar particles. Additionally, the proposed model introduces two hidden forces alongside the Standard Model interactions. We elucidate the origin of neutrino masses through the binding dynamics of a hidden force. The model offers a dynamical framework for explaining neutrino oscillation phenomena using these hidden forces. The basic framework of the standard model is preserved in this model.

21.1 Introduction

The quest of particle physics is to know nature at its most fundamental level. Currently, our understanding about the fundamental particles and forces of nature is contained in the standard model (SM) of particle physics, which says that there are twelve fundamental fermions of which six are leptons (e, μ, τ, ν_e, ν_μ, ν_τ) and six quarks (u, d, c, s, t, b). Besides, the SM includes three fundamental interactions, the electromagnetic (em) force which has a photon (γ) as its carrier, weak interaction which has three gauge bosons ($W^{\pm}$, Z^0) as force carriers, and strong force that has eight gluons (g) as force carriers.

In the present article, we are concerned with only the EM and weak interactions. The Glashow–Weinberg–Salam (GWS) model [1–4] relates the unified form of the electromagnetic (EM) and the weak forces, known as the electroweak (EW) force with the gauge group $SU(2)_L \otimes U(1)_Y$. When Higgs mechanism [5–7] takes place, the said symmetry spontaneously breaks down to $U(1)_{em}$, during which $W^{\pm}$ and Z^0

S. T. Goswami (✉) · S. Roy
Department of Physics, Gauhati University, Guwahati, India
e-mail: sagartirtha@gauhati.ac.in

J. W. F. Valle et al. (eds.), *Proceedings of the International Conference on Future Prospects in Neutrino and Astroparticle Physics*, Springer Proceedings in Physics 361,
https://doi.org/10.1007/978-981-97-7441-8_22

obtain masses, while γ remains massless. Moreover, all the fundamental fermions except the neutrinos (ν) get masses through the Higgs mechanism.

The ν is a very peculiar particle. According to the SM, it is massless. There is no fundamental symmetry that prevents it from having mass, but rather it is one of the axioms of the SM that forbids its mass term. Actually, the SM stipulates that every fundamental fermion is a Dirac particle and as such, one needs both left-handed (LH) and right-handed (RH) counterparts to form a mass term. The ν does not have mass because the SM doesn't recognize RH ν's. But experimental evidence [8–10] suggests that ν has a tiny but non-zero mass. Now, the question is how to form the mass term of ν's.

The first choice is obvious: we assume the existence of RH neutrinos. Then Dirac terms for them can easily be formed. But this step takes us beyond the SM (BSM). The other way is to use the electrical neutrality of ν's to consider them as Majorana fermions and form the Majorana mass term without the RH ν's. But this step is also a step in BSM area. Besides, experimental verification of either of the said two scenarios is not achieved yet.

In such a situation, there might be another interesting way of going forward, which is to review the basic principles of the SM and find a way to accommodate a mass term for ν's without violating the basic structure of GWS model. The structure of the paper is as follows: in Sect. 21.2, a description of the proposed model is provided, in Sect. 21.3, some important features of the model are discussed, and in Sect. 21.4, we give the summary of the model.

21.2 Description of the Model

As we review the basic principles of the SM, we make two important observations. The first observation is that Dirac or Majorana mass terms are needed provided the particle is a fundamental particle. If we have a particle that is not fundamental, then there would be no requirement for such mass terms, but such particles would obtain mass from other mechanisms. Perhaps, we could apply this concept to ν's. Now if we consider that ν's are composite particles, surely the charged lepton sectors would also be affected. Keeping this in mind, a model is put forward in which the first generation of leptons is fundamental, while the other two generations are composites. This proposition is also supported by the observation that μ and τ are unstable particles, while e is the only stable one. For more details, one can see [11]. We present the axioms of our model as follows:

- ν is a Dirac particle and it has no RH components.
- Among the leptons, only e and ν_e are fundamental, and ν_e is massless.
- In addition to the fundamental leptons, there are also two fundamental hidden scalar particles, $Dakshina(\chi)$ and $Vama(\xi)$.
- In addition to the SM EW interactions, there are two other fundamental hidden interactions, $Shupti(Sh)$ and $Sushupti(Ssh)$.

- *Sh* interaction is associated with the $U(1)_h$ group, while *Ssh* is associated with $SU(2)_h$ group.
- The bound states are formed in the following way: $\mu = (e\chi)$, $\nu_\mu = (\nu_e\,\chi)$, $\tau = (e\,\xi)$, $\nu_\tau = (\nu_e\,\xi)$.
- Apart from the SM electric charge (Q_{em}) and weak charge (T^3), fundamental particles are also endowed with two new hidden charges: $U(1)_h$ charge, Q_h and hidden isospin or $SU(2)_h$ charge, T_h^3. Bound states are formed such that $Q_h^{total} = 0$ and $T^3 = T_h^3 = \pm 1/2$.
- Every fundamental particle obtains masses through Higgs mechanism. But the composite particles obtain their masses through the intrinsic mass of the constituents and the composite nature.
- $M_\chi \lesssim M_\xi \sim 10^2$ GeV.

One of the consequences of the axioms is that now all the fermions have the same lepton number (electron number). The need for family lepton number is compensated by different compositions of the lepton families. The new gauge group under the extended model is $SU(2)_L \otimes U(1)_Y \otimes U(1)_h \otimes SU(2)_h$. The charges associated with different fields are given in Table 21.1

Now, from the table it is clear that the fundamental leptons, i.e., e and ν_e form a doublet($L = \begin{pmatrix} \nu_{eL} \\ e_L \end{pmatrix}$) under $SU(2)_L$ symmetry. The doublet also transforms under $U(1)_Y$ and $U(1)_h$, but is a singlet under $SU(2)_h$. The RH e, i.e., e_R transforms under both $U(1)_Y$ and $U(1)_h$, but remains a singlet under $SU(2)_L$ and $SU(2)_h$. The complex scalar fields (ϕ^+ and ϕ) form a doublet under $SU(2)_L$ ($\Phi = \begin{pmatrix} \phi^+ \\ \phi^0 \end{pmatrix}$) and transform under $U(1)_Y$, but remain invariant under the hidden interactions. On the other hand, the hidden scalar fields (χ and ξ) form a doublet under $SU(2)_h$ ($\phi_h = \begin{pmatrix} \chi \\ \xi \end{pmatrix}$) and also transform under $U(1)_h$, but it remains singlet under the EW interactions. The transformation rules of these fields are given in (21.1).

Table 21.1 The charges associated with different fields for the proposed model [11]

	Q_{em}	Y	T^3	Q_h	T_h^3
ν_{eL}	0	−1	$+\frac{1}{2}$	−1	0
e_L	−1	−1	$-\frac{1}{2}$	−1	0
e_R	−1	−2	0	−1	0
χ	0	0	0	+1	$+\frac{1}{2}$
ξ	0	0	0	+1	$-\frac{1}{2}$
ϕ^+	+1	+1	$+\frac{1}{2}$	0	0
ϕ^0	0	+1	$-\frac{1}{2}$	0	0

$$
\begin{aligned}
L &\xrightarrow{U(1)_Y} e^{-i\,\frac{Y}{2}\alpha(x)}\,L = e^{\frac{+i}{2}\alpha(x)}\,L,\\
L &\xrightarrow{U(1)_h} e^{-i\,Q_h\beta(x)}\,L = e^{+i\beta(x)}\,L,\\
L &\xrightarrow{SU(2)_L} e^{-i\,\frac{\tau^i}{2}\theta^i(x)}\,L,\\
L &\xrightarrow{SU(2)_h} L' = L,\\
e_R &\xrightarrow{U(1)_Y} e^{-i\,\frac{Y}{2}\alpha(x)}\,e_R = e^{+i\alpha(x)}\,e_R,\\
e_R &\xrightarrow{U(1)_h} e^{-i\,Q_h\beta(x)}\,e_R = e^{+i\beta(x)}\,e_R,\\
e_R &\xrightarrow{SU(2)_L} e'_R = e_R,\\
e_R &\xrightarrow{SU(2)_h} e'_R = e_R,\\
\Phi &\xrightarrow{U(1)_Y} e^{-i\,\frac{Y}{2}\alpha(x)}\,\Phi = e^{-\frac{i}{2}\alpha(x)}\,\Phi,\\
\Phi &\xrightarrow{SU(2)_L} e^{-i\,\frac{\tau^i}{2}\theta^i(x)}\,\Phi,\\
\Phi &\xrightarrow{U(1)_h} \Phi' = \Phi,\\
\Phi &\xrightarrow{SU(2)_h} \Phi' = \Phi,\\
\phi_h &\xrightarrow{U(1)_h} e^{-i\,Q_h\beta(x)}\,\phi_h = e^{-i\beta(x)}\,\phi_h,\\
\phi_h &\xrightarrow{SU(2)_h} e^{-i\,\frac{\tau^i}{2}\kappa^i(x)}\,\phi_h,\\
\phi_h &\xrightarrow{U(1)_Y} \phi'_h = \phi_h,\\
\phi_h &\xrightarrow{SU(2)_L} \phi'_h = \phi_h.
\end{aligned}
\tag{21.1}
$$

The covariant derivative of our model under $SU(2)_L \otimes U(1)_Y \otimes U(1)_h \otimes SU(2)_h$ symmetry is given by

$$
D_\mu = \partial_\mu - i\,g\frac{\tau^i}{2}W^i_\mu - i\,g'\frac{Y}{2}B_\mu - i\,g_h\frac{\tau^i}{2}G^i_\mu - i\,g'_h Q_h C_\mu,
\tag{21.2}
$$

where B_μ is the EM force carrier, W^i_μ are weak force carriers, C_μ is the Sh force carrier, and G^i_μ are Ssh force carriers.

Now, the unbroken Lagrangian before the Higgs mechanism can be written in the following parts:

$$
\mathcal{L} = \mathcal{L}_F + \mathcal{L}_H + \mathcal{L}_S + \mathcal{L}_G + \mathcal{L}_Y,
$$

where

$$
\begin{aligned}
\mathcal{L}_F &= i\,\bar{L}\gamma^\mu(\partial_\mu - i\,g\frac{\tau^i}{2}W^i_\mu + i\,g'B_\mu)L + i\,\bar{e}_R\gamma^\mu(\partial_\mu + i\,g'B_\mu)e_R \\
&\quad - g'_h(\bar{L}\gamma^\mu L + \bar{e}_R\gamma^\mu e_R)C_\mu, \\
\mathcal{L}_h &= \{(\partial^\mu - i\,g_h\frac{\tau^i}{2}G^{\mu^i} - i\,g'_h C^\mu)\phi_h\}^\dagger\{(\partial_\mu - i\,g_h\frac{\tau^i}{2}G^i_\mu \\
&\quad - i\,g'_h C_\mu)\phi_h\}, \\
\mathcal{L}_S &= \{(\partial^\mu - i\,g\frac{\tau^i}{2}W^{\mu^i} - i\,g'B^\mu)\Phi\}^\dagger\{(\partial_\mu - i\,g\frac{\tau^i}{2}W^i_\mu - i\,g'B_\mu)\Phi\} \\
&\quad + \mu^2(\Phi^\dagger\Phi) - \lambda(\Phi^\dagger\Phi)^2, \\
\mathcal{L}_G &= -\frac{1}{4}B_{\mu\nu}B^{\mu\nu} - \frac{1}{4}W^i_{\mu\nu}W^{\mu\nu^i} - \frac{1}{4}C_{\mu\nu}C^{\mu\nu} - \frac{1}{4}G^i_{\mu\nu}G^{\mu\nu^i}, \\
\mathcal{L}_Y &= -y_e(\bar{L}\Phi e_R) - y^2_\chi(\phi^\dagger_h\Phi)(\Phi^\dagger\phi_h) - y^2_\xi(\phi^\dagger_h\tilde{\Phi})(\tilde{\Phi}^\dagger\phi_h) + H.C.
\end{aligned}
\tag{21.3}
$$

where $\mu^2 > 0$, $\lambda > 0$. The $B_{\mu\nu}$, $W^i_{\mu\nu}$, $C_{\mu\nu}$, and $G^i_{\mu\nu}$ represent the field tensors for the B_μ, W^i_μ, C_μ, and G^i_μ, respectively.

Now, Higgs mechanism breaks the original group $SU(2)_L \otimes U(1)_Y \otimes U(1)_h \otimes SU(2)_h$ to $U(1)_{em} \otimes U(1)_h$. During Higgs mechanism, all the hidden gauge bosons along with the physical photon remain massless, while the gauge bosons for weak interaction acquire non-zero masses. Our axiom dictates that χ and ξ obtain unequal masses through higgs mechanism.

The Lagrangian after the Higgs mechanism can be separated into GWS and hidden parts as follows:

$$\mathcal{L} \to \mathcal{L}_{GWS} + \mathcal{L}_{hidden},$$

where

$$
\begin{aligned}
\mathcal{L}_{GWS} &= i\,\bar{e}\gamma^\mu\partial_\mu e + i\,\bar{\nu}_{eL}\gamma^\mu\partial_\mu\nu_{eL} - \frac{y_e v}{\sqrt{2}}\bar{e}\,e + \frac{g}{\sqrt{2}}\left(\bar{\nu}_{eL}\,\gamma^\mu\,e_L\,W^+_\mu\right. \\
&\quad \left.+\bar{e}_L\,\gamma^\mu\,\nu_{eL}\,W^-_\mu\right) + \frac{g}{2}\left(\bar{\nu}_{eL}\,\gamma^\mu\,\nu_{eL} - \bar{e}_L\,\gamma^\mu\,e_L\right)\left(Z_\mu\cos\theta_W\right. \\
&\quad \left.+A_\mu\sin\theta_W\right) - \frac{g'}{2}\left(\bar{\nu}_{eL}\,\gamma^\mu\,\nu_{eL} + \bar{e}_L\,\gamma^\mu\,e_L + 2\bar{e}_R\,\gamma^\mu\,e_R\right)\left(A_\mu\cos\theta_W\right. \\
&\quad \left.-Z_\mu\sin\theta_W\right) + \frac{g^2v^2}{4}W^+_\mu W^{\mu^-} + \frac{v^2}{8}(g^2 + g'^2)Z^\mu Z_\mu \\
&\quad + \partial_\mu H\,\partial^\mu H - \mu^2\,H^2 - \lambda\,v\,H^3 - \frac{\lambda}{4}H^4 \\
&\quad - \frac{y_e}{\sqrt{2}}H\,\bar{e}e + \frac{g^2}{8}(H^2 + 2\,Hv)\left(\frac{1}{\cos^2\theta_W}Z^\mu Z_\mu + 2W^+_\mu W^{\mu^-}\right),
\end{aligned}
\tag{21.4}
$$

and

$$
\begin{aligned}
\mathcal{L}_{hidden} = {} & \partial^\mu \chi^* \partial_\mu \chi - y_\chi^2 v^2 \chi^* \chi + \partial^\mu \xi^* \partial_\mu \xi - y_\xi^2 v^2 \xi^* \xi \\
& + \frac{g_h}{2} \{ i\,(\chi^* \partial^\mu \chi - \chi\, \partial^\mu \chi^*) - i\,(\xi^* \partial^\mu \xi - \xi\, \partial^\mu \xi^*) \}\, G_\mu^3 \\
& + \frac{g_h}{2}\, i\,(\xi^* \partial^\mu \chi - \chi\, \partial^\mu \xi^*)\, G_\mu^* + \frac{g_h}{2}\, i\,(\chi^* \partial^\mu \xi - \xi\, \partial^\mu \chi^*)\, G_\mu \\
& + g_h' \{ i\,(\chi^* \partial^\mu \chi - \chi\, \partial^\mu \chi^*) + i\,(\xi^* \partial^\mu \xi - \xi\, \partial^\mu \xi^*) \\
& - (\bar{e}\gamma^\mu e + \bar{\nu}_{eL}\gamma^\mu \nu_{eL}) \}\, C_\mu + \Big(\frac{g_h}{2}\,(\chi^* G^{\mu^3} + G^{\mu^*} \xi^*) \\
& + g_h' \chi^* C^\mu \Big) \Big(\frac{g_h}{2}\,(\chi\, G_\mu^3 + G_\mu\, \xi) + g_h' \chi\, C_\mu \Big) + \Big(\frac{g_h}{2}\,(\xi^* G^{\mu^3} \\
& - G^\mu \chi^*) - g_h' \xi^* C^\mu \Big) \Big(\frac{g_h}{2}\,(\xi\, G_\mu^3 + G_\mu^* \chi) - g_h' \xi\, C_\mu \Big) \\
& - (H^2 + 2\, v\, H)(y_\chi^2\, \chi^* \chi + y_\xi^2\, \xi^* \xi).
\end{aligned}
\tag{21.5}
$$

In (21.4), we have $W_\mu^+ = \frac{1}{\sqrt{2}}\,(W_\mu^1 - i\, W_\mu^2)$, $W_\mu^- = \frac{1}{\sqrt{2}}\,(W_\mu^1 + i\, W_\mu^2)$, $\theta_W = \tan^{-1}(g'/g)$ and, in (21.5), we have $G_\mu = (G_\mu^1 - i\, G_\mu^2)$, and $G_\mu^* = (G_\mu^1 + i\, G_\mu^2)$. Here, we would like to call the C_μ field as hidden photon (γ_h) and G_μ, G_μ^*, G_μ^3 as hidden gauge bosons.

Now, we turn to some important points on the two hidden forces.

- In the *Sh* force, we see that γ_h interacts with ν_e unlike the physical photon. And, both the hidden scalar fields interact with it. The primary vertices of *Sh* are given as follows (Fig. 21.1):

$$
\begin{aligned}
e^- &\rightarrow e^- + \gamma_h, \\
\nu_e &\rightarrow \nu_e + \gamma_h \\
\chi &\rightarrow \chi + \gamma_h, \\
\xi &\rightarrow \xi + \gamma_h.
\end{aligned}
\tag{21.6}
$$

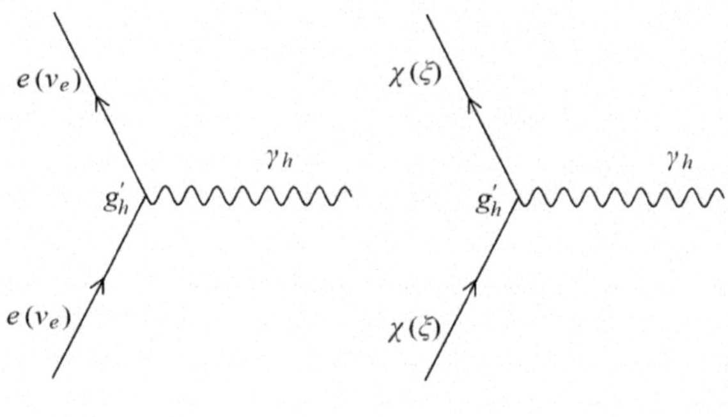

Fig. 21.1 The primary vertices for *Sh* interaction [11]

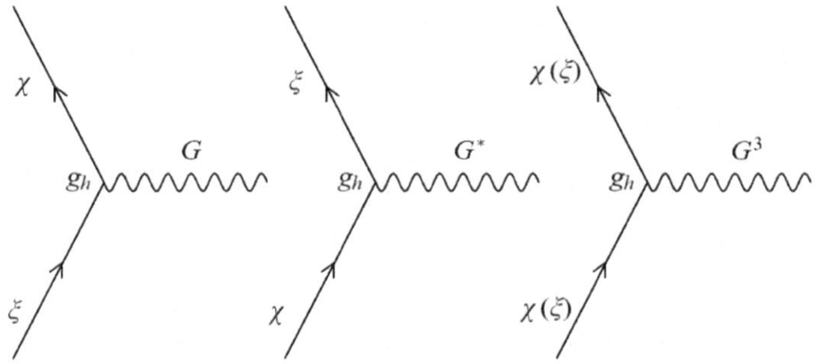

Fig. 21.2 The primary vertices for Ssh interaction [11]

- The Ssh force is related to hidden isospin lowering (HILC), raising (HIRC), and neutral (HINC) currents. The primary vertices are shown as follows (Fig. 21.2):

$$\begin{aligned} \xi &\rightarrow \chi + G, \quad \text{(HIRC)} \\ \chi &\rightarrow \xi + G^*, \quad \text{(HILC)} \\ \xi\,(\chi) &\rightarrow \xi\,(\chi) + G^3. \quad \text{(HINC)} \end{aligned} \tag{21.7}$$

21.3 Important Features of the Model

At this point, we would like to discuss some features of our model. It can, for example, give an alternative picture of μ decay and a dynamical explanation for ν oscillation as opposed to the explanation based on quantum mechanical interference. It also tries to explain the absence charged lepton flavour violation (CLFV) based on a dynamic consideration. We discuss a few such points as follows:

- In the SM, the μ decay process is represented by the following equation:

$$\mu \rightarrow e + \nu_\mu + \bar{\nu}_e. \tag{21.8}$$

  Here, only the weak force is at play. In our model, both the weak interaction and the Sh interaction govern the μ decay process. The process is depicted as follows:

$$\mu\,(e\,\chi) \rightarrow \nu_\mu\,(\nu_e\,\chi) + W^-; \quad W^- \rightarrow e + \bar{\nu}_e \tag{21.9}$$

  The corresponding diagram is given in Fig. 21.3.
- The model can also explain the phenomenon of ν oscillation in a dynamical framework. Here, we divide the scenario into two cases,

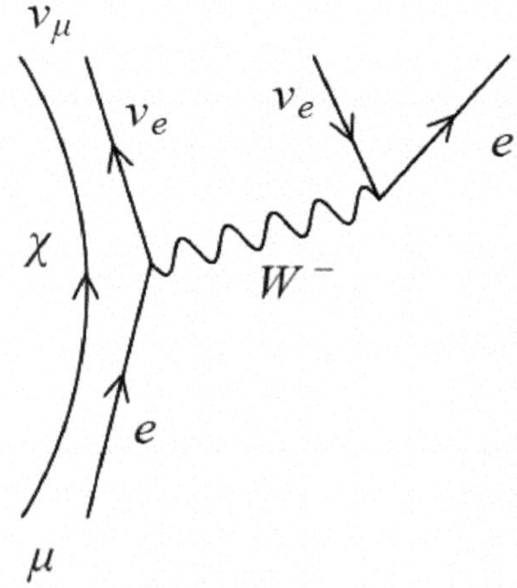

Fig. 21.3 The muon decay process in the light of the proposed model (time flows in the upward direction) [11]

1. Oscillation involving ν_e: Here only the Sh force is involved. The process consists of making and breaking down of composite neutrinos as shown:

$$\nu_e + \chi \longleftrightarrow \nu_\mu, \quad \nu_e + \xi \longleftrightarrow \nu_\tau \tag{21.10}$$

2. Oscillation not involving ν_e: In this case, both the Sh and Ssh forces are at play. The Ssh force facilitates the hidden isospin change, while the Sh force keeps the ν_e and the scalars together. The mediators ultimately decay again to the hidden scalars. The situation is depicted in the following diagram (Fig. 21.4).
   As can be seen from the figures, the relevant equations are given as

$$\nu_\mu \rightarrow \nu_\tau + G^*, \quad G^* \rightarrow \chi + \bar{\xi}, \quad \nu_\tau \rightarrow \nu_\mu + G, \quad G \rightarrow \bar{\chi} + \xi \tag{21.11}$$

It seems that in the case where ν_e is involved, the oscillation frequency might depend on the strength of the Sh interaction and also, on the density of the hidden scalars in space. On the other hand, in the case where ν_e is not involved, the oscillation time or frequency might depend on the lifetimes of the hidden scalars on the Ssh interaction.

– Now, we want to ask this question: The forces that allow ν oscillation can also stimulate oscillation in the charged lepton sector. Such oscillation would amount to CLFV. But it is not so far observed in nature. Would it be possible to explain the absence of CLFV based on our proposed model? It seems that the current picture of the model can't answer this. To answer this, we postulate the possibility of a new

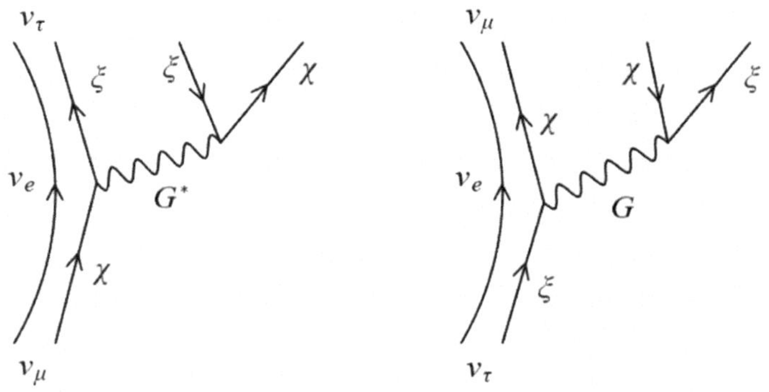

Fig. 21.4 The neutrino flavour oscillation between ν_μ and ν_τ as described in the proposed model (the time flows in the upward direction) [11]

force that could exist only in the composite charged leptons, but not in composite neutrinos. The minimum requirements of this new force are listed as follows:

1. The force exists in a composite state when both the constituents have masses.
2. This force is a repulsive force.
3. For bound states to be formed, a certain barrier called charged lepton fusion (CLF) scale needs to be overcome. Further, the repulsive power of the force in a bound state towards a third particle is far greater.
4. Selection rule: $\Delta T_h^3 \neq \pm 1$.
5. Once a bound state is formed, only $Q_h^{total} = 0$ is possible.

So far, no quantitative details are given about this force. There are some other important points about this force that deserve mention here. First, it allows weak decay to take place, but prevents hidden weak decay (Ssh). Second, in weak decays, it vanishes in the final product, e.g., in μ decay. It also prevents the formation of multi-particle bound states like $e\chi e\bar{e}$, $e\xi\xi\bar{\xi}$, etc. Finally, we do not yet know whether such a force can be a gauge interaction or not.

- Assuming that the new force exists, we write the mass of the composite leptons as follows:

$$\begin{aligned} M_\mu &= M_e + M_\chi + C + D_1, \\ M_{\nu_\mu} &= M_\chi + C, \\ M_\tau &= M_e + M_\xi + C + D_2, \\ M_{\nu_\tau} &= M_\xi + C, \end{aligned} \tag{21.12}$$

where C is the contribution from the $Shupti$ force and D_1 and D_2 are the contributions from the new force. Solving the above equations, we find

$$\begin{aligned} D_1 &= M_\mu - (M_{\nu_\mu} + M_e) \sim M_\mu \sim 10^2\ \text{MeV} \\ D_2 &= M_\tau - (M_{\nu_\tau} + M_e) \sim M_\tau \sim 10^3\ \text{MeV} \\ M_\xi - M_\chi &= M_{\nu_\tau} - M_{\nu_\mu} \\ C &= M_{\nu_\mu} - M_\chi \sim -M_\chi \end{aligned} \tag{21.13}$$

From (21.13), we see that in the composite charged leptons, the contributions from the new force are almost equal to the masses of those charged leptons themselves. Contribution from the Sh interaction is of the order of $M_\chi \sim 100\,\text{GeV}$. So, although it is not very clear from this stage, but there would be some cancellations to obtain the required masses for the particles.

21.4 Summary and Discussion

So far, we have discussed a composite model that we have proposed. Here we have considered neutrinos to be Dirac particles and introduced two new hidden forces and two hidden scalar particles. The work to obtain more details about the forces and the other two new hidden forces is underway. Unlike the leptons, the model does not talk about compositeness of quarks. An apparent limitation of our model is that it is not able to predict the common neutrino oscillation parameters such as the mass squared differences (Δm_{21}^2, Δm_{31}^2) or the mixing angles (θ_{12}, θ_{23}, θ_{13}). It is due to the fact such parameters arise if one considers the neutrino oscillation as a quantum mechanical phenomenon. In contrast, our model considers neutrino oscillation as a dynamical phenomenon and so, the above-mentioned parameters are expected to be replaced by some new dynamical parameters. Such a complete picture of the neutrino oscillation phenomenon is beyond the scope of this paper. The proposed model is merely an attempt to address the issue of neutrino mass by incorporating partial compositeness, while preserving the core GWS model framework.

References

1. S.L. Glashow, Nucl. Phys. **22**, 579 (1961). https://doi.org/10.1016/0029-5582(61)90469-2
2. S. Weinberg, Phys. Rev. Lett. **19**, 1264 (1967). https://doi.org/10.1103/PhysRevLett.19.1264
3. A. Salam, Conf. Proc. C **680519**, 367 (1968). https://doi.org/10.1142/9789812795915_0034
4. S.L. Glashow, J. Iliopoulos, L. Maiani, Phys. Rev. D **2**, 1285 (1970). https://doi.org/10.1103/PhysRevD.2.1285
5. F. Englert, R. Brout, Phys. Rev. Lett. **13**, 321 (1964). https://doi.org/10.1103/PhysRevLett.13.321
6. P.W. Higgs, Phys. Lett. **12**, 132 (1964). https://doi.org/10.1016/0031-9163(64)91136-9
7. P.W. Higgs, Phys. Rev. Lett. **13**, 508 (1964). https://doi.org/10.1103/PhysRevLett.13.508
8. R. Davis Jr., D.S. Harmer, K.C. Hoffman, Phys. Rev. Lett. **20**, 1205 (1968). https://doi.org/10.1103/PhysRevLett.20.1205

9. Q.R. Ahmad et al., Phys. Rev. Lett. **87**, 071301 (2001). https://doi.org/10.1103/PhysRevLett.87.071301
10. S. Fukuda et al., Phys. Rev. Lett. **86**, 5651 (2001). https://doi.org/10.1103/PhysRevLett.86.5651
11. S. Roy, S.T. Goswami, J. Phys. G **49**(11), 115005 (2022). https://doi.org/10.1088/1361-6471/ac94ea

Chapter 22
Study of One Zero Texture and One Vanishing Subtrace in Neutrino Mass Matrix

Sangeeta Dey and Mahadev Patgiri

Abstract In this work, we investigate the existence of one zero texture and one vanishing subtrace in the Majorana neutrino mass matrix. There are a total of thirty-six possible textures of neutrino mass matrix with the constrained conditions of one zero texture and one vanishing subtrace. Here we explore the phenomenological implications of only two classes of textures, i.e., Class A and Class B. We analyze the viability of all the textures using the ratio of the solar and atmospheric mass-squared differences (R_ν). Out of 12 textures of both the classes, only one texture (A_1) from class A and three textures (B_1, B_4 and B_5) from class B are found to be viable under the current neutrino oscillation data at 3σ confidence level. Finally, we study the ranges of the Majorana Phases (α and β) for the allowed range of the Dirac CP phase, δ for all the viable textures.

22.1 Introduction

Convincing evidence of neutrino oscillation experiments surely confirms neutrino mass and flavor mixing. We have accumulated the knowledge on the precision measurement of the three mixing angles (θ_{12}, θ_{23}, θ_{13}) and two mass-squared differences (Δm^2_{12}, Δm^2_{23}) from several neutrino oscillation experiments [1–4]. However, we have yet to determine the leptonic CP-violation, the nature of neutrinos, the absolute neutrino mass scale, neutrino mass hierarchy, and the origin of such small neutrino masses and large mixing. Although the absolute mass scale of neutrino is unknown, KATRIN [5], GERDA [6], EXO [7], and KamLAND-ZEN [8] experiments and the cosmological searches have reduced the upper bound and indicated the absolute scale of the neutrino in the sub-eV range. The ordering of the neutrino masses can be probed in the experiment viz., JUNO experiment [9], long-baseline experiment with Hyper-Kamiokande detector, J-PARK accelerator [10], and DUNE experiment [11].

S. Dey (✉) · M. Patgiri
Cotton University, Guwahati, India
e-mail: phy1891006_sangeeta@cottonuniversity.ac.in

J. W. F. Valle et al. (eds.), *Proceedings of the International Conference on Future Prospects in Neutrino and Astroparticle Physics*, Springer Proceedings in Physics 361,
https://doi.org/10.1007/978-981-97-7441-8_23

The 3×3 Majorana neutrino mass matrix, M_ν which is symmetric, is used to describe the neutrinos in three neutrino flavor schemes. M_ν is diagonalized by the 3×3 flavor mixing matrix V containing three CP violating phases (δ, α, β), where δ is the Dirac CP violating phase and α and β are the Majorana Phases. The current experimental data has however shed some light on the measurement of δ but α and β are still unconstrained. The texture study of the neutrino mass matrix has been carried out in the literature using various phenomenological models based on Abelian as well as non-Abelian flavor symmetries. These approaches are interesting as they reduce the number of free parameters and restrict the possible structures of the neutrino mass matrix. Some of the phenomenologically viable models according to the current experimental data studied in the literature are the vanishing minors [12–14], zero elements [15–20], equality of elements/minors [21–24], zero trace [25, 26], vanishing subtrace [27, 28], etc. The recent neutrino oscillation experiment gives us the data on two mass-squared differences, three mixing angles, Dirac CP phase, Jarlskog Invariant, J_{cp}, and neutrinoless double beta decay $|m_{ee}|$. Therefore, the texture study proves to be useful for investigating the other unknown parameters of neutrino mass matrix based on the available data.

In this work, we explore a particular texture of neutrino mass matrix by simultaneously imposing two constrained conditions, i.e., one zero texture and one vanishing subtrace. The following are the primary motivations of our work:

(a) Firstly from the phenomenological point of view, the study of neutrino mass matrix with one zero texture and one vanishing subtrace constrains M_ν. This class of texture is highly predictive as it reduces the number of free parameters of M_ν to five, leading to interesting relations among the parameters and predicting the observables.

(b) Our model proves to be useful for measuring the CP violating phases at 3σ level and also to study the neutrino mass hierarchy by carrying out a detailed phenomenological investigation.

We have thirty-six possible textures of M_ν with one zero texture and one vanishing subtrace. Each texture has two simultaneous constraint equations in two variables defined from the ratio of mass eigenvalues with CP phases. With the solutions, we plot R_ν versus δ to check the viability of the particular texture. The 3σ values of R_ν either restrict a texture to the sub-range of δ or completely rule out. Then we explore the range of Majorana CP phases by plotting α and β against the allowed range of δ for viable cases.

These textures of neutrino mass matrix can be obtained in the framework of Type II Seesaw Mechanism using S_3 symmetry. The symmetry realization of these textures deserves a detailed survey in the future. The paper is organized as follows: In Sects. 22.2 and 22.3, we present the formalism and framework of the neutrino mass matrix having one zero texture and one vanishing subtrace. In Sects. 22.4 and 22.5, we do the texture analysis. Finally, we present the results and discussion in Sect. 22.6.

22.2 Formalism

We have 36 possible textures of neutrino mass matrix with one zero texture and one vanishing subtrace. There are 6 one zero textures of neutrino mass matrix and 6 textures of one vanishing subtrace.

The textures of one vanishing subtrace are as follows:

$$s_{11} = 0, m_{22} + m_{33} = 0$$
$$s_{12} = 0, m_{21} + m_{33} = 0$$
$$s_{13} = 0, m_{21} + m_{32} = 0$$
$$s_{22} = 0, m_{11} + m_{33} = 0$$
$$s_{23} = 0, m_{11} + m_{32} = 0$$
$$s_{33} = 0, m_{11} + m_{22} = 0$$

Now we classify the textures of one zero texture and one vanishing subtrace of class A and class B as follows:

$$A_1 : m_{11} = 0, s_{11} = 0; \; A_2 : m_{11} = 0, s_{12} = 0; \; A_3 : m_{11} = 0, s_{13} = 0;$$

$$A_4 : m_{11} = 0, s_{22} = 0; \; A_5 : m_{11} = 0, s_{23} = 0; \; A_6 : m_{11} = 0, s_{33} = 0$$

$$B_1 : m_{12} = 0, s_{11} = 0; \; B_2 : m_{12} = 0, s_{12} = 0; \; B_3 : m_{12} = 0, s_{13} = 0;$$

$$B_4 : m_{12} = 0, s_{22} = 0; \; B_5 : m_{12} = 0, s_{23} = 0; \; B_6 : m_{12} = 0, s_{33} = 0$$

Out of these 12 possible textures, textures (A_4, A_5, A_6, B_3) are already ruled out by the study of two-zero textures [15]. The phenomenology of all the other textures is studied in the subsequent sections.

22.3 Neutrino Mass Matrix Framework

On the basis of the diagonal charged lepton mass matrix, we can construct the neutrino mass matrix M_ν using the three neutrino mass eigenvalues (m_1, m_2, m_3) along with the diagonalizing PMNS matrix V with the following parametrization [29]:

$$M_\nu = V \begin{pmatrix} m_1 & 0 & 0 \\ 0 & m_2 & 0 \\ 0 & 0 & m_3 \end{pmatrix} V^T, \tag{22.1}$$

$$V = UP_\nu$$

$$= \begin{pmatrix} c_{12}c_{13} & c_{13}s_{12} & s_{13}e^{-i\delta} \\ -s_{12}c_{23} - c_{12}s_{13}s_{23}e^{i\delta} & c_{12}c_{23} - s_{12}s_{13}s_{23}e^{i\delta} & c_{13}s_{23} \\ s_{23}s_{12} - c_{12}c_{23}s_{13}e^{i\delta} & -c_{12}s_{23} - c_{23}s_{12}s_{13}e^{i\delta} & c_{13}c_{23} \end{pmatrix} diag(1, e^{i\alpha}, e^{i(\beta+\delta)}), \tag{22.2}$$

where δ is the Dirac CP phase and α, β are the Majorana CP phases and (θ_{12}, θ_{23}, θ_{13}) are the solar, atmospheric, and reactor mixing angles, respectively.

Now the neutrino mass matrix can be rewritten in the following form with $\lambda_1 = m_1$, $\lambda_2 = m_2 e^{2i\alpha}$, $\lambda_3 = m_3 e^{2i(\beta+\delta)}$:

$$M_\nu = U \begin{pmatrix} \lambda_1 & 0 & 0 \\ 0 & \lambda_2 & 0 \\ 0 & 0 & \lambda_3 \end{pmatrix} U^T. \tag{22.3}$$

Now using (22.3) any element of the neutrino mass matrix M_ν can be expressed as

$$(M_\nu)_{mn} = \sum_{i=1}^{3} U_{mi} U_{ni} \lambda_i. \tag{22.4}$$

The zero texture for the neutrino mass matrix can be given by

$$(M_\nu)_{mn} = 0 \tag{22.5}$$

$$A_1\lambda_1 + A_2\lambda_2 + A_3\lambda_3 = 0 \tag{22.6}$$

$$A_i = U_{mi} U_{ni}. \tag{22.7}$$

The subtrace of the off-diagonal elements of the symmetric matrix M_ν can be written in the following form:

$$S_{mn} = (M_\nu)_{(m+1,m)} + (M_\nu)_{(m+2,2n-m)}, \tag{22.8}$$

and that for diagonal elements:

$$S_{mm} = (M_\nu)_{(m+1,m+1)} + (M_\nu)_{(m+2,m+2)}. \tag{22.9}$$

For $m + l > 3$, we have to take the values $(m + l) - 3$. Here m, n can take values (1, 2, 3) and $l = 1, 2$. Now we impose the condition for vanishing subtrace, i.e.,

$$S_{mn} = 0, \quad S_{mm} = 0. \tag{22.10}$$

Solving (22.10), we get

$$B_1\lambda_1 + B_2\lambda_2 + B_3\lambda_3 = 0, \tag{22.11}$$

where

$$B_i = U_{ai}U_{bi} - U_{ci}U_{di}, \tag{22.12}$$

here i takes the values (1, 2, 3). Therefore, the two constraint equations are

$$A_1\lambda_1 + A_2\lambda_2 + A_3\lambda_3 = 0, \tag{22.13}$$

$$B_1\lambda_1 + B_2\lambda_2 + B_3\lambda_3 = 0. \tag{22.14}$$

Considering $\lambda_1 > 0$ and defining $X = \frac{\lambda_2}{\lambda_1}$ and $Y = \frac{\lambda_3}{\lambda_1}$, (22.13) and (22.14) become

$$A_1 + A_2X + A_3Y = 0, \tag{22.15}$$

$$B_1 + B_2X + B_3Y = 0. \tag{22.16}$$

Solving (22.15) and (22.16), we get the following ratios:

$$X = \frac{A_1B_3 - A_3B_1}{A_3B_2 - A_2B_3}, \tag{22.17}$$

$$Y = \frac{B_1A_2 - A_1B_2}{A_3B_2 - A_2B_3}. \tag{22.18}$$

Now the Majorana phases can be expressed as

$$\alpha = \frac{1}{2}Arg\left[\frac{A_1B_3 - A_3B_1}{A_3B_2 - A_2B_3}\right], \tag{22.19}$$

$$\beta = \frac{1}{2}Arg\left[\frac{B_1A_2 - A_1B_2}{A_3B_2 - A_2B_3}e^{-2i\delta}\right]. \tag{22.20}$$

Now we define two parameters ρ and σ as

$$\rho = \left|\frac{m_2}{m_1}e^{2i\alpha}\right| \tag{22.21}$$

and

$$\sigma = \left|\frac{m_3}{m_1}e^{2i\beta}\right| \tag{22.22}$$

Therefore, the ratio of solar and atmospheric mass-squared differences can be expressed as

$$R_\nu = \frac{\delta m^2}{\Delta m^2} = \frac{2(\rho^2 - 1)}{2\sigma^2 - \rho^2 - 1}, \tag{22.23}$$

Table 22.1 Current neutrino oscillation parameters from global fits [30]. Here $\Delta m^2_{3l} = \Delta m^2_{31} > 0$ for normal hierarchy and $\Delta m^2_{3l} = \Delta m^2_{32} < 0$ for inverted hierarchy

Parameter	Normal ordering		Inverted ordering	
	Best fit $\pm 1\sigma$	3σ range	Best fit$\pm 1\sigma$	3σ range
θ°_{12}	$33.45^{+0.77}_{-0.75}$	31.27 – 35.86	$33.45^{+0.78}_{-0.75}$	31.27 – 35.87
θ°_{23}	$42.1^{+1.1}_{-0.9}$	39.7 – 50.9	$49.0^{+0.9}_{-1.3}$	39.8 – 51.6
θ°_{13}	$8.62^{+0.12}_{-0.12}$	8.25 – 8.98	$8.61^{+0.14}_{-0.12}$	8.24 – 9.02
δ°_{cp}	230^{+36}_{-25}	144 – 350	278^{+22}_{-30}	194 – 345
$\Delta m^2_{21}/10^{-5}eV^2$	$7.42^{+0.21}_{-0.20}$	6.82 – 8.04	$7.42^{+0.21}_{-0.20}$	6.82 – 8.04
$\lvert\Delta m^2_{3l}\rvert/10^{-3}eV^2$	$2.510^{+0.027}_{-0.027}$	2.430 – 2.593	$2.490^{+0.026}_{-0.028}$	−2.574 – 2.410

where $\delta m^2 = m_2^2 - m_1^2$ is the solar mass splitting and $\Delta m^2 = |m_3^2 - \frac{1}{2}(m_2^2 + m_1^2)|$ is the atmospheric mass splitting. For NH, we have $R_\nu = \frac{2\epsilon_1}{\sigma^2}$, if we consider $\rho = 1 + \epsilon$ for m_1 and m_2 being very close to each other with the values $0.013 < \epsilon < 0.017$ on 3σ. For IH, we have $R_\nu = \frac{2(\rho^2-1)}{\rho^2+1}$. From the 3σ values of Δm^2_{21} and $|\Delta m^2_{3l}|$, we obtain the range $R_\nu = (0.026 - 0.035)$ (Table 22.1).

In Sections 22.4 and 22.5, we discuss the numerical analysis of Class A and B textures.

22.4 Class A

Among the textures classified under class A, we show the analysis of only texture A_1.

22.4.1 Texture A_1

For this texture, we have

$$A_1 = c_{12}^2 c_{13}^2, \tag{22.24}$$

$$A_2 = c_{13}^2 s_{12}^2, \tag{22.25}$$

$$A_3 = s_{13}^2 e^{-2i\delta}. \tag{22.26}$$

$$B_1 = s_{12}^2 + c_{12}^2 s_{13}^2 e^{2i\delta}, \tag{22.27}$$

$$B_2 = c_{12}^2 + s_{12}^2 s_{13}^2 e^{2i\delta}, \tag{22.28}$$

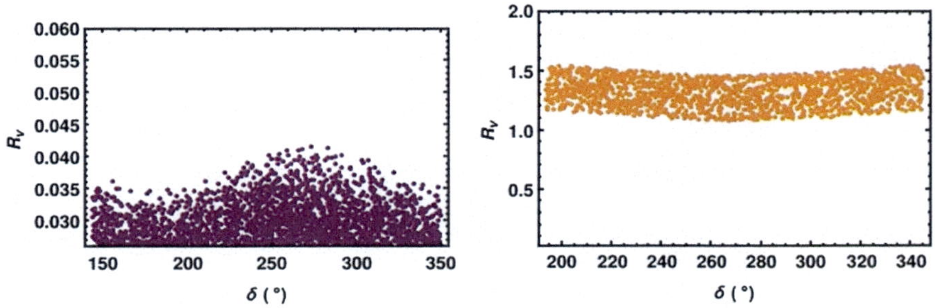

Fig. 22.1 R_ν plot for NH and IH

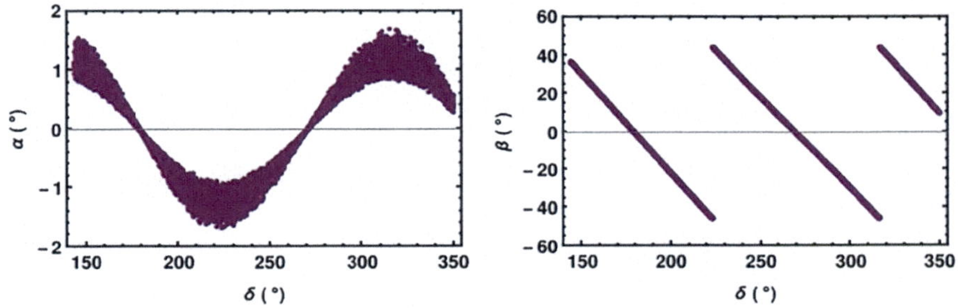

Fig. 22.2 α and β plots for NH

$$B_3 = c_{13}^2. \tag{22.29}$$

Now we plot R_ν (22.23) for both NH and IH (Fig. 22.1).

We show the scatter plot of α and β in Fig. 22.2.

On analyzing the rest of the textures of class A (not shown here), it is found that all the other textures are ruled out except texture A_1 at 3σ C.L.

22.5 Class B

Here we show the analysis of the textures B_1 and B_4 of class B.

22.5.1 *Texture* B_1

For this texture, we have

$$A_1 = -c_{12}c_{13}(c_{23}s_{12} + e^{i\delta}c_{12}s_{13}s_{23}), \tag{22.30}$$

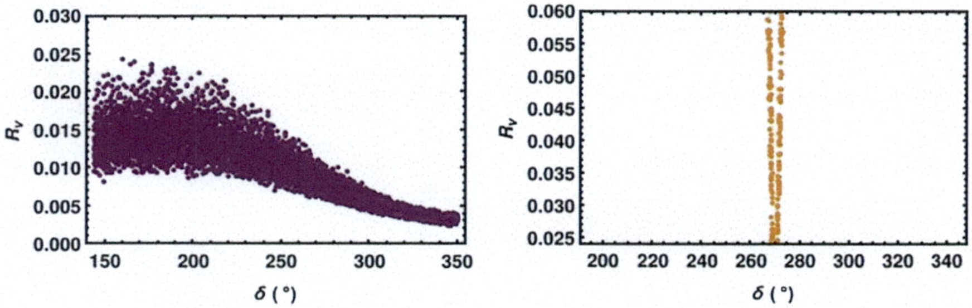

Fig. 22.3 R_ν plot for NH and IH

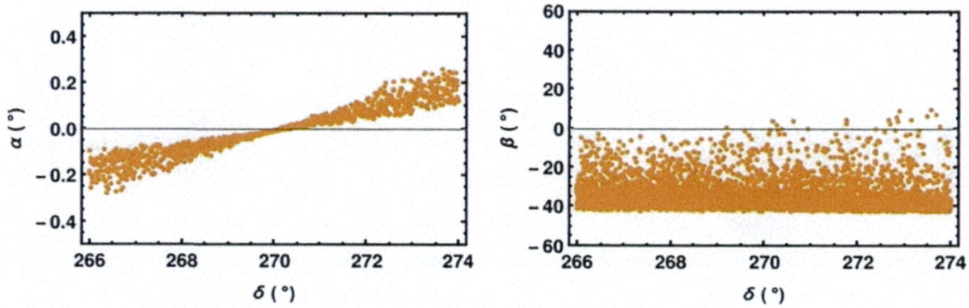

Fig. 22.4 α and β plots for IH

$$A_2 = c_{13}s_{12}(c_{12}c_{23} - e^{i\delta}s_{12}s_{13}s_{23}), \tag{22.31}$$

$$A_3 = e^{-i\delta}c_{13}s_{13}s_{23}. \tag{22.32}$$

$$B_1 = s_{12}^2 + e^{2i\delta}c_{12}^2s_{13}^2, \tag{22.33}$$

$$B_2 = c_{12}^2 + e^{2i\delta}s_{12}^2s_{13}^2, \tag{22.34}$$

$$B_3 = c_{13}^2. \tag{22.35}$$

Now we plot R_ν (22.23) for both NH and IH (Fig. 22.3).
We plot α and β in Fig. 22.4.

22.5.2 Texture B_4

For this texture, we have (Fig. 22.5)

$$A_1 = -c_{12}c_{13}(c_{23}s_{12} + e^{i\delta}c_{12}s_{13}s_{23}), \tag{22.36}$$

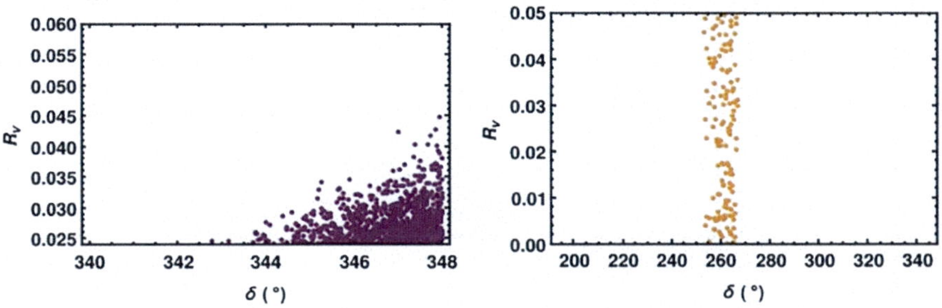

Fig. 22.5 R_ν plot for NH and IH

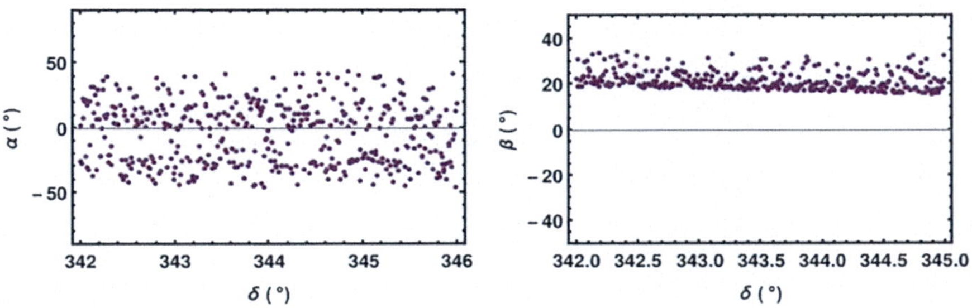

Fig. 22.6 α and β plots for NH

$$A_2 = c_{13}s_{12}(c_{12}c_{23} - e^{i\delta}s_{12}s_{13}s_{23}), \tag{22.37}$$

$$A_3 = e^{-i\delta}c_{13}s_{13}s_{23}. \tag{22.38}$$

$$B_1 = c_{12}^2c_{13}^2 + (e^{i\delta}c_{12}c_{23}s_{13} - s_{12}s_{13})^2, \tag{22.39}$$

$$B_2 = c_{13}^2s_{12}^2 + (e^{i\delta}c_{23}s_{12}s_{13} + c_{12}s_{23})^2, \tag{22.40}$$

$$B_3 = c_{13}^2c_{23}^2 + e^{-2i\delta}s_{13}^2. \tag{22.41}$$

Now we plot α and β as (Fig. 22.6).

All the other textures of class B are studied and analyzed and it is found that only textures B_1, B_4, and B_5 could survive the experimental data at 3σ level. The results of all the viable textures of these two classes are summarized in Table 22.2.

Table 22.2 Allowed ranges of CP phases δ, α, and β for the viable cases

Texture	δ	α	β
		NH	
A_1	$(144°, 350°)$	$\alpha = (-2°, 2°)$	$\beta = (-60°, 60°)$
		IH	
	x	x	x
		NH	
B_1	x	x	x
		IH	
	$\delta = (266°, 274°)$	$\alpha = (-0.2°, 0.2°)$	$\beta = (-45°, 1°)$
		NH	
B_4	$\delta = (342°, 346°)$	$\alpha = (-50°, 50°)$	$\beta = (10°, 40°)$
		IH	
	x	x	x
		NH	
B_5	$\delta = (160°, 250°)$	$\alpha = (-1°, 1°)$	$\beta = (-50°, 40°)$
		IH	
	x	x	x

22.6 Results and Discussion

In this work, we have observed a specific group of textures of neutrino mass matrix with ansatzes of one texture zero and one vanishing subtrace. Here we have carried out a detailed analysis of twelve out of thirty-six textures and summarized the results in Table 22.2. At the very outset we have checked the viability of the textures within their 3σ experimental range for both Normal Hierarchy and Inverted Hierarchy by calculating the parameter R_ν by plotting R_ν vs Dirac CP phase δ and only those values of δ are considered to be viable for which R_ν lies within the experimental range, otherwise the texture is ruled out. Again for the viable range of δ, we have measured the viable ranges of α and β by plotting α and β vs δ. In this work, we have considered the textures of only class A and class B.

We have summarized below some of the interesting observations of our texture study:

a. Out of twelve considered textures of class A and class B, textures A_4, A_5, A_6, and B_3 are already ruled out at 3σ level according to the study of two-zero textures.
b. Among the twelve textures only four textures, i.e., A_1, B_1, B_4, and B_5 are found to be phenomenologically viable considering the current neutrino oscillation data at 3σ level.
c. Only texture B_1 is found to be viable for IH while textures A_1, B_4, and B_5 are allowed for NH.

d. Textures B_1 and B_4 predict a highly constrained range of Dirac CP phase δ. Majorana phase α is measured to be highly constrained for the textures A_1, B_1, and B_5.

Our analysis gives a strong prediction for the Majorana CP phases which are still unknown in the neutrino sector. Our predictions may prove to be fruitful in describing the form of a neutrino mass matrix in the near future.

References

1. K. Abe et al., Indication of electron neutrino appearance from an accelerator-produced off-axis muon neutrino beam. Phys. Rev. Lett. **107**(4), 041801 (2011). https://doi.org/10.1007/11823285_121
2. D. Ayres, NOvA Collaboration, NOvA proposal to build a 30 kiloton off-axis detector to study neutrino oscillations in the Fermilab NuMI beamline (2005). arXiv:hep-ex/0503053
3. F.P. An et al., Observation of electron-antineutrino disappearance at Daya Bay. Phys. Rev. Lett. **108**(17), 171803 (2012)
4. J.K. Ahn, S. Chebotaryov, J. Choi, S. Choi, W. Choi, Y. Choi, H. Jang, J. Jang, E. Jeon, I. Jeong et al., Phys. Rev. Lett. **108**(19), 191802 (2012)
5. J. Wolf, K. Collaboration, The KATRIN neutrino mass experiment. Nucl. Instrum. Methods Phys. Res., Sect. A **623**(1), 442–444 (2010)
6. I. Abt, et al., A new 76Ge double beta decay experiment at LNGS (2004). arXiv:hep-ex/0404039
7. M. Danilov et al., Detection of very small neutrino masses in double-beta decay using laser tagging. Phys. Lett. B **480**(1–2), 12–18 (2000)
8. A. Gando, et al., Search for Majorana neutrinos near the inverted mass hierarchy region with KamLAND-Zen. Phys. Rev. Lett. **117**(8), 082503 (2016)
9. F. An, et al., Neutrino physics with JUNO. J. Phys. G: Nuclear Particle Phys. **43**(3), 030401 (2016)
10. K. Abe, et al., A long baseline neutrino oscillation experiment using J-PARC neutrino beam and Hyper-Kamiokande (2014). arXiv:1412.4673
11. B. Abi, et al., Volume I. introduction to DUNE. J. Instrum. **15**(08), T08008 (2020)
12. L. Lavoura, Zeros of the inverted neutrino mass matrix. Phys. Lett. B **609**(3–4), 317–322 (2005)
13. E.I. Lashin, N. Chamoun, Zero minors of the neutrino mass matrix. Phys. Rev. D **78**(7), 073002 (2008)
14. E.I. Lashin, N. Chamoun, One vanishing minor in the neutrino mass matrix. Phys. Rev. D **80**(9), 093004 (2009)
15. H. Fritzsch, Z. Xing, S. Zhou, Two-zero textures of the Majorana neutrino mass matrix and current experimental tests. J. High Energy Phys. **2011**(9), 1–43 (2011)
16. L. Lavoura, W. Rodejohann, A. Watanabe, Reproducing lepton mixing in a texture zero model. Phys. Lett. B **726**(1–3), 352–355 (2013)
17. R.R. Gautam, M. Singh, M. Gupta, Neutrino mass matrices with one texture zero and a vanishing neutrino mass. Phys. Rev. D **92**(1), 013006 (2015)
18. R.R. Gautam, Trimaximal mixing with a texture zero. Phys. Rev. D **97**(5), 055022 (2018)
19. P.H. Frampton, S.L. Glashow, D. Marfatia, Zeroes of the neutrino mass matrix. Phys. Lett. B **536**(1–2), 79–82 (2002)
20. Z. Xing, Texture zeros and Majorana phases of the neutrino mass matrix. Phys. Lett. B **530**(1–4), 159–166 (2002)
21. S. Dev, S. Verma, S. Gupta, Phenomenological analysis of hybrid textures of neutrinos. Phys. Lett. B **687**(1), 53–60 (2010)

22. S. Dev, et al., Neutrino mass matrices with two texture zeros and near maximal atmospheric neutrino mixing, in *XXI DAE-BRNS High Energy Physics Symposium: Proceedings, Guwahati, India, December 8–12, 2014* (Springer International Publishing, 2016)
23. S. Kaneko, H. Sawanaka, M. Tanimoto, Hybrid textures of neutrinos. J. High Energy Phys. **2005**(08), 073 (2005)
24. S. Dev, D. Raj, New hybrid textures for neutrino mass matrices. Nuclear Phys. B **957**, 115081 (2020)
25. X.-G. He, A. Zee, Neutrino masses with a "zero sum" condition: $m_{\upsilon 1} + m_{\upsilon 2} + m_{\upsilon 3} = 0$. Phys. Rev. D **68**(3), 037302 (2003)
26. S. Dey, M. Patgiri, Study of neutrino mass matrices with vanishing trace and one vanishing minor. Phys. Rev. D **107**(3), 035012 (2023)
27. A. Ismael, E.I. Lashin, N. Chamoun, Texture of two vanishing subtraces in neutrino mass matrix and current experimental tests. Phys. Rev. D **107**(3), 035017 (2023)
28. A. Ismael et al., Texture of single vanishing subtrace in neutrino mass matrix. Phys. Rev. D **103**(3), 035020 (2021)
29. W. Wang, Neutrino mass textures with one vanishing minor and two equal cofactors. Europ. Phys. J. C **73**, 1–8 (2013)
30. M.C. Gonzalez-Garcia, M. Maltoni, T. Schwetz, NuFIT: three-flavour global analyses of neutrino oscillation experiments. Universe **7**(12), 459 (2021)

Chapter 23
Light Majorons as Freeze-in Dark Matter

Soumen Kumar Manna and Arunansu Sil

Abstract We demonstrate that Majoron, a pseudo-Nambu-Goldstone boson emerging due to the spontaneous lepton number symmetry breaking, serves as potential freeze-in dark matter within the keV-GeV mass regime due to a newly introduced explicit lepton number symmetry breaking higher dimensional operator. The proposal establishes a link between dark matter and neutrino physics within a minimal extension of Standard model featuring a singlet complex scalar and two singlet right-handed neutrinos.

23.1 Introduction

The fundamental nature of neutrinos, whether they are Dirac or Majorana, remains uncertain despite evidence of their masses. The latter possibility, involving lepton number violation (LNV), may lead to the emergence of Majoron, a massless Nambu-Goldstone boson (pNGB) if lepton number symmetry (LNS) is promoted to a global $U(1)_L$ followed by spontaneous breaking [1–3]. Majoron, with a non-zero mass (m_χ) generated by an explicit soft LNS breaking term, may turn out to be a promising candidate of dark matter (DM) owing to its interactions suppressed by the $U(1)_L$ breaking scale (v_ϕ). Several studies have investigated Majoron (χ) being a freeze-out as well as freeze-in kind of dark matter, mostly utilizing an explicit $U(1)_L$ breaking interaction involving the standard model (SM) Higgs (H) portal, $\lambda_{\chi H}\chi^2 H^\dagger H$, determining Majoron mass (m_χ) and its interaction channels [4–6]. While the freeze-out scenario (with $\lambda_{\chi H} \sim \mathcal{O}(0.1)$) is mostly excluded below $\mathcal{O}$(TeV) mass scale [7], Majoron remains a natural choice for being a feebly interacting massive particle (FIMP) due to its suppressed interaction. However, Majoron production through

S. K. Manna (✉) · A. Sil
Department of Physics, Indian Institute of Technology Guwahati, Guwahati 781039, Assam, India
e-mail: skmanna2021@gmail.com

A. Sil
e-mail: asil@iitg.ac.in

J. W. F. Valle et al. (eds.), *Proceedings of the International Conference on Future Prospects in Neutrino and Astroparticle Physics*, Springer Proceedings in Physics 361,
https://doi.org/10.1007/978-981-97-7441-8_24

$h \to \chi\chi$ automatically necessitates $\lambda_{\chi H} \sim O(10^{-10})$, leading to a fine-tuned scenario where Majoron mass is confined around $m_\chi \sim 3$ MeV for the correct relic density [4, 6].

In this work, we propose a minimal and natural approach for FIMP type-Majoron production by incorporating a soft LNS breaking at dimension-5 level. Introducing a novel Majoron production channel through the annihilation of right-handed neutrinos (RHNs) extends the Majoron mass range towards lower values ($\mathcal{O}$(keV) $-$ $\mathcal{O}$(GeV)), offering detection prospects in various neutrino experiments and γ-ray observations.

23.2 Interactions of Majoron

Here, we have revisited the original Majoron model, extending the SM by two singlet right-handed neutrinos (RHNs) ($\mathcal{N}_{R_i}$, where $i = 1, 2$) and a singlet scalar field ($\Phi = (v_\phi + \phi + i\chi)/\sqrt{2}$), carrying $U(1)_L$ charges as $L = 1$ and -2, respectively. We write the renormalizable part of the Lagrangian as

$$-\mathcal{L} \supset \frac{f_i}{2}\Phi\overline{\mathcal{N}^c_{R_i}}\mathcal{N}_{R_i} + y^\nu_{\alpha i}\,\overline{L}_\alpha \tilde{H}\mathcal{N}_{R_i} + h.c., \tag{23.1}$$

where H refers to the SM Higgs doublet and $y^\nu_{\alpha i}$ is the neutrino Yukawa coupling. Upon the spontaneous breaking (SSB) of $U(1)_L$ symmetry, followed by the electroweak symmetry breaking (EWSB), neutrino masses are generated through minimal type-I seesaw mechanism [8] as $m_\nu = m_D M_R^{-1} m_D^T$ with $(m_D)_{\alpha i} = y^\nu_{\alpha i} v/\sqrt{2}$ ($v = 246$ GeV) and $M_R = f v_\phi/\sqrt{2}$ with $f = \text{diag}(f_1, f_2)$. While the details of the scalar potential is not necessary here, we mention the Majoron mass generating soft LNS breaking term as

$$-\mathcal{L}_{\text{LNB}} = -\frac{m^2}{4}(\Phi^2 + \Phi^{*2}), \tag{23.2}$$

leading to $m_\chi^2 = m^2$, with the consideration that m^2 is small in t'Hooft's sense [9] and can be linked to Planck scale (M_{Pl}) suppressed higher dimensional operators, expressed as $\Phi^n/M_{Pl}^{n-4} + h.c.$ with $n > 4$. Here, we neglect any portal coupling between Φ and H and further assume the CP-even heavy scalar ϕ to remain effectively decoupled from the plasma. The preliminary interaction of Majoron emerges after $U(1)_L$ gets spontaneously broken as

$$-\mathcal{L} \supset \frac{if_i}{2\sqrt{2}}\chi\overline{\mathcal{N}^c_{R_i}}\mathcal{N}_{R_i} + h.c.. \tag{23.3}$$

However, after EWSB, type-I seesaw mechanism results in a mixing between the active and sterile neutrinos through m_D, parametrized as $\theta = m_D^* M_R^{-1}$, which gives rise two additional interactions of Majoron written in terms of the Majorana mass basis of heavy (N_i) and light (ν_j) neutrinos as follows

$$-\mathcal{L}_{\chi\nu\nu} = -\sum_{j,k}\mathcal{L}_{\chi\nu_j\nu_k} = -\frac{\chi}{2\sqrt{2}}\sum_{j,k}\left(\sum_i i f_i \overline{\nu_j} P_R \nu_k V^T_{ji} V_{ik} + h.c.\right), \quad (23.4)$$

and

$$-\mathcal{L}_{\chi N\nu} = -\sum_{i,j}\mathcal{L}_{\chi N_i\nu_j} = \frac{\chi}{2\sqrt{2}}\sum_{i,j} f_i\left(\overline{\nu_j} P_R N_i V^T_{ji} + \overline{N_i} P_R \nu_j V_{ij}\right) + h.c. \quad (23.5)$$

where $V = \theta^\dagger U$ with U implying the neutrino mixing PMNS matrix. Hence, Majoron can be non-thermally produced after EWSB through the decay of RHNs as $N \to \chi\nu$ (due to the interaction in (23.5)) while its decay channel: $\chi \to \nu\nu$ remains active because of the interaction in (23.4). Utilizing the decay widths of $N_i \to \chi\nu$ and $\chi \to \nu\nu$ as evaluated in [10] and taking into account the stability condition of Majoron in order to be a viable DM as $\Gamma^{-1}_{\chi\to\nu\nu} > \tau_U$ (τ_U being the age of the universe $\sim \mathcal{O}(10^{19})$ s), one obtains

$$\Omega_\chi h^2 < 1.28 \times 10^{-9}\left(\frac{10^{19}\ \text{s}}{\tau_U}\right), \quad (23.6)$$

which turns out to be very negligible in comparison to the correct relic density for cold DM ($\Omega_{\text{CDM}} h^2 \simeq 0.120 \pm 0.001$).

23.3 Our Framework and DM Phenomenology

As the minimal setup lacks the required relic density, we incorporate a soft LNS breaking term in dimension-5 involving RHNs and Φ as

$$-\mathcal{L}_{\text{d}_5} = \frac{\alpha_i}{2\Lambda}\left[\Phi^2 + (\Phi^*)^2\right]\overline{N^c_{R_i}} N_{R_i} + h.c.. \quad (23.7)$$

Here, the cut-off scale Λ ($\leq M_{Pl}$) represents the scale associated to explicit $U(1)_L$ breaking. After Φ obtains its vev (v_ϕ), the consequences of such a term are two fold: first, it contributes to the masses of RHNs

$$M_i = v_\phi\left(\frac{f_i}{\sqrt{2}} + \frac{v_\phi}{\Lambda}\right) \quad (23.8)$$

and secondly, a new production channel of Majoron opens up after the SSB of $U(1)_L$ through $NN \to \chi\chi$ in presence of the interaction

$$-\mathcal{L}_{int} = \frac{1}{2\Lambda}\chi^2 \overline{N^c_i} N_i, \quad (23.9)$$

emerges from (23.7). Considering the existence of Majoron is connected to $U(1)_L$ breaking at a temperature $T_L \sim v_\phi$, we examine the evolution of Majoron DM, taking T_L as the initial temperature (instead of the reheating temperature) for two situations: (a) Case A: RHNs are assumed to be present in thermal bath at T_L and (b) Case B: RHNs are not absent in thermal bath at T_L.

23.3.1 Case A: RHNs are Initially Present in Thermal Bath (at T_L)

In this case, we assume the RHNs to follow a thermal distribution at the temperature T_L, expressed in terms of co-moving number density $Y_{N_i}^{eq} = n_{N_i}^{eq}/\mathcal{S}$ ($\mathcal{S}$ being the entropy density of the universe) as

$$Y_{N_i}^{eq}(T) = \frac{45 g_i}{4\pi^4 g_\star^S(T)} \left(\frac{M_i}{T}\right)^2 K_2\left(\frac{M_i}{T}\right), \tag{23.10}$$

where K_2 denotes the modified Bessel's function of second kind. The yield of Majorons (Y_χ) can be evaluated by solving the following Boltzmann equation for χ as

$$\frac{dY_\chi}{dT} \simeq -\frac{2\mathcal{S}}{\mathcal{H}T} \sum_i \langle \sigma v \rangle_{N_i N_i \to \chi\chi} Y_{N_i}^{eq^2} \left[1 - \frac{Y_\chi^2}{Y_\chi^{eq^2}}\right], \tag{23.11}$$

where $\langle \sigma v \rangle_{N_i N_i \to \chi\chi}$ denotes the thermal average of the cross-section for the process $N_i N_i \to \chi\chi$ and $\mathcal{H}$ ($\approx T^2/M_{Pl}$) is Hubble rate of the universe expansion. The second term in the right hand side corresponds to a less significant back-reaction process due to negligible initial abundance of χ. The other production channel ($N_i \to \chi \nu_j$) contributes negligibly as we discussed earlier and hence omitted in (23.11). Considering degenerate RHNs ($f_1 = f_2 = 0.01$), we illustrate the evolution of χ and N_i by using a benchmark point (BP) in Fig. 23.1 (left panel), where the parameters are

$$\{\Lambda = 1.3 \times 10^{13}\ \text{GeV},\ v_\phi = 1.5 \times 10^8\ \text{GeV},\ m_\chi = 10^{-5}\ \text{GeV}\} \tag{23.12}$$

As showcased in Fig. 23.1 (left panel), in this case, Majoron production is sensitive to the initial temperature T_L, where maximum production takes place. This observation carries a signature of UV-freeze in scenario [11].

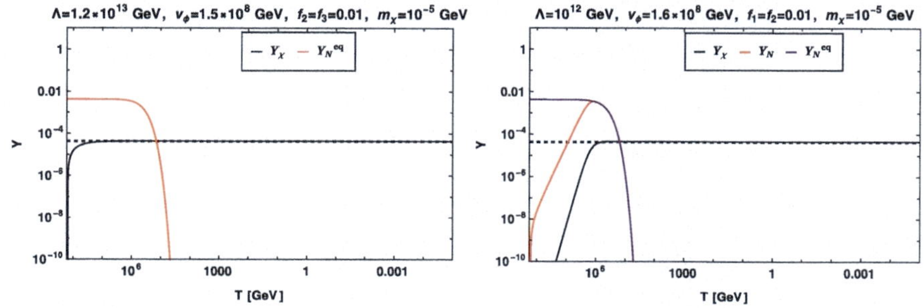

Fig. 23.1 *Left panel*: Evolution of Y_χ (solid black) and $Y_{N_i} = Y_{N_i}^{eq}$ (solid red) against T (GeV) (for case A) and *right panel*: evolution of Y_χ (solid black), Y_{N_i} (solid red) and $Y_{N_i}^{eq}$ (solid purple) against T (GeV) (for case B) with degenerate RHNs are shown corresponding to the respective BPs. The black dotted line denotes the correct value of yield required for satisfying the relic condition

23.3.2 Case B: RHNs are Initially Absent in Thermal Bath at T_L

On the contrary, if RHNs do not remain in thermal equilibrium initially, they will be produced out of the inverse decay of L and H due to the presence of Yukawa interaction and subsequently annihilate into Majorons. Hence, in this case, one needs to solve coupled Boltzmann equations indicating the evolution of χ and N_i as follows

$$\frac{dY_{N_i}}{dT} = \frac{2\mathcal{S}}{\mathcal{H}T}\langle\sigma v\rangle_{N_i N_i \to \chi\chi}\left(Y_{N_i}^2 - \frac{(Y_{N_i}^{eq})^2}{(Y_\chi^{eq})^2}Y_\chi^2\right) + \frac{1}{\mathcal{H}T}\langle\Gamma_{N_i \to LH}\rangle\left(Y_{N_i} - Y_{N_i}^{eq}\right),$$

$$\frac{dY_\chi}{dT} = -\frac{2\mathcal{S}}{\mathcal{H}T}\sum_i \langle\sigma v\rangle_{N_i N_i \to \chi\chi}\left(Y_{N_i}^2 - \frac{(Y_{N_i}^{eq})^2}{(Y_\chi^{eq})^2}Y_\chi^2\right), \tag{23.13}$$

where $\langle\Gamma_{N_i \to LH}\rangle$ denotes the thermal average of the decay width of the process $N_i \to LH$ which is expressed as

$$\Gamma_{N_i \to LH} = \sum_\alpha \frac{|y_{\alpha i}^\nu|^2 M_i}{8\pi}. \tag{23.14}$$

Therefore, the neutrino Yukawa coupling $y_{\alpha i}^\nu$ plays a critical role in this case in producing RHNs and consequently, Majorons. The structure of $y_{\alpha i}^\nu$ is determined utilizing the Casas-Ibarra (CI) parametrization [12] as

$$y^\nu = i\frac{\sqrt{2}}{v}U^\dagger D_{\sqrt{m_\nu}} R D_{\sqrt{M_R}} \quad \text{with } R = \begin{pmatrix} 0 & 0 \\ \cos\theta_R & \sin\theta_R \\ -\sin\theta_R & \cos\theta_R \end{pmatrix}. \tag{23.15}$$

Here $D_{\sqrt{m_\nu}}$ and $D_{\sqrt{M_R}}$ denote the square roots of the diagonal active neutrinos and RHNs mass matrices, respectively, and R is a complex orthogonal matrix with θ_R being a complex angle. Setting the lightest active neutrino to be massless, D_{m_ν} is defined employing the best fitted values from neutrino oscillation data [13]. The mixing angles and CP phase are determined using these fits. We have shown the evolution of χ and N_i for degenerate RHNs in Fig. 23.1 (right panel) following the benchmark point

$$\left\{\Lambda = 10^{12}\ \text{GeV},\ v_\phi = 1.6\times 10^8\ \text{GeV},\ m_\chi = 10^{-5}\ \text{GeV},\ \theta_R = \frac{\pi}{4}\right\} \tag{23.16}$$

As illustrated in Fig. 23.1 (right panel), starting from a negligible initial abundance, N_i is gradually being produced through $LH \to N_i$ process and subsequently enters into thermal equilibrium while annihilating into Majorons. The production of χ finally freezes as N_i catches the equilibrium distribution.

In case of non-degenerate RHNs, freeze-in of χ takes place when the lightest RHN enters thermal equilibrium, altering the parameter space compared to scenario of degenerate RHNs. Therefore, the RHN masses (or f_i) are important parameters in case B, unlike case A, where their masses do not play much significant role. Another crucial parameter in this case is θ_R, which significantly influences the neutrino Yukawa interaction. Furthermore, if θ_R is assumed to be complex, the scenario can be tied to the production of observed Baryon asymmetry of the universe [10].

23.4 Parameter Scan and Constraints

Employing the ingredients as stated in the preceding section, we performed a parameter scan in the keV-GeV Majoron mass regime for both case A and case B, as shown in Fig. 23.2, considering degenerate RHNs. In both cases, we obtain an extensive region of parameter space (the coloured region in Fig. 23.2) achieving the correct relic density in the $\mathcal{O}$(keV)-$\mathcal{O}$(GeV) mass range. In both cases, the lower boundary sets the stability condition of Majoron DM ($\Gamma^{-1}_{\chi\to\nu\nu} < \tau_U$), whereas the excluded region in the top left follows from the condition: $v_\phi/\Lambda > f/\sqrt{2}$ to keep the RHN mass contribution from the dimension-5 operator sub-dominant. However, for a given set of (m_χ, v_ϕ), case B demands a smaller value of Λ compared to case A. The distinction is evident in the parameter space where the top-right excluded region for case A (based on the condition: $\Lambda < M_{Pl}$) does not appear in case B.

The main signature of Majoron DM emerges from its tree-level decay channel into the light neutrinos which leads to monochromatic neutrino fluxes observable in experiments like Borexino, KamLAND, Super-Kamiokande and IceCube. Stringent

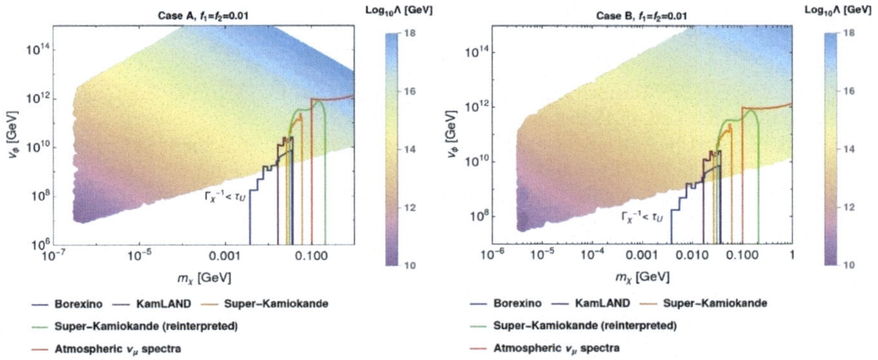

Fig. 23.2 Parameter space ensuring relic satisfaction (the coloured region) in $m_\chi - v_\phi$ plane for case A (*left panel*) and case B (*right panel*) with degenerate RHNs, where values of Λ are in the colour-bar (in log-scale). The lower bounds on v_ϕ are indicated by the solid lines corresponding to different neutrino search experiments

constraints from such experiments, as interpreted in [6], places lower limits on v_ϕ relative to m_χ, which are superimposed on the parameter spaces (in Fig. 23.2) in both cases. These constraints are effective for Majoron mass above MeV. Another decay channel of Majorons into two photons establishes exclusion limits from various X-ray and γ-ray telescopes [14]. However, this decay channel is two-loops suppressed and dependent on model parameters, rendering it less robust. Furthermore, we carefully explore the keV mass regime of Majoron, considering warm dark matter (WDM) constraints to avoid any violations of structure-formation constraints. Following the prescriptions given in [15, 16], our analysis infers an approximate limit on the Majoron mass, setting it at $m_\chi \geq 5.3$ keV [10]. This limit is translated from the Lyman-α forest observations for thermal relics employing a reference value of $m_{\text{WDM}} = 3$ keV.

23.5 Conclusion

This work enhances the minimal Majoron model addressing its viability as Freeze-in dark matter (FIMP) through a dimension-5 soft lepton number breaking operator. Compared to the previous studies, the inclusion of such an explicit LNS violating interaction broadens the parameter space for Majoron mass in the keV-GeV range, thanks to a novel Majoron production channel. Despite stringent experimental bounds (specially in the above-MeV mass regime), our region of parameter space exhibits an extensive dark matter phenomenology, offering compelling evidences for exploration through various ongoing and future neutrino as well as gamma-ray experiments.

References

1. Y. Chikashige, R.N. Mohapatra, R.D. Peccei, Phys. Rev. Lett. **45**, 1926 (1980). https://doi.org/10.1103/PhysRevLett.45.1926
2. Y. Chikashige, R.N. Mohapatra, R.D. Peccei, Phys. Lett. B **98**, 265–268 (1981). https://doi.org/10.1016/0370-2693(81)90011-3
3. G.B. Gelmini, M. Roncadelli, Phys. Lett. B **99**, 411–415 (1981). https://doi.org/10.1016/0370-2693(81)90559-1
4. M. Frigerio, T. Hambye, E. Masso, Phys. Rev. X **1**, 021026 (2011). https://doi.org/10.1103/PhysRevX.1.021026. arXiv:1107.4564 [hep-ph]
5. F.S. Queiroz, K. Sinha, Phys. Lett. B **735**, 69–74 (2014). arXiv:1404.1400 [hep-ph]. https://doi.org/10.1016/j.physletb.2014.06.016
6. C. Garcia-Cely, J. Heeck, JHEP **05**, 102 (2017). https://doi.org/10.1007/JHEP05(2017)102. arXiv:1701.07209 [hep-ph]
7. J.M. Cline, K. Kainulainen, P. Scott, C. Weniger, Phys. Rev. D **88**, 055025 (2013). [erratum: Phys. Rev. D **92**(3), 039906 (2015)]. arXiv:1306.4710 [hep-ph]. https://doi.org/10.1103/PhysRevD.88.055025
8. A. Ibarra, G.G. Ross, Phys. Lett. B **591**, 285–296 (2004). arXiv:hep-ph/0312138 [hep-ph]. https://doi.org/10.1016/j.physletb.2004.04.037
9. G. 't Hooft, NATO Sci. Ser. B **59**, 135–157 (1980). https://doi.org/10.1007/978-1-4684-7571-5_9
10. S.K. Manna, A. Sil, Phys. Rev. D **108**(7), 075026 (2023). arXiv:2212.08404 [hep-ph]. https://doi.org/10.1103/PhysRevD.108.075026
11. F. Elahi, C. Kolda, J. Unwin, JHEP **03**, 048 (2015). https://doi.org/10.1007/JHEP03(2015)048. arXiv:1410.6157 [hep-ph]
12. J.A. Casas, A. Ibarra, Nucl. Phys. B **618**, 171–204 (2001). arXiv:hep-ph/0103065 [hep-ph]. https://doi.org/10.1016/S0550-3213(01)00475-8
13. I. Esteban, M.C. Gonzalez-Garcia, M. Maltoni, T. Schwetz, A. Zhou, JHEP **09**, 178 (2020). https://doi.org/10.1007/JHEP09(2020)178. arXiv:2007.14792 [hep-ph]
14. M. Lattanzi, S. Riemer-Sorensen, M. Tortola, J.W.F. Valle, Phys. Rev. D **88**(6), 063528 (2013). arXiv:1303.4685 [astro-ph.HE]. https://doi.org/10.1103/PhysRevD.88.063528
15. G. Ballesteros, M.A.G. Garcia, M. Pierre, JCAP **03**, 101 (2021). https://doi.org/10.1088/1475-7516/2021/03/101. arXiv:2011.13458 [hep-ph]
16. A. Kamada, K. Yanagi, JCAP **11**, 029 (2019). https://doi.org/10.1088/1475-7516/2019/11/029. arXiv:1907.04558 [hep-ph]

Chapter 24
An Approach to Dirac Neutrino Mass with Singlet-Doublet Dark Matter and ΔNeff

Sujit Kumar Sahoo

Abstract We propose a scotogenic scenario where the singlet-doublet dark matter plays an important role in generating Dirac neutrino mass. It is possible to address muon anomalous magnetic moment with charged doublet in the loop. Due to the Dirac nature of neutrino, the additional right-handed neutrino (RHN) can contribute to the additional relativistic degrees of freedom, ΔNeff, in the Universe. We show that the tiny Dirac neutrino mass correlates the parameter space of muon $(g-2)$, ΔNeff and plays a non-trivial role in deciding the relic density of dark matter. We analyse the thermal dark matter and its detection prospects. We also explored thermal and non-thermal origin of ΔNeff and the parameter space, explored in this work, is verifiable at future CMB experiments like CMB-S4 and SPT-3G.

24.1 Introduction

Despite having gravitational effects on astrophysical bodies and cosmological observations, it is very challenging to get any signature in DM detection experiments. This DM accounts almost 27% of the total energy density of the Universe and often denoted with a dimensionless parameter $\Omega_{DM}h^2 = 0.120 \pm 0.001$ at a 68% C.L [1]. Among the different class of particles, weakly interacting massive particle (WIMP) is a potential candidate for DM, which has thermal origin. The neutrino oscillation experiments have shown the requirement of non-zero mass for neutrino with almost null results from neutrinoless double beta decay experiments motivate the Dirac nature of neutrinos. Because of Dirac neutrino mass, RHNs presence is also required, which share same mass as the SM neutrinos. Few CMB experiments also estimate a parameter, $\Delta N_{\rm eff}$, which is the presence of extra relativistic particle present in the universe other than photon and active neutrinos. The contribution to $N_{\rm eff}$ from SM active neutrinos has been calculated precisely to be 3.045 [2]. The current observation from Planck satellite gives $N_{\rm eff} = 2.99^{0.34}_{-0.33}$ at 95% CL [1]. Apart from these,

S. K. Sahoo (✉)
Department of Physics, Indian Institute of Technology Hyderabad, Kandi, Sangareddy, 502285 Telangana, India
e-mail: ph21resch11008@iith.ac.in

J. W. F. Valle et al. (eds.), *Proceedings of the International Conference on Future Prospects in Neutrino and Astroparticle Physics*, Springer Proceedings in Physics 361,
https://doi.org/10.1007/978-981-97-7441-8_25

recent observations on muon anomalous magnetic moment show deviation from the standard model value for muon magnetic moment [6]. This suggests some new physics interactions with muon.

24.2 The Model

In this work, we extend the standard model (SM) particle content by adding one singlet scalar (ϕ), three generations of right-handed neutrinos (ν_R), and three generations of singlet-doublet vector-like fermions (χ and $\Psi = (\psi^- \psi^0)^T$)(with hypercharge $Y = -1$, where we use $Q = T_3 + Y/2$). Three generations of singlet-doublet fermion are needed to produce three non-zero neutrino mass eigenstates. We have imposed two additional symmetries Z_2 and Z_3. Z_2 guarantees the stability of the lightest Z_2 odd particle. Z_3 symmetry is to restrict the Majorana mass term of neutrino along with the generation of Dirac neutrino mass at tree level (Table 24.1).

The relevant Lagrangian term is given by

$$\begin{aligned} \mathcal{L} \supset\, & i\overline{\Psi}\gamma^\mu D_\mu \Psi + i\overline{\chi}\gamma^\mu \partial_\mu \chi - M_\Psi \overline{\Psi}\Psi - M_\chi \overline{\chi}\chi \\ & - y\overline{\Psi}\tilde{H}\chi - \lambda_\psi \overline{L}\phi\Psi - \lambda_\chi \overline{\nu_R}\phi\chi + h.c. \end{aligned} \tag{24.1}$$

where L is the SM lepton doublet and H is the Higg's doublet. The generation index of Z_2-odd singlet-doublet fermions are suppressed in the above equation. The Z_2 symmetry remains unbroken for the stability of DM, whereas the Z_3 symmetry softly breaks because of the Lagrangian term $\lambda_\psi \overline{L}\phi\psi$. It is necessary to break Z_3 symmetry for light neutrino mass generation at one-loop lavel.

The scalar potential is given by

$$V_{H,\phi} = -\mu_H^2(H^\dagger H) + \lambda_H (H^\dagger H)^2 + \frac{M_\phi^2}{2}\phi^2 + \frac{\lambda_\phi}{4!}\phi^4 + \lambda_{\phi H}\phi^2 (H^\dagger H) \tag{24.2}$$

As we are considering one Z_2 odd scalar in this model, it does not get vev. After EWSB, Higg's particle gets vev and induces mixing between singlet fermion (χ) and neutral component of doublet fermion (ψ^0) through the term $y\overline{\Psi}\tilde{H}\chi$. Transformation from their flavour eigenstates, $\begin{pmatrix}\psi^0 & \chi\end{pmatrix}^T$ to mass eigenstates,$(\chi_1 \quad \chi_2)^T$ is given by

Table 24.1 Particles and their charge assignments in our setup

Gauge group	Fermion fields				Scalar
	L	Ψ	χ	ν_R	ϕ
$SU(2)_L$	2	2	1	1	1
$U(1)_Y$	-1	-1	0	0	0
Z_3	ω	ω^2	ω^2	ω^2	1
Z_2	$+1$	-1	-1	$+1$	-1

$$\begin{pmatrix} \psi^0 \\ \chi \end{pmatrix} = U \begin{pmatrix} \chi_1 \\ \chi_2 \end{pmatrix} = \begin{pmatrix} \cos\theta & -\sin\theta \\ \sin\theta & \cos\theta \end{pmatrix} \begin{pmatrix} \chi_1 \\ \chi_2 \end{pmatrix}, \tag{24.3}$$

where θ is the mixing angle, given by

$$\tan 2\theta = \frac{\sqrt{2}yv}{M_\psi - M_\chi}. \tag{24.4}$$

Here we have considered the mass hierarchy as $M_{\chi_1} > M_\phi > M_{\chi_2}$, where M_{χ_2} is dominantly the singlet fermion χ and DM candidate. As M_ϕ lies in between M_{χ_1} and M_{χ_2}, its interaction with χ_2 plays an important role in deciding DM relic and it relaxes the constraint on singlet-doublet mass splitting ($\Delta M = M_{\chi_1} - M_{\chi_2}$) from direct detection bounds as pointed out in [12]. The presence of ψ^- and its interaction with SM leptons allows to explain the muon $(g-2)$. The addition of 3 generations of singlet-doublet fermion can be motivated in context of W-mass anomaly. In the current work, we will consider the higher generation singlet-doublet fermions to be very massive, so that they won't contribute to the DM phenomenology other than neutrino mass generation. Detailed analysis can be found in [13].

24.3 Dirac Neutrino Mass and $(g-2)_\mu$

In this work, the Dirac neutrino mass term $\overline{L}\tilde{H}\nu_R$ can be realised at one loop level as shown in Fig. 24.1 (left) through radiative seesaw mechanism The neutrino mass expression [5] is given is (24.5). After imposing the stringent bounds on neutrino mass from cosmological data and mass squared differences from neutrino oscillation data, we found the product of Yukawa couplings involved in the loop $\lambda_\psi \lambda_\chi \sin\theta < 10^{-10}$, if we consider the loop particle masses in electroweak range ($\mathcal{O}(100 \text{ GeV})$).

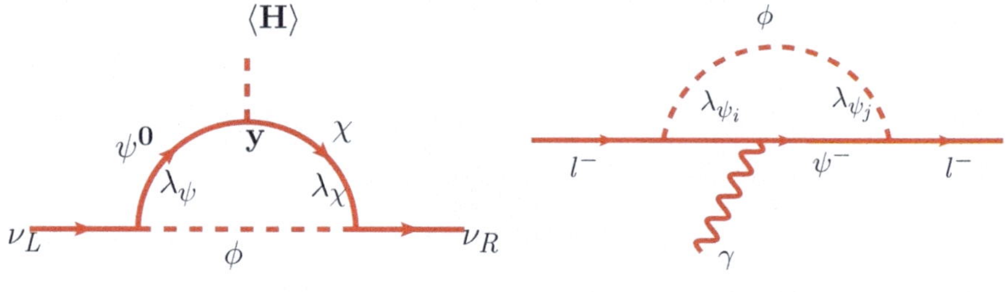

Fig. 24.1 (Left) Dirac Neutrino Mass generation at one-loop level, (Right) Feynman diagram for the dark sector contribution to $(g-2)_\mu$ and LFV

$$(m_\nu)_{\alpha\beta} = \frac{v}{4\pi^2} \sum_{i=1,2} (\lambda_\psi)_{\alpha i} y_i (\lambda_\chi)_{i\beta} F(M_{\chi_{1i}}, M_{\chi_{2i}}, M_\phi) \tag{24.5}$$

where

$$F = \left[M_{\chi_1} M_{\chi_2} I_3(M_\phi, M_{\chi_1}, M_{\chi_2}) + J_3(M_\phi, M_{\chi_1}, M_{\chi_2})\right].$$

Here I_3 and J_3 are the loop integrals given as

$$I_3(M_\phi, M_{\chi_1}, M_{\chi_2}) = \left[\frac{M_\phi^2 \ln \frac{M_{\chi_2}^2}{M_\phi^2}}{(M_\phi^2 - M_{\chi_1}^2)(M_\phi^2 - M_{\chi_2}^2)} + \frac{M_{\chi_1}^2 \ln \frac{M_{\chi_2}^2}{M_{\chi_1}^2}}{(M_{\chi_1}^2 - M_\phi^2)(M_{\chi_1}^2 - M_{\chi_2}^2)}\right]$$

$$J_3(M_\phi, M_{\chi_1}, M_{\chi_2}) = \left[A - \ln\left(\frac{M_{\chi_2}^2}{\mathrm{GeV}^2}\right) + \frac{M_\phi^4 \ln \frac{M_{\chi_2}^2}{M_\phi^2}}{(M_\phi^2 - M_{\chi_1}^2)(M_\phi^2 - M_{\chi_2}^2)} + \frac{M_{\chi_1}^4 \ln \frac{M_{\chi_2}^2}{M_{\chi_1}^2}}{(M_{\chi_1}^2 - M_\phi^2)(M_{\chi_1}^2 - M_{\chi_2}^2)}\right] \tag{24.6}$$

where $A = \frac{2}{\varepsilon} - \gamma_E + 1 + \ln 4\pi$.

In this framework, we have positive contribution from the loop shown in Fig. 24.1 (right) to muon $(g-2)$ and the loop contribution can be calculated using the expression [7] given in (24.7). Figure 24.2 shows the allowed region in the parameter space of loop particle mass (M_{ψ^-}) and the Yukawa coupling, λ_{ψ_μ}. We can see λ_{ψ_μ} coupling is constrained in the range of 0.01–0.1 for $M_{\psi^-} \in [100, 1800]$ GeV.

Fig. 24.2 The allowed region for muon $(g-2)$ is shown in the parameter space of M_{ψ^-} and λ_{ψ_μ}

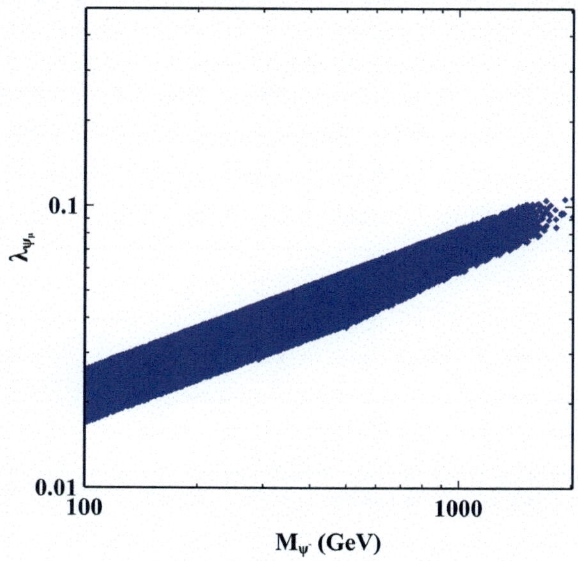

$$\Delta a_\mu = \frac{\lambda_{\psi_\mu}^2}{8\pi^2} \frac{m_\mu^2}{M_\phi^2} \int_0^1 dx \frac{x^2(1 - x - \frac{M_\Psi}{m_\mu})}{(1-x)(1 - x\frac{m_\mu^2}{M_\phi^2}) + x\frac{M_\Psi^2}{M_\phi^2}} \tag{24.7}$$

24.4 Dark Matter Phenomenology

We have considered χ_2 as our lightest Z_2 odd particle and the possible DM candidate in this model. As it is an admixture of both singlet and doublet fermion, the gauge interactions and the other Yukawa interactions are responsible for its thermalization in the early universe and thereafter freeze-out from the thermal bath. The required DM relic density can be attained in our model through number changing processes such as χ_2 annihilation, and its coannihilation with χ_1, ψ^- and ϕ to SM bath particle and ν_R.

Our main focus in this work is to calculate the contribution to $\Delta N_{\rm eff}$ from the extra RHNs introduced to generate Dirac neutrino mass. This depends on the production mechanism of ν_R, which can be of thermal origin or non-thermal origin. ν_R is coupled to SM through dark sector coupling λ_χ. We have compared the interaction rate of ν_R for the elastic scattering processes with χ_2 and ϕ with Hubble Rate, $H(T)$, and found that λ_χ coupling needs to be larger than 10^{-3} for M_ϕ and M_{χ_2} in order of $\mathcal{O}(100\,\text{GeV})$ in order to keep ν_R in equilibrium with thermal bath. We have considered two scenarios, (i) $10^{-3} < \lambda_\chi < \sqrt{4\pi}$ (**Case-1**) and (ii) $10^{-11} < \lambda_\chi \leq 10^{-3}$ (**Case-2**).

In case-1, λ_χ is large and to satisfy neutrino mass bound, we are keeping $\sin\theta$ small. So the dominating process would be $\chi_2\chi_2 \to \nu_R\nu_R$. Other than this, coannihilation of χ_2 with χ_1 and ϕ would be there depending on their respective mass differences. The left Fig. 24.3 shows the relic satisfying points in the plane of $\Delta M_1 (= M_\phi - M_{\chi_2})$ and M_{χ_2}, where the colour code is for λ_χ. We can see as the value of λ_χ is decreasing, the mass difference (ΔM_1) is decreasing, i.e. the coannihilation processes are required to get correct relic density. In case-2, λ_χ is small, but $\sin\theta$ is large. The coannihilation processes along with the gauge mediated processes will determine the relic density. The relic satisfying points are shown in right Fig. 24.3 in the plane of ΔM_1 and M_{χ_2} with $\lambda_{\phi H}$ coupling in colour code. The arc formed by the black coloured points are dominated by χ_2 and ϕ coannihilation process, where λ_ϕ value is maximum. As the $\lambda_{\phi H}$ value decreases, the mass difference (ΔM_1) also decreases. But the colour pattern shown in this figure is distorted, this is because of the contribution coming from gauge processes. Large $\sin\theta$ value gives significant contribution from gauge mediated processes to effective thermal averaged cross section ($\langle \sigma v \rangle_{\rm eff}$). The grey coloured points in both of the plots are falling under LEP bounds, though they satisfy relic density. The relic density is calculated using MicrOmega [8].

In case-1, $\sin\theta$ will be small. Hence, Higg's mediated and Z mediated processes will be suppressed. The loop diagram shown in Fig. 24.4 gives the dominant contribution to direct detection cross section. In case-2, because of large $\sin\theta$, the usual gauge mediated and Higg's mediated processes contribute to direct detection cross

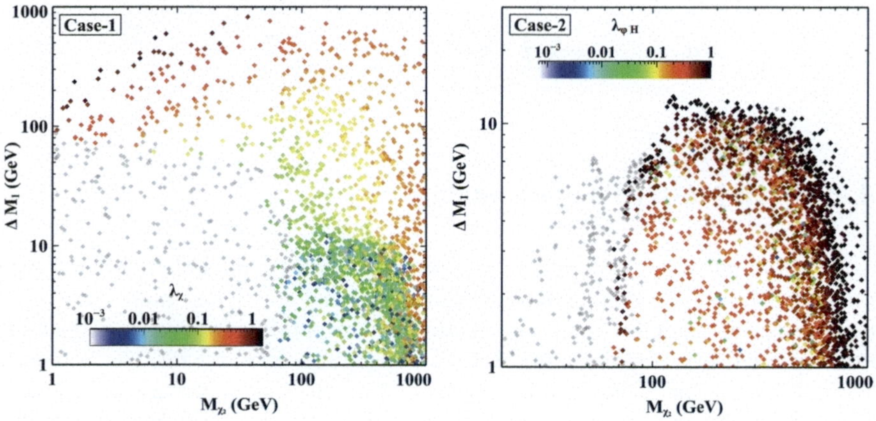

Fig. 24.3 Relic satisfying points for case-1(left) and case-2(right) in the plane of $\Delta M_1 (= M_\phi - M_{\chi 2})$ and $M_{\chi 2}$. The colour code is for λ_χ coupling in left plot and $\lambda_{\phi H}$ in the right plot

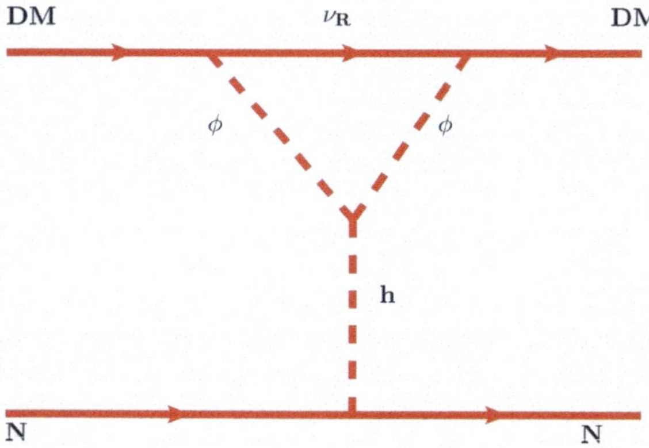

Fig. 24.4 Spin-independent elastic DM-nucleon scattering arising at one loop

section. In Fig. 24.5, current experimental bounds and future sensitivity is shown and we found that most of the parameter space is safe from direct detection cross section [9–11].

24.5 ΔN_{eff}

As we have mentioned in the previous section, case-1 corresponds to thermal production of ν_R through annihilation of χ_2 and ϕ. This annihilation process depends on the λ_χ coupling and masses of the particles (χ_2 and ϕ). Earlier the decoupling from thermal bath, less contribution to ΔN_{eff}. Using entropy conservation relation, we have calculated today's temperature of the decoupled ν_R followed by its energy

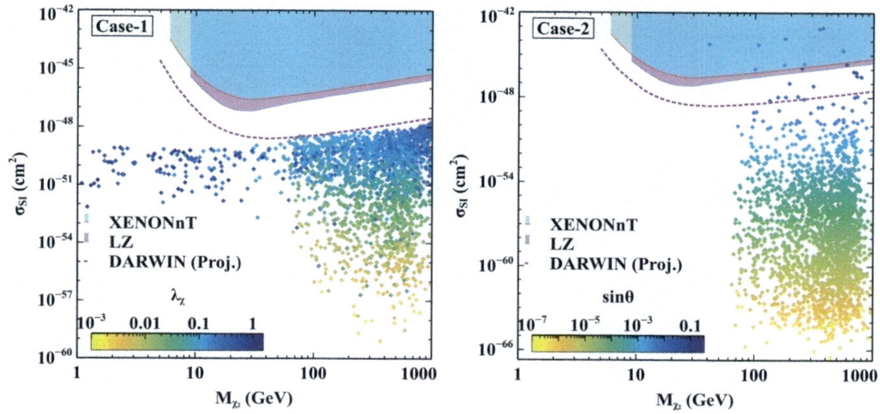

Fig. 24.5 Spin-independent DM-nucleon scattering cross section as a function of DM mass for case-1(left) and case-2(right). The colour code is for λ_χ coupling in left plot and $\sin\theta$ in the right plot

density and ΔN_{eff} using (24.8). In left Fig. 24.6, we have shown the ΔN_{eff} contribution. We can see, as the mass (M_{χ_2}) increases, interaction rate of ν_R with thermal bath decreases and hence the contribution to ΔN_{eff} decreases. Similarly, as the λ_χ coupling decreases, interaction rate also decreases, following less contribution to ΔN_{eff}. In case-2, though it is difficult to produce ν_R through thermal processes, it is still possible to produce enough radiation density through freeze-in mechanism. In this model, ϕ being Z_2 odd particle and heavier than χ_2, it can decay to ν_R and χ_2. We have used the evolution equation of ν_R written in (24.9) to get the abundance of ν_R.

$$\Delta N_{\text{eff}} = N_{\nu_R} \times \left.\frac{\rho_{\nu_R}}{\rho_{\nu_L}}\right|_{\mathrm{T=T_{CMB}}}, \tag{24.8}$$

The evolution equation of ϕ and ν_R is given as

$$\frac{dY_\phi}{dx} = \frac{\beta s}{Hx}\left[-\langle\sigma v\rangle_{\phi\phi\to X\bar{X}}(Y_\phi^2 - (Y_\phi^{\text{eq}})^2) - \frac{\Gamma_\phi}{s}\frac{K_1(x)}{K_2(x)}Y_\phi\right]$$

$$\frac{d\tilde{Y}_{\nu_R}}{dx} = \frac{\beta}{Hs^{1/3}x}\langle E\Gamma\rangle Y_\phi, \tag{24.9}$$

where the dimensionless parameters $Y_\phi = n_\phi/s$ and $\tilde{Y}_{\nu_R} = \rho_{\nu_R}/s^{4/3}$. In the above equation Γ_ϕ and $\langle E\Gamma\rangle$ can be expressed as

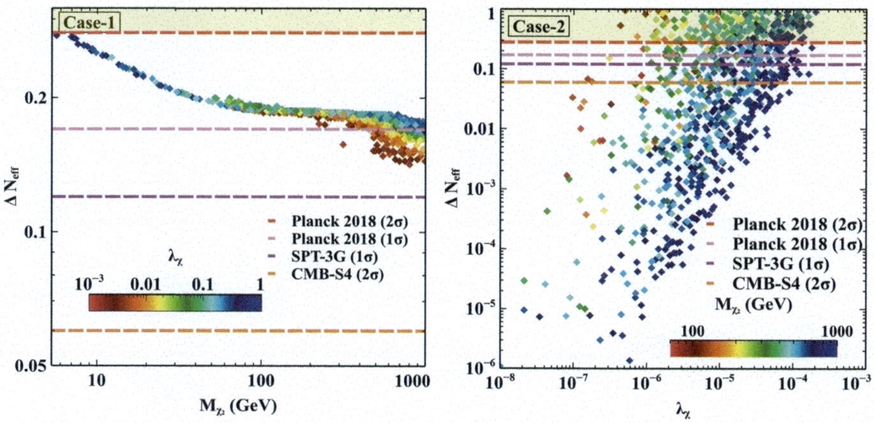

Fig. 24.6 The contribution to $\Delta N_{\rm eff}$ for case-1 (left) and case-2 (right) is shown. The yellow shaded regions are already ruled out from experiments. The orange and magenta colour dashed lines are the future sensitivities

$$\Gamma(\phi \to \chi \nu_R) = \frac{1}{8\pi}\lambda_\chi^2 \cos\theta^2 M_\phi \left(1 - \frac{M_{\chi_2}^2}{M_\phi^2}\right)^2 \tag{24.10}$$

$$\langle E\Gamma(\phi \to \chi \nu_R)\rangle = \frac{1}{16\pi}\lambda_\chi^2 \cos\theta^2 M_\phi^2 \left(1 - \frac{M_{\chi_2}^2}{M_\phi^2}\right)^3 . \tag{24.11}$$

After solving the above Boltzmann equations, $\Delta N_{\rm eff}$ produced by ρ_{ν_R} can be calculated as:

$$\begin{aligned}\Delta N_{\rm eff} &= 3 \times 2 \left(\frac{\rho_{\nu_R}}{\rho_{\nu_L}}\right)_{T_D(\nu_L)} \\ &= 6 \left(\frac{s^{4/3}\tilde{Y}_{\nu_R}}{\rho_{\nu_L}}\right)_{T_D(\nu_L)}\end{aligned} \tag{24.12}$$

The right Fig. 24.6 shows the $\Delta N_{\rm eff}$ dependency on λ_χ and M_{χ_2} is shown in colour code. We can see as the λ_χ increases or M_{χ_2} decreases, the decay width increases and hence more contribution to $\Delta N_{\rm eff}$. In both the plots shown in Fig. 24.6, yellow shaded region is ruled out from Planck 2018 data at 2σ [1]. We have shown the projected sensitivities of future experiments such as SPT-3G [3] and CMB-S4 [4].

24.6 Conclusion

In this work, we studied the singlet-doublet DM model by extending with a singlet scalar (ϕ) and three generations of ν_R to explain Dirac neutrino mass in a scotogenic setup. The addition of extra scalar in model plays an important role in deciding relic, and it relaxes the bound on the singlet-doublet mass splitting ($M_{\chi_1} - M_{\chi_2}$) from direct detection as in case of usual singlet-doublet model. We also checked the allowed parameter space from muon ($g - 2$). Because of additional relativistic particles in the model, we tried to estimate their contribution to ΔN_{eff} and potential detection prospects in CMB.

Acknowledgements We would like to thank Narendra Sahu, Debasish Borah, Satyabrata Mohapatra and Dibyendu Nanda for useful discussion.

References

1. N. Aghanim, et al. (Planck), Astron. Astrophys. **641**, A6 (2020). [Erratum: Astron. Astrophys. **652**, C4 (2021)]. https://arxiv.org/abs/1807.06209
2. G. Mangano, G. Miele, S. Pastor, T. Pinto, O. Pisanti, P.D. Serpico (2005). https://arxiv.org/abs/hep-ph/0506164
3. J.S. Avva, et al. (SPT-3G), J. Phys. Conf. Ser. **1468**, 012008 (2020). https://arxiv.org/abs/1911.08047
4. K. Abazajian, et al. (2019). https://arxiv.org/abs/1907.04473
5. E. Ma, O. Popov, Phys. Lett. B **764** 142 (2017). https://arxiv.org/abs/1609.02538
6. D.P. Aguillard, et al. (Muon g-2) (2023). https://arxiv.org/abs/2308.06230
7. P. Lindner, Q.M. Lindner, M. Platscher, F.S. Queiroz, Phys. Rept. **731**, 1 (2018). https://arxiv.org/abs/1610.06587
8. Bélanger, et al.: micrOMEGAs4.1: two dark matter candidates. https://arxiv.org/abs/1407.6129
9. J. Aalbers, et al. (LZ), Phys. Rev. Lett. **131**, 041002 (2023). https://arxiv.org/abs/2207.03764
10. E. Aprile, et al. (XENON), Phys. Rev. Lett. **131**, 041003 (2023). https://arxiv.org/abs/2303.14729
11. J. Aalbers, et al. (DARWIN), JCAP **11**, 017 (2016). https://arxiv.org/abs/1606.07001
12. S. Bhattacharya, P. Ghosh, N. Sahoo, N. Sahu, Front. Phys. **7**, 80 (2019). https://arxiv.org/abs/1812.06505
13. D. Borah, S. Mahap- atra, D. Nanda, S.K. Sahoo, N. Sahu (2023). https://arxiv.org/abs/2310.03721

Chapter 25
Unveiling the Impact of Generalized CP Symmetry on 2HDM and Neutrino Physics

Tapender, Sanjeev Kumar, and Surender Verma

Abstract We have proposed a model that incorporates the idea of generalized CP symmetries (GCP) in two-Higgs-doublet model (2HDM), with Type-I seesaw mechanism, in the leptonic sector. The imposition of GCP results in simplified scalar and Yukawa sectors leading to reduction in the number of free parameters. The framework results in stringent constraints on the Yukawa couplings yielding a model characterized by only 12 independent real parameters in the Yukawa coupling matrices. Furthermore, under the GCP symmetry, the right-handed Majorana neutrino mass matrix becomes diagonal featuring two degenerate masses. The second Higgs doublet acquires a complex vacuum expectation value (VEV) after soft breaking of the GCP symmetry which seeds the CP-violation in the model. We have investigated the phenomenology of the model considering both, normal (NH) as well as inverted (IH), hierarchies of neutrino masses. There exists an interesting interplay between possible neutrino mass hierarchy and atmospheric mixing angle, θ_{23}. For normal (inverted) hierarchy θ_{23} is found to be in the lower (higher) octant. Further, for NH, the Dirac CP phase δ is constrained to a range of $-10°$ to $10°$. On the other hand, when considering IH, the model becomes inherently CP-violating with CP-violating phase δ confined to small regions near $\pm 40°$.

Tapender (✉) · S. Verma
Department of Physics and Astronomical Science, Central University of Himachal Pradesh, Dharamshala 176215, India
e-mail: tapenderphy@gmail.com

S. Verma
e-mail: s_7verma@hpcu.ac.in

S. Kumar
Department of Physics and Astrophysics, University of Delhi, Delhi 110007, India
e-mail: skverma@physics.du.ac.in

J. W. F. Valle et al. (eds.), *Proceedings of the International Conference on Future Prospects in Neutrino and Astroparticle Physics*, Springer Proceedings in Physics 361,
https://doi.org/10.1007/978-981-97-7441-8_26

25.1 Introduction

The Standard Model (SM), which describes strong and electroweak interactions, has been highly successful in explaining various aspects of natural phenomena. It explains the interactions between particles and how massive particles acquire their mass. Despite its success, the SM falls short in explaining certain observed phenomena, such as neutrino oscillation and matter-antimatter asymmetry. Consequently, various extensions of the SM are under development to address these gaps in our understanding of nature. Among various extensions to explain the origin of neutrino masses observed in neutrino oscillations, a straightforward method is to augment the field content of the SM by introducing right-handed neutrino singlets. Then neutrinos can possess the Dirac mass term, and a lepton number violating Majorana mass term. The Type-I seesaw mechanism, famous for achieving small neutrino masses by setting the right-handed neutrinos mass at a high-energy scale, emerges from this extension [1–5]. For explaining the observed matter-antimatter asymmetry in the universe, CP-violation is a necessary condition [6]. As an attempt to explain this required CP-violation, SM scalar sector is extended by adding the second Higgs doublet whose vacuum expectation value (VEV) breaks the CP symmetry, as proposed by Lee [7]. However, such extensions often result in models with a large number of free parameters, diminishing their predictability. To address this issue, models are often supplemented with additional symmetries. Generalized CP symmetry (GCP) applied to the 2HDM scalar potential has been explored in previous works [8, 9], classifying GCP into three classes: CP1, CP2, and CP3. Further from [9], we know that apart from standard CP symmetry (CP1), CP3 symmetry of 2HDM scalar sector can be extended to Yukawa sector and they applied this to the quark sector. Based on these findings, in our model, we combine the concepts of 2HDM and the Type-I seesaw mechanism. We utilize the GCP symmetry of 2HDM, extending it to the leptonic Yukawa sector, enabling predictions related to neutrinos and CP-violation [10]. As we will adopt the most general 2HDM (without forbidding any Higgs-fermion Yukawa couplings), there exists a potential risk of large flavor changing neutral currents (FCNCs) [11]. However, the primary focus of our study is to investigate the consequences of GCP symmetry on the most general Yukawa couplings without imposing additional symmetry constraints and explore their corresponding neutrino phenomenology. To mitigate the risk of FCNCs, we consider a simpler approach by assuming that the new scalar particles are all very heavy, and thus FCNCs are suppressed.

The paper is organized as follows. In Sect. 25.2, we introduce the general 2HDM and discuss the GCP symmetries associated with it. Section 25.3 is dedicated to discussing the impact of the CP3 symmetry on the leptonic Yukawa sector within the Type-I seesaw framework. Numerical analysis is conducted in Sect. 25.4. At last, we have summarized our conclusions in Sect. 25.5.

25.2 Symmetry of Two-Higgs-Doublet Model: Generalized CP Symmetry

We will extend the scalar sector of the SM by adding a second Higgs doublet, ϕ_2, which has the same charge assignment as the SM Higgs, ϕ_1. Therefore, the most general scalar potential with two Higgs doublets can be written as:

$$\begin{aligned}V_H &= m_{11}^2\phi_1^\dagger\phi_1 + m_{22}^2\phi_2^\dagger\phi_2 - [m_{12}^2\phi_1^\dagger\phi_2 + \text{H.c.}] + \frac{1}{2}\lambda_1(\phi_1^\dagger\phi_1)^2 + \frac{1}{2}\lambda_2(\phi_2^\dagger\phi_2)^2 \\ &+\lambda_3(\phi_1^\dagger\phi_1)(\phi_2^\dagger\phi_2) + \lambda_4(\phi_1^\dagger\phi_2)(\phi_2^\dagger\phi_1) \\ &+[\frac{1}{2}\lambda_5(\phi_1^\dagger\phi_2)^2 + \lambda_6(\phi_1^\dagger\phi_1)(\phi_1^\dagger\phi_2) + \lambda_7(\phi_2^\dagger\phi_2)(\phi_1^\dagger\phi_2) + \text{H.c.}],\end{aligned} \tag{25.1}$$

which has fourteen free parameters. Among these fourteen parameters, m_{12}^2, λ_5, λ_6, and λ_7 are generally complex, while m_{11}^2, m_{22}^2, and λ_1 through λ_4 are real. As we can see, there is an increase in the number of free parameters in the model. To reduce the number of free parameters, additional symmetries can be imposed on the 2HDM. These symmetries can be either Higgs family symmetry or GCP symmetry. In our model, we will be employing the GCP symmetry on the 2HDM. Under GCP, fields transform as [8]

$$\phi_a \xrightarrow{\text{GCP}} \phi_a' = S_{ab}\phi_b^*, \tag{25.2}$$

where S is an arbitrary unitary matrix. According to Ecker, Grimus, and Neufeld [12], there always exists a basis of scalar fields for which we can bring the GCP transformation matrix S to the form

$$S = \begin{pmatrix} \cos\theta & \sin\theta \\ -\sin\theta & \cos\theta \end{pmatrix}, \tag{25.3}$$

where $0 \leq \theta \leq \pi/2$. From [8], we know that we can classify these GCP transformations into three different classes: (i) CP1 when $\theta = 0$, (ii) CP2 when $\theta = \pi/2$, (iii) CP3 when $0 < \theta < \pi/2$. When we impose any of these GCP symmetries, our scalar potential parameters get reduced. However, since we have to include the masses of the charged leptons and neutrinos, we must keep in mind whether the GCP symmetry can be extended to the Yukawa sector or not. From the work done by Ferreira and Silva [9], we know that, apart from standard CP (CP1), only CP3 can be extended to the Yukawa sector. On imposing CP3 on the scalar sector of the 2HDM, the requirement for invariance of the scalar potential under CP3 transformations dictates that the coefficients must satisfy $m_{11}^2 = m_{22}^2$, $m_{12}^2 = 0$, $\lambda_2 = \lambda_1$, $\lambda_6 = 0$, $\lambda_7 = 0$, and $\lambda_5 = \lambda_1 - \lambda_3 - \lambda_4$, all of which must be real parameters. However, since CP3 is a continuous symmetry, its spontaneous breaking would result in the emergence of Goldstone bosons, which need to be eliminated. To achieve this, we need to softly break the CP3 symmetry in the scalar sector, by setting $m_{11}^2 \neq m_{22}^2$ and $\Re[m_{12}^2] \neq 0$. This softly broken scalar potential also makes it possible to have a complex VEV for the second Higgs doublet, serving as a source for CP-violation. In the next section, we will extend the CP3 symmetry to the leptonic Yukawa sector.

25.3 Extending CP3 to Yukawa Sector within the Type-I Seesaw Framework

To generate neutrino mass via the Type-I seesaw mechanism, we have added three right-handed neutrinos N_{Ri} $(i = 1, 2, 3)$ to the SM field content in addition to the second Higgs doublet ϕ_2. The leptonic part of the Yukawa Lagrangian within the Type-I seesaw framework can be written as

$$-\mathcal{L}_Y = \overline{L}_L \Delta_a \phi_a l_R + \overline{L}_L \Lambda_a \tilde{\phi}_a N_R + \frac{1}{2}\overline{N^c_R} M_R N_R + \text{H.c.}, \tag{25.4}$$

where to simplify notation, we omitted flavor indices, ϕ_a $(a = 1, 2)$ denotes Higgs doublets, and N_R denotes right-handed neutrino singlets. L_L and l_R denotes the left-handed lepton doublets and right-handed lepton singlets within the framework of the SM. Yukawa coupling matrices for charged leptons and neutrinos are denoted by Δ_a and Λ_a, respectively. The lepton number violating Majorana mass term for right-handed neutrinos is denoted by M_R. Also, $\tilde{\phi}_a = i\sigma_2\phi_a^*$, where σ_2 is a Pauli spin matrix, and $N^c_R = C\overline{N_R}^T$, where C is the charge conjugation matrix. We want to use the GCP symmetry of 2HDM in the leptonic Yukawa sector.[1] The fields under GCP transform as

$$\left.\begin{aligned}
\phi_a &\xrightarrow{\text{GCP}} \phi_a^{'} = S_{ab}\phi_b^*, \\
\tilde{\phi}_a &\xrightarrow{\text{GCP}} \tilde{\phi}_a^{'} = S_{ab}^*(\tilde{\phi}_b^{\dagger})^T, \\
L_L &\xrightarrow{\text{GCP}} L_L^{'} = iS_\zeta\gamma^0 C\overline{L}_L^T, \\
l_R &\xrightarrow{\text{GCP}} l_R^{'} = iS_\beta\gamma^0 C\overline{l}_R^T, \\
N_R &\xrightarrow{\text{GCP}} N_R^{'} = iS_\eta\gamma^0 C\overline{N}_R^T,
\end{aligned}\right\} \tag{25.5}$$

where γ^0 is the Dirac gamma matrix, S is given by (25.3), and S_ζ, S_β, and S_η are CP transformation matrices which are given by

$$S_\kappa = \begin{pmatrix} \cos\kappa & \sin\kappa & 0 \\ -\sin\kappa & \cos\kappa & 0 \\ 0 & 0 & 1 \end{pmatrix} \text{ where } \kappa = \zeta,\ \beta,\ \eta. \tag{25.6}$$

The invariance under these transformations requires that Yukawa coupling matrices have to satisfy the following constraint

[1] See [9] for the quark Yukawa sector.

$$\Delta_b^* = S_\zeta^\dagger \Delta_a S_\beta S_{ab}, \tag{25.7}$$
$$\Lambda_b^* = S_\zeta^\dagger \Lambda_a S_\eta S_{ab}^*, \tag{25.8}$$

and the right-handed neutrinos mass matrix M_R has to satisfy the following constraint

$$M_R^* = S_\eta^T M_R S_\eta, \tag{25.9}$$

the matrix M_R is a complex symmetric matrix, which, in general, can be written as

$$M_R = \begin{pmatrix} (M_R)_{11} & (M_R)_{12} & (M_R)_{13} \\ - & (M_R)_{22} & (M_R)_{23} \\ - & - & (M_R)_{33} \end{pmatrix}. \tag{25.10}$$

Since, under GCP transformations, fields in the leptonic sector (as shown in (25.5)) undergo the same transformations as those in the quark sector of [9], the results derived in the quark sector are applicable to the leptonic sector. In determining the correct quark masses in [9], it was established that, apart from the standard CP (CP1), only CP3 symmetry with $\theta = \pi/3$ ($\zeta = \beta = \eta = \pi/3$) can be extended to the quark Yukawa sector. Consequently, in the leptonic Yukawa sector, we similarly extend CP3 symmetry with $\theta = \pi/3$ ($\zeta = \beta = \eta = \pi/3$). Thus, the solution to (25.7) and (25.8) is given by

$$\Delta_1 = \begin{pmatrix} ix_{11} & ix_{12} & x_{13} \\ ix_{12} & -ix_{11} & x_{23} \\ x_{31} & x_{32} & 0 \end{pmatrix}, \quad \Delta_2 = \begin{pmatrix} ix_{12} & -ix_{11} & -x_{23} \\ -ix_{11} & -ix_{12} & x_{13} \\ -x_{32} & x_{31} & 0 \end{pmatrix}, \tag{25.11}$$

$$\Lambda_1 = \begin{pmatrix} iy_{11} & iy_{12} & y_{13} \\ iy_{12} & -iy_{11} & y_{23} \\ y_{31} & y_{32} & 0 \end{pmatrix}, \quad \Lambda_2 = \begin{pmatrix} iy_{12} & -iy_{11} & -y_{23} \\ -iy_{11} & -iy_{12} & y_{13} \\ -y_{32} & y_{31} & 0 \end{pmatrix}, \tag{25.12}$$

respectively, where all x_{ij} and y_{ij} are real parameters.

Now, our objective is to solve the constraint provided by (25.9). By substituting (25.6) (with $\kappa = \eta$) and (25.10) into (25.9), we arrive at the set of constraints

$$(M_R)_{11}^* - (M_R)_{11}\cos^2\eta - (M_R)_{22}\sin^2\eta + (M_R)_{12}\sin 2\eta = 0, \tag{25.13}$$
$$(M_R)_{12}^* - (M_R)_{12}\cos 2\eta + (-(M_R)_{11} + (M_R)_{22})\sin\eta\cos\eta = 0, \tag{25.14}$$
$$(M_R)_{22}^* - (M_R)_{22}\cos^2\eta - 2(M_R)_{12}\cos\eta\sin\eta - (M_R)_{11}\sin^2\eta = 0, \tag{25.15}$$
$$(M_R)_{13}^* - (M_R)_{13}\cos\eta + (M_R)_{23}\sin\eta = 0, \tag{25.16}$$
$$(M_R)_{23}^* - (M_R)_{23}\cos\eta - (M_R)_{13}\sin\eta = 0, \tag{25.17}$$
$$(M_R)_{33}^* - (M_R)_{33} = 0. \tag{25.18}$$

We have two sets of simultaneous linear equations. The first set consists of (25.13), (25.14), and (25.15). The second set includes (25.16) and (25.17). As these equations are complex, we can solve these linear equations by separating the real and imaginary parts and requiring both to vanish.

For real part of (25.13), (25.14), and (25.15), we have

$$\begin{pmatrix} 1-\cos^2\eta & \sin 2\eta & -\sin^2\eta \\ -\cos\eta\sin\eta & 1-\cos 2\eta & \cos\eta\sin\eta \\ -\sin^2\eta & -2\cos\eta\sin\eta & 1-\cos^2\eta \end{pmatrix} \begin{pmatrix} \Re[(M_R)_{11}] \\ \Re[(M_R)_{12}] \\ \Re[(M_R)_{22}] \end{pmatrix} = 0, \tag{25.19}$$

and for the imaginary part, we have

$$\begin{pmatrix} -1-\cos^2\eta & \sin 2\eta & -\sin^2\eta \\ -\cos\eta\sin\eta & -1-\cos 2\eta & \cos\eta\sin\eta \\ -\sin^2\eta & -2\cos\eta\sin\eta & -1-\cos^2\eta \end{pmatrix} \begin{pmatrix} \Im[(M_R)_{11}] \\ \Im[(M_R)_{12}] \\ \Im[(M_R)_{22}] \end{pmatrix} = 0. \tag{25.20}$$

Further, the real part of (25.16) and (25.17) is given by

$$\begin{pmatrix} 1-\cos\eta & \sin\eta \\ -\sin\eta & 1-\cos\eta \end{pmatrix} \begin{pmatrix} \Re[(M_R)_{13}] \\ \Re[(M_R)_{23}] \end{pmatrix} = 0, \tag{25.21}$$

and for the imaginary part, we have

$$\begin{pmatrix} -1-\cos\eta & \sin\eta \\ -\sin\eta & -1-\cos\eta \end{pmatrix} \begin{pmatrix} \Im[(M_R)_{13}] \\ \Im[(M_R)_{23}] \end{pmatrix} = 0. \tag{25.22}$$

Since $\eta = \pi/3$, we have

$$\begin{pmatrix} \frac{3}{4} & \frac{\sqrt{3}}{2} & -\frac{3}{4} \\ -\frac{\sqrt{3}}{4} & \frac{3}{2} & \frac{\sqrt{3}}{4} \\ -\frac{3}{4} & -\frac{\sqrt{3}}{2} & \frac{3}{4} \end{pmatrix} \begin{pmatrix} \Re[(M_R)_{11}] \\ \Re[(M_R)_{12}] \\ \Re[(M_R)_{22}] \end{pmatrix} = 0, \tag{25.23}$$

$$\begin{pmatrix} -\frac{5}{4} & \frac{\sqrt{3}}{2} & -\frac{3}{4} \\ -\frac{\sqrt{3}}{4} & -\frac{1}{2} & \frac{\sqrt{3}}{4} \\ -\frac{3}{4} & -\frac{\sqrt{3}}{2} & -\frac{5}{4} \end{pmatrix} \begin{pmatrix} \Im[(M_R)_{11}] \\ \Im[(M_R)_{12}] \\ \Im[(M_R)_{22}] \end{pmatrix} = 0, \tag{25.24}$$

and

$$\begin{pmatrix} \frac{1}{2} & \frac{\sqrt{3}}{2} \\ -\frac{\sqrt{3}}{2} & \frac{1}{2} \end{pmatrix} \begin{pmatrix} \Re[(M_R)_{13}] \\ \Re[(M_R)_{23}] \end{pmatrix} = 0, \tag{25.25}$$

$$\begin{pmatrix} -\frac{3}{2} & \frac{\sqrt{3}}{2} \\ -\frac{\sqrt{3}}{2} & -\frac{3}{2} \end{pmatrix} \begin{pmatrix} \Im[(M_R)_{13}] \\ \Im[(M_R)_{23}] \end{pmatrix} = 0. \tag{25.26}$$

Only those parts for which the coefficient matrix determinant is zero will yield non-trivial solutions. Thus, we are left with only (25.23), the determinant of whose coefficient matrix is zero. Consequently, we have arbitrary solutions, with $\Re[(M_R)_{11}] = \Re[(M_R)_{22}] \equiv M_{R1}$ and $\Re[(M_R)_{12}]$ equal to zero. Additionally, the solution to (25.18) leads to $\Im[(M_R)_{33}]$ being zero, and we are left with only one relevant parameter $\Re[(M_R)_{33}] \equiv M_{R3}$.

We can clearly see the impact of GCP symmetry on our right-handed neutrinos mass matrix M_R, which now becomes

$$M_R = \begin{pmatrix} M_{R1} & 0 & 0 \\ 0 & M_{R1} & 0 \\ 0 & 0 & M_{R3} \end{pmatrix}. \tag{25.27}$$

To proceed further, we need to consider spontaneous symmetry breaking (SSB). After SSB, both Higgs doublets acquire vacuum expectation values (VEVs), which can be written as

$$\langle \phi_1 \rangle = \frac{1}{\sqrt{2}} \begin{pmatrix} 0 \\ v_1 \end{pmatrix}, \quad \langle \phi_2 \rangle = \frac{1}{\sqrt{2}} \begin{pmatrix} 0 \\ v_2 e^{i\gamma} \end{pmatrix}, \tag{25.28}$$

with the constraint that $v = \sqrt{v_1^2 + v_2^2} \approx 245$ GeV, where v is the Standard Model VEV. Consequently, now the charged leptons can get their mass, and the mass matrix for them becomes

$$M_l = \frac{1}{\sqrt{2}} (v_1 \Delta_1 + e^{i\gamma} v_2 \Delta_2), \tag{25.29}$$

and Dirac mass matrix for neutrinos is given by

$$M_D = \frac{1}{\sqrt{2}} (v_1 \Lambda_1 + e^{-i\gamma} v_2 \Lambda_2), \tag{25.30}$$

The charged lepton mass matrix M_l is a complex matrix which can be diagonalized by the biunitary transformation by using two 3×3 unitary matrices, denoted by U_L and U_R as

$$M_l = U_L m_d U_R^\dagger, \tag{25.31}$$

where $m_d = diag(m_e, m_\mu, m_\tau)$ is diagonal matrix with entries giving mass eigenvalues of electron, muon, and tau, respectively. Using this we can have

$$U_L^\dagger M_l M_l^\dagger U_L = m_d^2. \tag{25.32}$$

For convenience, we want to work in the basis where charged lepton mass matrix is diagonal, we can use the matrix U_L to rotate M_D into the required basis, i.e.,

$$M_D^{'} = U_L^\dagger M_D. \tag{25.33}$$

We can use this rotated Dirac mass matrix and right-handed neutrinos mass matrix M_R to obtain effective light neutrino mass matrix M_ν by using the Type-I seesaw mechanism. The effective light neutrino mass matrix can be written as

$$M_\nu = -M_D^{'} M_R^{-1} (M_D^{'})^T, \tag{25.34}$$

which can be diagonalized by 3×3 unitary matrix U as

$$U^\dagger M_\nu U^* = m, \tag{25.35}$$

where $m_{ik} = m_i \delta_{ik}$, $m_i > 0$ $(i, k = 1, 2, 3)$.

Now, we will proceed to numerically study the masses of charged leptons and neutrino phenomenology by varying various parameters to obtain the permissible parameter space within this model.

25.4 Numerical Analysis and Discussion

For the numerical study, we have varied the free parameters using a uniform random number generator. The VEV-phase γ is varied in the range from 0 to 2π, while the VEV v_1 is varied in the range $(10 - 17)$ GeV (note that $v_2 = \sqrt{v^2 - v_1^2}$), and the masses of right-handed neutrinos M_{R1} and M_{R3} are varied in the range 10^{11} to 10^{13} GeV and 1.1×10^{13} to 10^{15} GeV, respectively.

25.4.1 Charged Lepton Masses

To determine the masses of charged leptons, we varied the Yukawa coupling parameters associated with the Δ_1 and Δ_2 Yukawa matrices within the ranges shown in the second column of Table 25.1. The squared masses of charged leptons $(m_e^2, m_\mu^2, m_\tau^2)$ were obtained by calculating the eigenvalues of the mass matrix $M_l M_l^\dagger$. The corresponding eigenvectors were then used to construct the charged lepton mixing matrix U_L. We explored the parameter space of charged leptons to get their masses, for the normal $(m_1 < m_2 < m_3)$ and inverted $(m_3 < m_1 < m_2)$ hierarchies of neutrinos. The benchmark points obtained are shown in second and third columns of Table 25.2, and corresponding masses of charged leptons (m_e, m_μ, m_τ) obtained are $(0.511, 105.66, 1776.84)$ eV for both NH and IH of neutrinos.

Table 25.1 Ranges of Yukawa coupling parameters used in numerical analysis for charged leptons (second column) and for neutrinos (columns four and five)

Charged leptons		Neutrinos		
Yukawa coupling parameter	Range	Yukawa coupling parameter	Range	
			NH	IH
x_{11}	$(2-3)\times 10^{-5}$	y_{11}	$(6-8)\times 10^{-2}$	$(8-9)\times 10^{-2}$
x_{12}	$(6-8)\times 10^{-5}$	y_{12}	$(2-3)\times 10^{-2}$	$(2-3)\times 10^{-3}$
x_{13}	$(6-7)\times 10^{-3}$	y_{13}	$(6-8)\times 10^{-1}$	$(1-2)\times 10^{-1}$
x_{23}	$(8-9)\times 10^{-3}$	y_{23}	$(1-2)\times 10^{-3}$	$(4-6)\times 10^{-1}$
x_{31}	$(4-6)\times 10^{-4}$	y_{31}	$(1-2)\times 10^{-2}$	$(3-5)\times 10^{-2}$
x_{32}	$(1-2)\times 10^{-4}$	y_{32}	$(4-6)\times 10^{-2}$	$(6-7)\times 10^{-2}$

Table 25.2 The benchmark points yielding correct charged lepton masses (columns two and three) and corresponding parameters in neutrino sector yielding neutrino oscillation parameters within 3σ experimental range [13] (columns five and six)

Parameters (Units)	NH	IH	Parameters (Units)	NH	IH
x_{11}	2.6924×10^{-5}	2.6924×10^{-5}	y_{11}	6.0098×10^{-2}	8.0066×10^{-2}
x_{12}	7.9938×10^{-5}	7.9938×10^{-5}	y_{12}	2.3348×10^{-2}	2.6014×10^{-3}
x_{13}	6.3827×10^{-3}	6.3827×10^{-3}	y_{13}	6.9357×10^{-1}	1.5228×10^{-1}
x_{23}	8.0280×10^{-3}	8.0280×10^{-3}	y_{23}	1.7912×10^{-3}	4.5311×10^{-1}
x_{31}	5.9456×10^{-4}	5.9456×10^{-4}	y_{31}	1.6556×10^{-2}	3.9092×10^{-2}
x_{32}	1.0655×10^{-4}	1.0656×10^{-4}	y_{32}	4.8802×10^{-2}	6.1573×10^{-2}
v_1 (GeV)	11.00	15.30	M_{R1} (GeV)	3.06×10^{12}	4.38×10^{12}
v_2 (GeV)	244.75	244.52	M_{R3} (GeV)	2.75×10^{14}	6.14×10^{14}
γ (°)	182.27	189.80			

25.4.2 Neutrino Oscillation Parameters

For neutrino oscillation parameters, we varied the Yukawa coupling parameters of the Yukawa matrices Λ_1 and Λ_2 in the ranges shown in the fourth and fifth columns of Table 25.1 for normal and inverted hierarchies, respectively. To obtain neutrino masses, we first rotated the Dirac mass matrix of neutrinos, M_D, using U_L to transform it to the basis where charged lepton mass matrix is diagonal. Subsequently, we obtained the eigenvalues of the matrix M_ν given in (25.34). These eigenvalues provide the masses of the neutrinos, and the neutrino mixing matrix U is determined using the corresponding eigenvectors. Utilizing this mixing matrix, we further calculated the neutrino oscillation parameters. The relevant formulas used are (i) for neutrino mixing angles we have

$$\sin^2\theta_{12} = \frac{|U_{12}|^2}{1-|U_{13}|^2},\ \sin^2\theta_{13} = |U_{13}|^2,\ \sin^2\theta_{23} = \frac{|U_{23}|^2}{1-|U_{13}|^2},$$

(ii) for Jarlskog CP invariant $J_{\rm CP}$ [14–16], we have

$$J_{\rm CP} = \Im[U_{11}U_{12}^*U_{21}^*U_{22}],$$

and for Dirac phase δ we have

$$\delta = \arcsin\left[J_{\rm CP}/(\sin\theta_{12}\cos\theta_{12}\sin\theta_{23}\cos\theta_{23}\sin\theta_{13}\cos^2\theta_{13})\right],$$

(iii) for effective Majorana mass we have, $|m_{\beta\beta}| = |(M_\nu)_{11}| = |\sum_{i=1}^{3} U_{ei}^2 m_i|$.

The benchmark points obtained from the parameter scan, satisfying the neutrino oscillation data at 3σ experimental range [13] for NH and IH, are presented in columns five and six of Table 25.2, respectively. The corresponding values of neutrino oscillation parameters $(\theta_{13}, \theta_{12}, \theta_{23}, \Delta m_{21}^2, \Delta m_{31}^2)$ are $(8.82°, 32.1°, 42.96°, 7.00\times10^{-5}\,\text{eV}^2, 2.50\times10^{-3}\,\text{eV}^2)$ for NH and $(8.21°, 31.9°, 48.12°, 7.44\times10^{-5}\,\text{eV}^2, 2.47\times10^{-3}\,\text{eV}^2)$ for IH. The values of $|m_{\beta\beta}|$ corresponding to these benchmark points are 0.01201 eV and 0.04766 eV for NH and IH, respectively.

The correlation plots obtained are shown in Fig. 25.1, where the sharp predictions from this model are evident. In Fig. 25.1a and b, plots for the effective Majorana mass $|m_{\beta\beta}|$ with the sum of neutrino masses $\sum_i m_i$ $(i = 1, 2, 3)$ are presented for NH and IH of neutrinos, respectively. Both exhibit a linear correlation, but IH displays a better correlation compared to NH. The predicted range of $|m_{\beta\beta}|$ lies in the sensitivity range of the current and future neutrinoless double beta decay $(0\nu\beta\beta)$ experiments (SuperNEMO [17], KamLAND-Zen [18], NEXT [19, 20], nEXO [21]) and can be tested. As per the recent results obtained from KamLAND-Zen [18] experiment, the IH from this model may be ruled out; however, additional confirmation from other experiments is required. Furthermore, the sum of neutrino masses, $\sum_i m_i$, remains below the stringent upper limit established by Planck data [TT, TE, EE+lowE+lensing+BAO]. This data imposes a stringent constraint of 0.12 eV at the 95% confidence level (CL) [22] for the sum of neutrino masses $(\sum_i m_i)$. The Fig. 25.1c shows that for NH we have values of Dirac CP phase δ approximately from $(-10°$ to $10^0)$, allowing for solutions of both types, i.e., CP-violating and CP-conserving. Conversely, for IH, the values of δ are restricted to two small regions around $\pm40°$, indicating that only CP-violating solutions are available (see Fig. 25.1d). Another prediction from this model pertains to the octant of the neutrino mixing angle θ_{23}. In the case of NH, θ_{23}, resides in the lower octant with values approximately 43° (Fig. 25.1e), while for IH, θ_{23} is situated in the upper octant with values approximately 48° (Fig. 25.1f). For NH, the VEV-phase γ is allowed to take values in the range from 177° to 183° (Fig. 25.1e), while for IH, γ is constrained to only two values, approximately 170° and 190° (Fig. 25.1f).

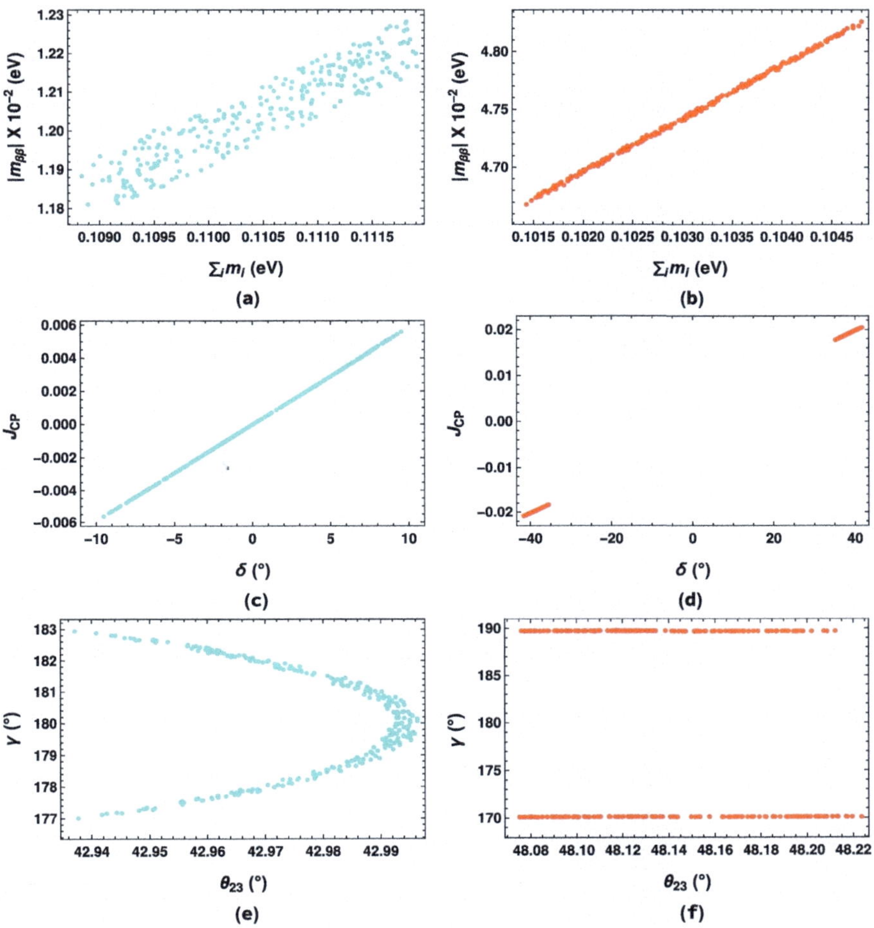

Fig. 25.1 Correlation plots are presented depicting the relationships between the effective Majorana mass $|m_{\beta\beta}|$ and the sum of neutrino masses $\sum_i m_i$, the Jarlskog invariant J_{CP} and the CP-violating phase δ as well as for the VEV-phase γ and the neutrino mixing angle θ_{23} for both normal (first column) and inverted (second column) hierarchies

25.5 Conclusions

We have developed a model that utilizes the concept of GCP symmetry in the 2HDM and extends it to the leptonic Yukawa sector within the Type-I seesaw framework. As a result of the GCP symmetry, there is a substantial reduction in the number of free parameters in this model. In the softly broken CP3 symmetric scalar potential of this model, we have only six real parameters. In the leptonic Yukawa coupling matrices, we now have 12 independent real parameters, six in each of the charged lepton and neutrino sectors. This reduction in parameters enables us to obtain precise predictions from the model regarding neutrino phenomenology. For the effective Majorana mass, we find $|m_{\beta\beta}| \in (1.18 - 1.23) \times 10^{-2}$ eV for NH and $|m_{\beta\beta}| \in (4.67 - 4.83) \times 10^{-2}$ eV for IH, which can be tested in current and future $0\nu\beta\beta$ decay experiments. Furthermore, this model predicts distinct outcomes for CP-violation and the atmospheric mixing angle θ_{23} based on the neutrino mass hierarchies. For NH, both CP-violating and CP-conserving solutions are allowed, with δ approximately ranging from ($-10°$ to $10°$) and $\theta_{23} \approx 43°$ in the lower octant. In the case of IH, CP-violation is inevitable with δ around $\pm 40°$ and $\theta_{23} \approx 48°$ in the upper octant. Thus, this model has revealed a rich and unique neutrino phenomenology.

Acknowledgements Tapender acknowledges the financial support provided by Central University of Himachal Pradesh. The authors, also, acknowledge Department of Physics and Astronomical Science for providing necessary facility to carry out this work.

References

1. P. Minkowski, Phys. Lett. B **67**, 421–428 (1977)
2. T. Yanagida, Conf. Proc. C **7902131**, 95–99 (1979)
3. S.L. Glashow, N.A.T.O. Sci, Ser. B **61**, 687 (1980)
4. M. Gell-Mann, P. Ramond, R. Slansky, Conf. Proc. C **790927**, 315–321 (1979)
5. R.N. Mohapatra, G. Senjanovic, Phys. Rev. Lett. **44**, 912 (1980)
6. A.D. Sakharov, Pisma. Zh. Eksp. Teor. Fiz. **5**, 32–35 (1967)
7. T.D. Lee, Phys. Rev. D **8**, 1226–1239 (1973)
8. P.M. Ferreira, H.E. Haber, J.P. Silva, Phys. Rev. D **79**, 116004 (2009)
9. P.M. Ferreira, J.P. Silva, Eur. Phys. J. C **69**, 45–52 (2010)
10. Tapender, S. Kumar, S. Verma, Phys. Rev. D **109**(1), 015004 (2024)
11. A. Crivellin, A. Kokulu, C. Greub, Phys. Rev. D **87**(9), 094031 (2013)
12. G. Ecker, W. Grimus, H. Neufeld, J. Phys. A **20**, L807 (1987)
13. P.F. de Salas, D.V. Forero, S. Gariazzo, P. Martínez-Miravé, O. Mena, C.A. Ternes, M. Tórtola, J.W.F. Valle, JHEP **02**, 071 (2021)
14. C. Jarlskog, Phys. Rev. Lett. **55**, 1039 (1985)
15. S.M. Bilenky, S.T. Petcov, Rev. Mod. Phys. **59**, 671 (1987) [erratum: Rev. Mod. Phys. **61**, 169 (1989); erratum: Rev. Mod. Phys. **60**, 575–575 (1988)]
16. P.I. Krastev, S.T. Petcov, Phys. Lett. B **205**, 84–92 (1988)
17. A.S. Barabash, J. Phys. Conf. Ser. **375**, 042012 (2012)
18. S. Abe, et al. [KamLAND-Zen], Phys. Rev. Lett. **130**(5), 051801 (2023)
19. F. Granena, et al. [NEXT]. arXiv:0907.4054 [hep-ex]
20. J.J. Gomez-Cadenas et al., NEXT. Adv. High Energy Phys. **2014**, 907067 (2014)
21. C. Licciardi [nEXO], J. Phys. Conf. Ser. **888**(1), 012237 (2017)
22. N. Aghanim, et al. [Planck], Astron. Astrophys. **641**, A6 (2020) [erratum: Astron. Astrophys. **652**, C4 (2021)]

Chapter 26
Perturbation to Golden Ratio Mixing and Charged Lepton Flavor Violation

Victoria Puyam and Ngangkham Nimai Singh

Abstract A_5 discrete symmetry group is used to construct a neutrino mass model that can reproduce deviation from the exact golden ratio mixing through the contribution from the charged lepton sector in the linear seesaw framework. We can obtain all the values of the neutrino oscillation parameters, including the Dirac CP-violating phase, within the current experimental bounds. The model predicts only normal neutrino mass ordering. Furthermore, we also study the charged lepton flavor violation (cLFV) process $\mu \to e + \gamma$. When the quasi-Dirac neutrino masses are in the (1–10) TeV range, this cLFV process is within reach of future planned experiments.

26.1 Introduction

Present knowledge on neutrino physics comes from neutrino oscillation experiments [1–7], neutrinoless double beta decay experiments [8–10], the end point of tritium beta decay experiment [11] and cosmological observations [12]. The precision measurement of the leptonic mixing angles in the Pontecorvo-Maki-Nakagawa-Sakata (PMNS) matrix U_{PMNS} shows that θ_{23} and θ_{12} are large and θ_{13} is small but non-zero. The upper bound on the sum of the three neutrino masses $\sum m_i \leq 0.12$ eV is also observed from the Planck cosmological observations [12]. The current global fit on neutrino oscillation data is given in [13]. In light of current experimental results, there is a need for serious consideration in the lepton mixing sector as the presently available exact lepton mixing forms that predict vanishing θ_{13} like Tri-Bimaximal (TBM) [14, 15], Golden ratio (GR) mixings [16, 17] are excluded by experiments.

These neutrino experiment data clearly hints at physics beyond the Standard Model (BSM). However, they still have shortcomings like octant of θ_{23}, neutrino mass ordering, CP violation in the lepton sector, baryon asymmetry of the universe,

V. Puyam (✉) · N. N. Singh
Department of Physics, Manipur University, Imphal 795003, India
e-mail: victoriapuyam@gmail.com

N. N. Singh
Research Institute of Science and Technology, Manipur University, Imphal 795003, India

J. W. F. Valle et al. (eds.), *Proceedings of the International Conference on Future Prospects in Neutrino and Astroparticle Physics*, Springer Proceedings in Physics 361,
https://doi.org/10.1007/978-981-97-7441-8_27

etc. So, many new theories and experiments are pursued to study BSM. One such exciting approach is the charged lepton flavor violation (cLFV). These processes are highly suppressed in the Standard Model (SM). In the neutrino mass model based on the type I seesaw framework, the cLFV processes are often suppressed by the heavy masses of right-handed neutrino introduced to account for the smallness of active neutrino masses. However, sizeable structures of electroweak currents that contribute to the cLFV process, can be explained in a low-scale seesaw-like linear or inverse seesaw. There are several presently available and planned experiments for cLFV decays ($l_i \rightarrow l_j\gamma$) [18], and three-body decays ($l_i \rightarrow l_j l_k l_k$) [19]. In this work, we study the most sensitive cLFV decay $\mu \rightarrow e + \gamma$ with a neutrino mass model based on a linear seesaw in the basis where the charged lepton mass matrix is non-diagonal.

On the other hand, the lepton masses and mixing patterns are often investigated by using several discrete symmetry groups such as A_4, A_5, S_3, S_4, $\Delta(96)$, etc. [16, 17, 20–31]. Out of these groups, the A_5 symmetry group is used to accommodate golden ratio (GR) mixing in a neutrino mass matrix. Here, it is interesting to note that in the GR mixing scenario, θ_{23} is maximal and $\theta_{13} = 0$, while the solar mixing angle, θ_{12} is governed by the golden ratio, $\phi_g = (1 + \sqrt{5})/2$ as $\tan\theta_{12} = 1/\phi_g$ with $\sin^2\theta_{12} \approx 0.276$ in GR1 mixing or $\cos\theta_{12} = \phi_g/2$ with $\sin^2\theta_{12} \approx 0.345$ in GR2 mixing [17, 32]. Both the GR mixing patterns are excluded by current data. As a result, the neutrino mass models based on GR mixing need modifications. These deviations are obtained either by adding the contribution from the charged lepton sector [21, 33–35] or by considering the vacuum alignment correction of the A_5 triplets or A_5 quintuplet [16, 17] or by including the extra contribution in the neutrino sector [33, 36–40].

In this work, we aim to construct a lepton mass model based on $A_5 \times Z_5 \times Z_3$ symmetry group that can produce the necessary deviation from GR1 mixing to accommodate the current neutrino masses and mixing data. Here, the light neutrino masses are generated through the linear seesaw mechanism, and its mass matrix retains the GR1 mixing pattern. On the other hand, a specific pattern of charged lepton mass matrix can perturb GR1 mixing and control the neutrino mixing pattern. It is interesting to note that there are earlier models with different structures of charged lepton mass matrix [21, 34, 35]. In the present work, we employ charged lepton correction in 12 sector, which is quite different from the charged lepton correction adopted in other works mentioned above. The way the charged lepton mass matrix is structured has a significant impact on the lepton mixing matrix [21, 34]. Here, unlike the earlier model based on A_5 symmetry, our model predicts only the normal ordering of neutrino masses. Again, our model's specific structure of the charged lepton mass matrix allows us to study a distinct lepton mixing matrix and its related phenomenology. One such interesting prediction of the model is that the values of θ_{23} are found in the lower octant region for normal hierarchy (NH). In addition, the model also shows a strong correlation between θ_{13} and other mixing angles and Dirac CP-violating phase δ_{CP}, as they can be obtained in their current bounds by taking the experimental values of the θ_{13} as the input value. Furthermore, we also discuss the implication of cLFV ($\mu \rightarrow e + \gamma$) process to our model.

The paper is organized as follows: In Sect. 26.2, we describe the model and its particle contents in the underlying symmetry group and lepton mixing pattern arising from the model. Section 26.3 deals with numerical analysis, diagonalization, and some details of cLFV. The results of our numerical analysis are given in Sect. 26.4. In Sect. 26.5, we give the conclusion of the present work.

26.2 Description of the Model

The lepton flavor model considered in the present work is based on A_5 symmetry group. Here, the standard model (SM) scalar sectors are extended by introducing two Higgs doublet h_u and h_d and seven Higgs singlets ϕ_1, ϕ_2, ϕ_3, ϕ_5, χ_1, ξ_1 and ξ_2. In the lepton sector, the right-handed neutrino fields N_R and S_R are considered along with the SM left-handed lepton doublet field l and the right-handed lepton singlet fields e_R, μ_R and τ_R. These right-handed neutrino fields contribute to the masses of light-active neutrinos through the linear seesaw mechanism. In this model, the A_5 symmetry group is further supplemented by the $Z_5 \times Z_3$ group to achieve the desired pattern of the lepton mass matrix. Inclusion of this additional symmetry restricts the introduction of the unwanted terms in the charged lepton mass matrix, which are otherwise allowed by A_5 symmetry, and hence necessary to obtain the specific pattern of the charged lepton mass matrix. The transformation properties of the fields used in our model under the symmetry group so far considered are provided in Table 26.1. It is important to note that the Z_5 and Z_3 charges used in our model are in additive notation.

The Yukawa Langrangian terms for the charged leptons and neutrino which are invariant under $A_5 \times Z_5 \times Z_3$ transformation are given below

$$
\begin{aligned}
-L_y = {} & y_1(\bar{l}\,\frac{\phi_1}{\lambda^3} e_R) h_d \xi_1^2 + y_2(\bar{l}\,\frac{\phi_3}{\lambda^2}\mu_R) h_d \xi_1 + y_3(\bar{l}\,\frac{\phi_2}{\lambda}\tau_R) h_d + y_4(\bar{l}\,\frac{\phi_2}{\lambda^3} e_R) h_d \xi_1 \xi_2 \\
& + y_5(\bar{l}\,\frac{\phi_1}{\lambda^2}\mu_R) h_d \xi_2 + \frac{y_a}{\lambda}(\bar{l}\, N_R \phi_5)_1 h_u + \frac{y_b}{\lambda}\xi_2(\bar{l}\, S_R)_1 h_u + M(\bar{N}_R^c \chi_1 S_R).
\end{aligned}
\tag{26.1}
$$

Table 26.1 Transformation properties of various fields under symmetry group considered in the model

Fields	l	e_R	μ_R	τ_R	N_R	S_R	$H_{u,d}$	ϕ_1	ϕ_2	ϕ_3	ϕ_5	ξ_1	ξ_2	χ_1	η_1	η_2
A_4	3	1	1	1	3	3	1	3	3	3	5	1	1	1	1	1
Z_5	−1	−1	1	0	1	1	0	0	1	4	0	1	0	−2	0	0
Z_3	2	0	−1	1	0	−1	0	0	0	0	1	−1	−1	1	1	2

The vacuum expectation value (VEV) pattern for the A_5 scalar triplets and the A_5 scalar quintuplet used in the model are similar with [16, 21]. The VEV patterns for the A_5 scalar triplets and quintuplets considered in our model, can be obtained from the superpotential given below

$$W = M_o\eta_1\eta_2 + M_1\phi_1^2 + M_2(\phi_2\phi_3) + g(\phi_1\phi_2\phi_3) + g_1\eta_1(\phi_5^2) + g_2(\phi_5^3)_1 + g_3(\phi_5^3)_2 + \frac{g_4}{3}\eta_1^3 + \frac{g_5}{3}\eta_2^3 + \cdots \tag{26.2}$$

where η_1 and η_2 are the driving field.

It is to be noted that the suffices 1 and 2 in the terms $g_2(\phi_5^3)_1$ and $g_2(\phi_5^3)_2$ which appear in W represent two independent cubic invariants that can be built from a quintuplet in A_5 multiplication. Again, terms like $((\phi_1\phi_1)_3\phi_1)_1$ are omitted in the superpotential since the contraction $(\phi_1\phi_1)_3$ vanishes. The detailed calculations are shown in [16, 21].

Since the VEVs of the scalars are realized, the mass matrix for the leptons can be calculated after spontaneous symmetry breaking. The charged lepton mass matrix obtained has a unique structure as shown below

$$M_l = \begin{pmatrix} \frac{1}{\lambda^3}y_1v_dv_1u_1^2 & \frac{1}{\lambda^2}y_5v_dv_1u_2 & 0 \\ \frac{1}{\lambda^3}y_4v_dv_3u_2u_1 & \frac{1}{\lambda^2}y_2v_dv_3u_1 & 0 \\ 0 & 0 & \frac{1}{\lambda}y_3v_dv_2 \end{pmatrix} = \begin{pmatrix} m_{11} & m_{12} & 0 \\ m_{21} & m_{22} & 0 \\ 0 & 0 & m_{33} \end{pmatrix}. \tag{26.3}$$

In the neutrino sector, after the electroweak symmetry breaking, the neutrino mass matrix in the basis (ν_L, N_R^c, S_R) has the structure

$$M_\nu = \begin{pmatrix} 0 & M_D & M_L \\ M_D^T & 0 & M \\ M_L^T & M^T & 0 \end{pmatrix} \tag{26.4}$$

After block diagonalization, the light active neutrino mass matrix arising from the linear seesaw mechanism [41, 42] has the form

$$m_\nu = M_D(M_LM^{-1})^T + M_LM^{-1}M_D^T \tag{26.5}$$

where sub-matrices M_L, M_D and M_R have the following patterns,

$$M_L = m\begin{pmatrix} 1 & 0 & 0 \\ 0 & 0 & 1 \\ 0 & 1 & 0 \end{pmatrix}, \; M_d = \begin{pmatrix} \frac{2}{3}(b+c) & \frac{b}{\sqrt{2}} & \frac{b}{\sqrt{2}} \\ \frac{b}{\sqrt{2}} & c & -\frac{1}{3}(b+c) \\ \frac{b}{\sqrt{2}} & -\frac{1}{3}(b+c) & c \end{pmatrix} \text{ and } M_R = \begin{pmatrix} M & 0 & 0 \\ 0 & 0 & M \\ 0 & M & 0 \end{pmatrix}. \tag{26.6}$$

where $\frac{y_bu_2}{\lambda} = m$, $\frac{y_buv_u}{\lambda} = a$, $\frac{y_apv_u}{\lambda} = b$ and $\frac{y_aqv_u}{\lambda} = c$, and.

Using (26.6) in (26.5), the light active neutrino mass matrix has the following resultant structure

$$m_\nu = \begin{pmatrix} (4(b+c)m)/(3M) & (\sqrt{2}bm)/M & (\sqrt{2}bm)/M \\ (\sqrt{2}bm)/M & (2cm)/M & (2(-b-c)m)/(3M) \\ (\sqrt{2}bm)/M & (2(-b-c)m)/(3M) & (2cm)/M \end{pmatrix} \quad (26.7)$$

On the other hand, the six doubly degenerate quasi-Dirac heavy neutrinos have the mass matrix structure

$$M_{N^-} = -M_R - M_D^T M_D M_R^{-1} + \frac{1}{2}[M_D M_R^{-1} M_L^T + M_L^T M_R^{-1} M_D^T] \quad (26.8)$$

$$M_{N^+} = M_R + M_D^T M_D M_R^{-1} + \frac{1}{2}[M_D M_R^{-1} M_L^T + M_L^T M_R^{-1} M_D^T] \quad (26.9)$$

So, the physical neutrino spectrum consists of three active neutrinos and six nearly degenerate quasi-Dirac heavy sterile neutrinos. Here, it can be seen that in the limit $M_L \to 0$, the active light neutrino masses become massless. In the model, light active neutrino masses are explained by considering $m \sim 40$ eV and $M \sim 10$ TeV, consistent with cLFV.

26.3 Numerical Analysis and Diagonalisation

The charged leptons mass matrix obtained in (26.3) can be diagonalized through the use of a rotation matrix R_{12}, which diagonalizes the charge lepton mass matrix as

$$R_{12}^\dagger M_L M_L^\dagger R_{12} = R_{12}^\dagger h R_{12} = \begin{pmatrix} m_e^2 & 0 & 0 \\ 0 & m_\mu^2 & 0 \\ 0 & 0 & m_\tau^2 \end{pmatrix},$$

$$\text{where, } R_{12} = \begin{pmatrix} \cos\theta & \sin\theta e^{i\psi} & 0 \\ -\sin\theta e^{-i\psi} & \cos\theta & 0 \\ 0 & 0 & 1 \end{pmatrix} \text{ and } \tan\theta \approx \frac{m_{12}}{m_{22}} \quad (26.10)$$

On the other hand, the resultant light active neutrino mass matrix given in (26.7) can be diagonalized by U_{GR1} as

$$U_{GR1}^T m_\nu U_{GR1} = \text{diag}(m_1, m_2, m_3) \quad (26.11)$$

where m_i when $i = 1, 2, 3$ are neutrino mass eigenvalues, and the mixing matrix U_{GR1} is given by

$$U_{GR1} = \begin{pmatrix} \frac{\sqrt{\phi_g}}{\sqrt[4]{5}} & -\frac{\sqrt{\frac{1}{\phi_g}}}{\sqrt[4]{5}} & 0 \\ \frac{\sqrt{\frac{1}{\phi_g}}}{\sqrt{2}\sqrt[4]{5}} & \frac{\sqrt{\phi_g}}{\sqrt{2}\sqrt[4]{5}} & -\frac{1}{\sqrt{2}} \\ \frac{\sqrt{\frac{1}{\phi_g}}}{\sqrt{2}\sqrt[4]{5}} & \frac{\sqrt{\phi_g}}{\sqrt{2}\sqrt[4]{5}} & \frac{1}{\sqrt{2}} \end{pmatrix}, \quad \phi_g = \frac{1+\sqrt{5}}{2} \tag{26.12}$$

The PMNS leptonic mixing matrix U for this scenario can be constructed as

$$U = R_{12}^{\dagger} U_{GR1} = \begin{pmatrix} \frac{-\sqrt{2}e^{-i\psi}\sin\theta\sqrt{\frac{1}{\phi_g}} - 2\cos\theta\sqrt{\phi_g}}{\sqrt{2}\sqrt[4]{5}} & \frac{2\cos\theta\sqrt{\frac{1}{\phi_g}} - \sqrt{2}e^{-i\psi}\sin\theta\sqrt{\phi_g}}{\sqrt{2}\sqrt[4]{5}} & \frac{e^{i\psi}\sin\theta}{\sqrt{2}} \\ -\frac{\sqrt{2}\cos\theta\sqrt{\frac{1}{\phi_g}} + 2e^{i\psi}\sin\theta\sqrt{\phi_g}}{\sqrt{2}\sqrt[4]{5}} & \frac{2e^{i\psi}\sin\theta\sqrt{\frac{1}{\phi_g}} - \sqrt{2}\cos\theta\sqrt{\phi_g}}{\sqrt{2}\sqrt[4]{5}} & \frac{\cos\theta}{\sqrt{2}} \\ -\frac{\sqrt{\frac{1}{\phi_g}}}{\sqrt{2}\sqrt[4]{5}} & -\frac{\sqrt{\phi_g}}{\sqrt{2}\sqrt[4]{5}} & -\frac{1}{\sqrt{2}} \end{pmatrix} \tag{26.13}$$

The model parameters θ and ψ can be expressed in terms of lepton mixing angles as

$$s_{13}^2 = \frac{1}{2}\sin^2\theta, \quad s_{23}^2 = \frac{\cos^2\theta}{2-\sin^2\theta}, \quad s_{12}^2 = \frac{\sqrt{2}\cos\psi\sin 2\theta + \phi_g\sin^2\theta + \frac{2\cos^2\theta}{\phi_g}}{\sqrt{5}(2-\sin^2\theta)} \tag{26.14}$$

The mixing angles θ_{23} and θ_{12} are then related to reactor angleθ_{13} as

$$s_{23}^2 = \frac{1-2\sin^2\theta_{13}}{2(1-\sin^2\theta_{13})} \text{ and } s_{12}^2 = \frac{2\cos\psi\sin\theta_{13}\sqrt{1-2\sin^2\theta_{13}} + \phi_g\sin^2\theta_{13} + \frac{1-2\sin^2\theta_{13}}{\phi_g}}{\sqrt{5}(2-\sin^2\theta_{13})} \tag{26.15}$$

Furthermore, the Jarlskorg invariant J_{CP} for this scenario can be extracted from the relation

$$J_{CP} = \mathrm{Im}(U_{11}^* U_{23}^* U_{13} U_{21}) = \frac{\cos\theta\sin\theta\sin\psi}{2\sqrt{10}} = \frac{\sin\theta_{13}\sqrt{1-2\sin^2\theta_{13}}\sin\psi}{2\sqrt{5}} \tag{26.16}$$

While the Dirac CP violating δ can be obtained by comparing (26.16) with the definition of J_{CP} in the standard parametrisation [43],

$$J_{CP} = \frac{1}{8}\sin 2\theta_{12}\sin 2\theta_{23}\sin 2\theta_{13}\cos\theta_{12}\sin\delta \tag{26.17}$$

As shown in (26.13), the contribution from the charged lepton sector plays a crucial role in explaining leptonic mixing parameters; in contrast, the contribution from the neutrino mass model parameters b and c to the lepton mixing matrix is trivial. However, the information for light neutrino masses can be explained by the neutrino mass model parameters by adjusting their values to fix the mass square differences (Δm_{21}^2 and Δm_{31}^2) to their current experimental bound and the sum of the absolute neutrino masses $\sum m_i \leq 0.12$ eV. Here, we also found that for the resultant pattern of m_ν, no data points can simultaneously explain lepton mixing angles and masses for an inverted hierarchy case. As a result, our model only supports NH.

26.3.1 Charged Lepton Flavor Violation (cLFV)

This section discusses the charged lepton flavor violation (cLFV) process, especially $\mu \to e\gamma$ decay, and its implication through our model. Here, in the linear seesaw scenario, the decay process is induced through the exchange of the nine neutrinos (consisting of three active light neutrinos and six heavy neutrinos) coupled to the charged leptons in the charged current [41]. So, the resultant decay ratio for the $\mu \to e\gamma$ is given by [41]

$$Br\ (\mu \to e\gamma) = \frac{\alpha_W^3 s_W^2}{256\pi^2} \frac{m_\mu^5}{M_W^4} \frac{1}{\Gamma_\mu} |G_{ij}^W|^2, \tag{26.18}$$

$$\text{where,} \quad G_{ij}^W = \sum_{k=1}^{9} K_{ik}^* K_{jk} G_\gamma^W \left(\frac{m_{N_K}}{M_W^2}\right), \quad \text{and}$$

$$G_\gamma^W(x) = \frac{1}{12(1-x)^4}(10 - 43x + 78x^2 - 49x^3 + 18x^3 \ln x + 4x^4). \tag{26.19}$$

Here, K is the effective lepton mixing matrix that describes the CC weak interaction. It can be expressed in rectangular form and has dimensions 3×9, including the exchange of the three light active neutrinos with k = 1, 2, 3 as well as the six mediators, with k = 4, 5, . . . 9. The matrix K in complete form is written as

$$K = (K_L, K_H) \tag{26.20}$$

where K_L and K_H are 3×3 and 3×6 matrices, respectively. The sub-matrices have the forms [42]:

$$K_L = \left(1_{3\times3} - \frac{1}{2} M_D (M^{-1})^T M^{-1} M_D^\dagger\right) U = \left(1_{3\times3} - \frac{1}{2} V V^\dagger\right) U \tag{26.21}$$

$$K_H = \left(-\frac{i}{\sqrt{2}} V, \frac{i}{\sqrt{2}} V\right), \qquad V = M_D (M^{-1})^T \tag{26.22}$$

The $\mu \rightarrow e\gamma$ decay process is the most sensitive of cLFV processes. Here, the CC contribution is sizeable even in the absence of supersymmetry due to the exchange of the six sub-dominantly coupled heavy quasi-Dirac states having masses in the 1–10 TeV range.

26.4 Result

In Sect. 26.2, it is shown in (26.15) that there is correlation between the reactor angle θ_{13} and other two mixing angles (θ_{23} and θ_{12}). So, the values of the θ_{23} and θ_{12} can be predicted by using the values of θ_{13} in 3σ range of the current global fit data as input. Additionally, the rotational matrix parameter ψ which is necessary for explaining the mixing angle θ_{12} is taken in the range (0, 360°). After evaluating and selecting the values of θ_{12} that lie in the current experimental bound, the values of ψ are found to be in a definitive range, whereas the range for the rotational matrix parameter θ can be extracted from (26.14). The allowed region of ψ and mass parameter b and c that can satisfy current data are shown in Fig. 26.1(a) and Fig. 26.1(b). The model predictions for θ_{23}, θ_{12}, and δ in normal hierarchy (NH) case are shown in the correlation plots Fig. 26.2. From Fig. 26.2, it is evident that the present model successfully predicts θ_{23} values in the lower octant range (<45°), which lie well within the 3σ range of the current oscillation data. This outcome is a significant characteristic of our lepton mass model. However, the values of θ_{12} are dispersed throughout the 3σ range. This is due to the relaxation on the constraint imposed on the parameter ψ. The Dirac CP-violating phase δ and the mass squared differences Δm^2_{21} and Δm^2_{31} are also obtained in 3σ range as shown in Fig. 26.2. Thus, the contribution from the specific charged lepton mass matrix predicted by our model can give necessary deviations to exact GR1 mixing, and these deviations are required to explain current lepton mixing data.

The model can also explain current charged lepton masses data. For the charged lepton sector, the current experimental values of SM charged lepton masses

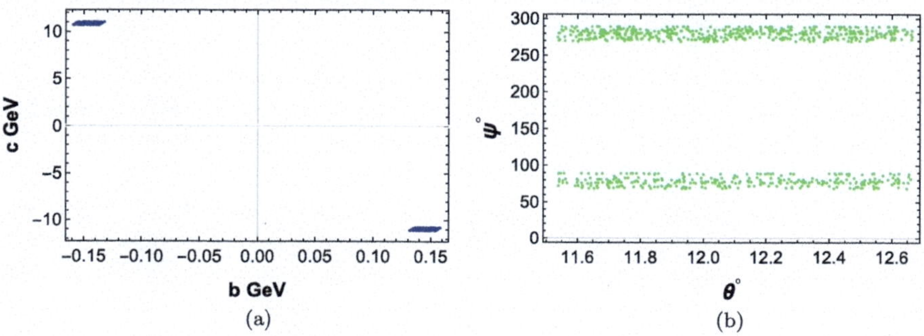

Fig. 26.1 Allowed parameter space of model parameters **a** b and c and **b** θ and ψ

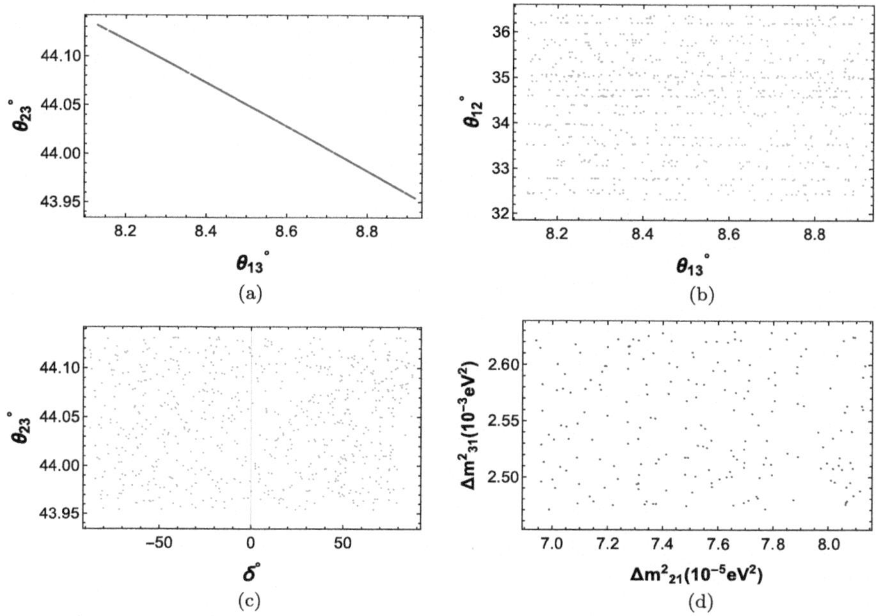

Fig. 26.2 Correlation plot between different neutrino oscillation parameters for NH

$m_e(m_z) = 0.4883266 \pm 0.0000017$ MeV, $m_\mu(m_z) = 102.87267 \pm 0.00021$ MeV and $m_\tau(m_z) = 1747.43 \pm 0.12$ MeV. To discuss this, we choose a representative value of the lepton mass parameter given in (26.8) as

$$h = \begin{pmatrix} 504.756 + 0.0235407i & 2254.9 + 0.049973i & 0 \\ 2254.9 + 0.049973i & 10078.3 - 0.0235407i & 0 \\ 0 & 0 & 3.0532 \times 10^6 \end{pmatrix} \quad (26.23)$$

They yield mass spectrum of $m_e = 0.488327$ MeV, $m_\mu = 102.87272$ MeV and $m_\tau = 1747.34$ MeV. Again, we also analyze the implication of our model on $\mu \to e\gamma$ decay. Here, the model predicts $Br(\mu \to e\gamma)$ in the range $(3.39055 \times 10^{-14} - 3.91148 \times 10^{-14})$ when we consider $m \sim 40$ eV and $M \sim 10$ TeV. In this scenario, the heavy quasi-Dirac neutrinos have masses of 1–10 TeV. This predicted range for the $\mu \to e\gamma$ decay process is in the projected future sensitivity for MEG II (6×10^{-14}) [44].

26.5 Summary and Conclusion

We have presented a lepton mass model which can generate suitable deviation to the GR1 mixing pattern. The model is based on A_5 symmetry, and it is supplemented by additional symmetry $Z_5 \times Z_3$. This model can give deviation from the GR1 mixing pattern and explain the tiny masses of neutrinos through the linear seesaw mechanism. The charged lepton sector of the model predicts a specific pattern of a mass matrix that can perturb the GR1 mixing pattern, resulting in a unique pattern of PMNS matrix U that accurately explains current leptonic mixing data. The specific pattern of m_ν considered in our model predicts NH as the mass pattern of neutrino masses. With this model, we can also express the leptonic mixing angles θ_{12}, θ_{23}, and δ in terms of the model parameter ψ and the mixing angle θ_{13}. The model predicts a lower octant region for θ_{23} while θ_{12} values lie within the 3σ range of current data for the NH case. It can also explain the charged lepton masses, neutrino masses, and mass square differences for neutrinos in their current bounds. Again, the model can explain $Br(\mu \rightarrow e\gamma)$ in the projected future sensitivity for MEG II (6×10^{-14}).

References

1. K. Eguchi et al., First results from KamLAND: evidence for reactor anti-neutrino disappearance. Phys. Rev. Lett. **90**, 021802 (2003)
2. Q.R. Ahmad et al., Direct evidence for neutrino flavor transformation from neutral current interactions in the Sudbury Neutrino Observatory. Phys. Rev. Lett. **89**, 011301 (2002)
3. Y. Fukuda et al., Evidence for oscillation of atmospheric neutrinos. Phys. Rev. Lett. **81**, 1562–1567 (1998)
4. Y. Abe et al., Indication of reactor $\bar{\nu}_e$ disappearance in the double chooz experiment. Phys. Rev. Lett. **108**, 131801 (2012)
5. T. Lasserre, G. Mention, M. Cribier, A. Collin, V. Durand, V. Fischer, J. Gaffiot, D. Lhuillier, A. Letourneau, M. Vivier, Comment on Phys. Rev. Lett. **108**, 191802 (2012); Observation of Reactor Electron Antineutrino Disappearance in the RENO Experiment, 5 (2012)
6. J. Ling, Observation of electron-antineutrino disappearance at Daya Bay. AIP Conf. Proc. **1560**(1), 206–210 (2013)
7. A.B. McDonald, Nobel lecture: the sudbury neutrino observatory: observation of flavor change for solar neutrinos. Rev. Mod. Phys. **88**(3), 030502 (2016)
8. M. Agostini et al., Probing Majorana neutrinos with double-β decay. Science **365**, 1445 (2019)
9. D.Q. Adams et al., Improved limit on neutrinoless double-beta decay in ^{130}Te with CUORE. Phys. Rev. Lett. **124**(12), 122501 (2020)
10. A. Gando et al., Search for majorana neutrinos near the inverted mass hierarchy region with KamLAND-Zen. Phys. Rev. Lett. **117**(8), 082503 (2016). [Addendum: Phys.Rev.Lett. 117, 109903 (2016)]
11. M. Aker et al., Improved upper limit on the neutrino mass from a direct kinematic method by KATRIN. Phys. Rev. Lett. **123**(22), 221802 (2019)
12. N. Aghanim et al., Planck 2018 results. I. Overview and the cosmological legacy of Planck. Astron. Astrophys. **641**, A1 (2020)
13. P.F. de Salas, D.V. Forero, S. Gariazzo, P. Martínez-Miravé, O. Mena, C.A. Ternes, M. Tórtola, J.W.F. Valle, 2020 global reassessment of the neutrino oscillation picture. JHEP **02**, 071 (2021)
14. P.F. Harrison, D.H. Perkins, W.G. Scott, Tri-bimaximal mixing and the neutrino oscillation data. Phys. Lett. B **530**, 167 (2002)
15. Z.-Z. Xing, Nearly tri bimaximal neutrino mixing and CP violation. Phys. Lett. B **533**, 85–93 (2002)
16. F. Feruglio, A. Paris, The golden ratio prediction for the solar angle from a natural model with a5 flavour symmetry. JHEP **03**, 101 (2011)

17. G.-J. Ding, L.L. Everett, A.J. Stuart, Golden ratio neutrino mixing and A_5 flavor symmetry. Nucl. Phys. B **857**, 219–253 (2012)
18. R. Bernstein, P.S. Cooper, Charged lepton flavor violation: an experimenter's guide. Phys. Rep. **532**(2), 27–64 (2013). Charged Lepton Flavor Violation: An Experimenter's Guide
19. S. Mihara, J. Miller, P. Paradisi, G. Piredda, Charged lepton flavor-violation experiments. Ann. Rev. Nuclear Particle Sci. **63**(1), 531–552 (2013)
20. L.L. Everett, A.J. Stuart, Icosahedral (A(5)) family symmetry and the golden ratio prediction for solar neutrino mixing. Phys. Rev. D **79**, 085005 (2009)
21. A.E. Cárcamo Hernández, I. de Medeiros Varzielas, An A5 inverse seesaw model with perturbed golden ratio mixing. Nucl. Phys. B **985**, 116031 (2022)
22. H. Ishimori, T. Kobayashi, H. Ohki, Y. Shimizu, H. Okada, M. Tanimoto, Non-Abelian discrete symmetries in particle physics. Prog. Theor. Phys. Suppl. **183**, 1–163 (2010)
23. G. Altarelli, F. Feruglio, Discrete flavor symmetries and models of neutrino mixing. Rev. Mod. Phys. **82**, 2701–2729 (2010)
24. V. Puyam, S.R. Singh, N.N. Singh, Deviation from tribimaximal mixing using A4 flavour model with five extra scalars. Nucl. Phys. B **983**, 115932 (2022)
25. V. Puyam, Nonzero θ_{13} with $A_4 \times Z_4$ flavor symmetry group. LHEP **2022**, 340 (2022)
26. S. Gupta, A.S. Joshipura, K.M. Patel, Minimal extension of tri-bimaximal mixing and generalized $Z_2 \to Z_2$ symmetries. Phys. Rev. D **85**, 031903 (2012)
27. N. Gautam, R. Krishnan, M.K. Das, Effect of sterile neutrino on low-energy processes in minimal extended seesaw with $\Delta(96)$ symmetry and TM1 mixing. Front. Phys. **0**, 417 (2021)
28. V.V. Vien, H.N. Long, D.P. Khoi, Neutrino mixing with Non-Zero θ_{13} and CP violation in the 3-3-1 model based on S_4 flavor symmetry. Int. J. Mod. Phys. A **30**(17), 1550102 (2015)
29. A.E. Cárcamo Hernández, S.F. King, Littlest inverse seesaw model. Nucl. Phys. B **953**, 114950 (2020)
30. T. Kobayashi, J. Kubo, H. Terao, Exact S(3) symmetry solving the supersymmetric flavor problem. Phys. Lett. B **568**, 83–91 (2003)
31. S. Morisi, K.M. Patel, E. Peinado, Model for T2K indication with maximal atmospheric angle and tri-maximal solar angle. Phys. Rev. D **84**, 053002 (2011)
32. W. Rodejohann, Unified parametrization for quark and lepton mixing angles. Phys. Lett. B **671**(2), 267–271 (2009)
33. I.K. Cooper, S.F. King, A.J. Stuart, A golden A_5 model of leptons with a minimal NLO correction. Nucl. Phys. B **875**, 650–677 (2013)
34. I. de Medeiros Varzielas, L. Lavoura, olden ratio lepton mixing and nonzero reactor angle with A_5. J. Phys. G **41**, 055005 (2014)
35. J. Gehrlein, J.P. Oppermann, D. Schäfer, M. Spinrath, An SU(5) $\times$ A_5 golden ratio flavour model. Nucl. Phys. B **890**, 539–568 (2014)
36. A. Di Iura, C. Hagedorn, D. Meloni, Lepton mixing from the interplay of the alternating group a5 and cp. J. High Energy Phys. **2015**(8), 1–27 (2015)
37. P. Ballett, S. Pascoli, J. Turner, Mixing angle and phase correlations from A_5 with generalized *cp* and their prospects for discovery. Phys. Rev. D **92**, 093008 (2015)
38. J. Turner, Predictions for leptonic mixing angle correlations and nontrivial dirac *cp* violation from A_5 with generalized *cp* symmetry. Phys. Rev. D **92**, 116007 (2015)
39. C.-C. Li, G.-J. Ding, Lepton mixing in a 5 family symmetry and generalized cp. J. High Energy Phys. **2015**(5), 1–56 (2015)
40. M.L. López-Ibáñez, A. Melis, D. Meloni, O. Vives, Lepton flavor violation and neutrino masses from a5 and cp in the non-universal mssm. J. High Energy Phys. **2019**(6), 1–34 (2019)
41. D.V. Forero, S. Morisi, M. Tortola, J.W.F. Valle, Lepton flavor violation and non-unitary lepton mixing in low-scale type-I seesaw. JHEP **09**, 142 (2011)
42. A.E.C. Hernández, K.N. Vishnudath, J.W.F. Valle, Linear seesaw mechanism from dark sector. J. High Energy Phys. **2023** (2023)
43. P.I. Krastev, S.T. Petcov, Resonance amplification and t violation effects in three neutrino oscillations in the earth. Phys. Lett. B **205**, 84–92 (1988)
44. M. Meucci, Meg II Experiment Status and Prospect (2022)

Chapter 27
Neutrino Masses and Mixing in an Inverse Seesaw Model with S_4 Modular Flavor Symmetry

Raktima Kalita

Abstract In our work, we have used modular invariance approach to construct a neutrino mass model in the framework of the inverse seesaw mechanism with modular S_4 flavor symmetry which requires less number of flavon fields. The phenomenological study of the neutrino mass matrix is carried out and correlation plots are drawn using the current 3σ ranges of neutrino oscillation data to test the compatibility of the model with experiments. Again the use of inverse seesaw mechanism allows the right-handed neutrinos to have masses in the TeV scale which may be feasible at the current large hadron collider experiment.

27.1 Introduction

The conclusive experimental evidence for existence of massive neutrinos indicates a new physics beyond Standard Model (SM) of particle physics. Currently the neutrino experiments are being engaged in precision measurements for neutrino data viz, the solar and atmospheric neutrino mass splittings (Δm_{ij}^2) and solar, atmospheric and reactor angles (θ_{ij}) of neutrino flavor mixings. However, some parameters, e.g., CP violating Dirac and Majorana phases δ, α, and β, octant degeneracy of the atmospheric mixing (θ_{23}) and neutrino mass ordering are yet to measure experimentally. To generate the non-zero masses of neutrinos, the scenarios beyond standard model (BSM) are very crucial. One of the main features of BSM scenarios is the existence of sterile neutrinos, which are SM gauge singlets. These are considered as right-handed neutrinos which are coupled to the standard model active neutrinos through Yukawa interactions. In canonical seesaw framework, to explain the eV scale light neutrinos, the mass of right-handed neutrinos are considered to be of the scale of 10^{15} GeV, which is beyond the reach of current neutrino oscillation experiments. But, some

R. Kalita (✉)
Department of Physics, Cotton University, Guwahati 781001, Assam, India
e-mail: phy2091007_raktima@cottonuniversity.ac.in

J. W. F. Valle et al. (eds.), *Proceedings of the International Conference on Future Prospects in Neutrino and Astroparticle Physics*, Springer Proceedings in Physics 361,
https://doi.org/10.1007/978-981-97-7441-8_28

low scale mechanisms like inverse seesaw [3, 4], linear seesaw [5], extended seesaw [6], etc., where the heavy neutrino mass is taken in the range of TeV scale makes them experimentally testable.

On the other hand, the understanding of flavor structure of fermions is one of the challenging issues in SM. So, to describe the flavor structure, symmetry can play a crucial role and some non-abelian discrete flavor symmetry groups S_4, A_4, A_5, etc., are found useful for model building in the literature [7–11]. In these approaches, extra flavon fields are required for spontaneous breaking of the discrete flavor symmetry [12] and the use of higher dimensional operators leads to less predictability of the model. To bypass such complicacies, a modular invariance approach has been introduced [13] in literature, which has a few appealing features. In this approach, the flavon fields other than the modulus are not needed, which enhances the predictability of the model [14]. This motivates us to study the present model.

In the present work, we construct a inverse seesaw model (2, 2) based on modular S_4 symmetry. In our scenario, the desired neutrino mass matrix is constructed with the proposed symmetry and a numerical analysis is carried out for various parameters to fit in the current neutrino data.

The paper is organized as follows. In Sect. 27.2, the formalism and the neutrino mass matrix from inverse seesaw mechanism with modular S_4 symmetry are discussed. In Sect. 27.3, numerical analysis for correlations between observables of neutrino sector and the input model parameters along with the prediction for lepton flavor violation is presented. Finally, in Sect. 27.4 we conclude our results.

27.2 Model Framework

In this section, we discuss our model framework for inverse seesaw mechanism introducing modular S_4 symmetry. Also, a Z_4 symmetry is introduced to remove the undesired interactions for the model. The importance of S_4 modular symmetry is the requirement of less number of flavons compared to usual S_4 group. Here, the Yukawa couplings have non-trivial group transformation. In the model, we introduce right-handed $SU(2)$ singlet fermion N_R which is transformed as doublet under modular S_4 symmetry and left-handed sterile fermions transformed as $\{1, 1^{'}\}$ under the same symmetry. Furthermore, two Higgs doublet H_u, H_d has charge 1 under modular S_4 symmetry (Table 27.1).

The particle content and the group charges under $S_4 \times Z_4 \times k_I$ for the model is given in Table 27.2. Here, k_I is denoted as the number of modular weights.

Based on the above charge assignments and symmetries, the Lagrangian of the model for the leptonic sector is given by

$$\mathcal{L}_{lepton} = \mathcal{L}_L + \mathcal{L}_D + \mathcal{L}_{NS} + \mathcal{L}_S \tag{27.1}$$

where $\mathcal{L}_L$ is the mass term for charged leptons, $\mathcal{L}_D$ is the Dirac mass term which connects left-handed (ν_L) and right-handed (N_R) components of neutrinos, $\mathcal{L}_{NS}$

Table 27.1 Particle content of the model and their charges

Fields	L	l_R	N_R	S_i	H_u	H_d	ϕ
$SU(2)_L$	2	1	1	1	2	2	1
S_4	3	$1, 1'$	2	$1, 1'$	1	1	1
Z_4	-1	i	$-i$	1	i	$-i$	-1
k_I	-2	-2	-2	-2	0	0	0

Table 27.2 Current neutrino oscillation parameters from global fits [15]

Oscillation parameters	Normal ordering		Inverted ordering	
	Best fit$\pm 1\sigma$	3σ range	Best fit$\pm 1\sigma$	3σ range
$\Delta m^2_{21}/10^{-5} eV^2$	$7.42^{+0.21}_{-0.20}$	6.82–8.04	$7.42^{+0.21}_{-0.20}$	6.82–8.04
$\lvert\Delta m^2_{3l}\rvert/10^{-3} eV^2$	$2.510^{+0.027}_{-0.027}$	2.430–2.593	$-2.490^{+0.026}_{-0.028}$	$-2.574 - 2.410$
θ°_{12}	$33.45^{+0.77}_{-0.75}$	31.27–35.87	$33.45^{+0.78}_{-0.75}$	31.27–35.87
θ°_{23}	$42.1^{+1.1}_{-0.9}$	39.7–50.9	$49.0^{+0.9}_{-1.9}$	39.8–51.6
θ°_{13}	$8.62^{+0.12}_{-0.12}$	8.25–8.98	$8.61^{+0.14}_{-0.12}$	8.24–9.02
δ°_{CP}	230^{+36}_{-25}	144–350	278^{+22}_{-30}	194–345

reprsents the mixing term for right-handed neutrinos (N_R) and sterile fermions (S_i) and $\mathcal{L}_S$ is the majorana mass term for gauge singlet sterile fermions (S_i).

27.2.1 Charged Lepton Mass Matrix

The Lagrangian for the charged leptons is given by

$$\mathcal{L}_L = \alpha(LE_1^c)_{3'}(H_d)_1 Y_{3'} + \beta(LE_2^c)_3(H_d)_1 Y_3^{(4)} + \gamma(LE_3^c)_{3'}(H_d)_1 Y_{3'}^{(4)} \tag{27.2}$$

where α, β, γ are the free parameters. The charged lepton mass matrix can be written as

$$M_L = v_d \begin{pmatrix} \alpha Y_3 & -2\beta Y_2 Y_3 & 2\gamma Y_1 Y_3 \\ \alpha Y_5 & \beta(\sqrt{3}Y_1 Y_4 + Y_2 Y_5) & \gamma(\sqrt{3}Y_2 Y_4 + Y_1 Y_5) \\ \alpha Y_2 & \beta(\sqrt{3}Y_1 Y_5 + Y_2 Y_4) & \gamma(\sqrt{3}Y_2 Y_5 - Y_1 Y_4) \end{pmatrix} \tag{27.3}$$

where v_d is the vev for down type Higgs doublet.

27.2.2 Dirac Mass Matrix for the Model

The Lagrangian for Dirac mass term for the model can be written as

$$\mathcal{L}_D = (LN_R)_3(H_u)_1 Y_3^{(4)} + (LN_R)_{3'}(H_u)_1 Y_{3'}^{(4)} \tag{27.4}$$

Now, the Dirac mass matrix for the neutrinos is given by

$$M_D = v_u \begin{pmatrix} -2\alpha_D a & -2\beta_D b \\ -\alpha_D(\frac{\sqrt{3}}{2}c + \frac{1}{2}d) + \beta_D(\frac{3}{2}d - \frac{\sqrt{3}}{2}e) & \alpha_D(\frac{3}{2}f + \frac{\sqrt{3}}{2}c) + \beta_D(\frac{3}{2}f + \frac{\sqrt{3}}{2}c) \\ -\alpha_D(\frac{\sqrt{3}}{2}f + \frac{1}{2}c) + \beta_D(\frac{3}{2}c - \frac{\sqrt{3}}{2}f) & \alpha_D(\frac{3}{2}e + \frac{\sqrt{3}}{2}d) + \beta_D(\frac{3}{2}e + \frac{\sqrt{3}}{2}d) \end{pmatrix} \tag{27.5}$$

where v_u is the vev for the up type Higgs doublet and α_D, β_D are the free parameters.

Here, we denote $a = Y_2Y_3$, $b = Y_1Y_3$, $c = Y_2Y_4$, $d = Y_2Y_5$, $e = Y_1Y_4$ and $f = Y_1Y_5$.

27.2.3 Mixing Mass Term Between the Right-Handed Neutrinos and Gauge Singlet Neutral Fermions

The Lagrangian for this mixing is given by

$$\mathcal{L}_{NS} = \alpha_{NS}(Y_2^{(4)} N_R)S_1\phi + \beta_{NS}(Y_2^{(4)} N_R)S_2\phi \tag{27.6}$$

The resultant mass matrix is found to be

$$M_{NS} = v_{NS} \begin{pmatrix} Y_2^2 - Y_1^2 & 2Y_1Y_2 \\ 2Y_1Y_2 & Y_1^2 - Y_2^2 \end{pmatrix} \begin{pmatrix} \alpha_{NS} & 0 \\ 0 & \beta_{NS} \end{pmatrix} \tag{27.7}$$

where α_{NS}, β_{NS} are the free parameters and v_ϕ is the vev for scalar field ϕ.

27.2.4 Majorana Mass Term for Gauge Singlet Fermions

The Lagrangian for Majorana mass term can be written as

$$\mathcal{L}_S = \mu_0 SSY_1^{(4)} \tag{27.8}$$

The Majorana mass matrix is found to be

$$M_S = \mu_0 \begin{pmatrix} 0 & Y_1^2 + Y_2^2 \\ Y_1^2 + Y_2^2 & 0 \end{pmatrix} \tag{27.9}$$

27.2.5 *Inverse Seesaw Mechanism for Light Neutrino Masses*

Within the present model, the complete neutrino mass matrix is given by

$$M_\nu = \begin{pmatrix} 0 & M_D & 0 \\ M_D^T & 0 & M_{NS} \\ 0 & M_{NS}^T & M_S \end{pmatrix} \tag{27.10}$$

Here, M_D is at electroweak scale, M_{NS} at TeV scale and M_S at keV scale to get SM neutrinos at sub-eV scale [16]. Now, the inverse seesaw formula for light neutrinos is given by

$$M_\nu = M_D M_{NS}^{-1} M_S (M_{NS}^{-1})^T M_D^T \tag{27.11}$$

27.3 Numerical Analysis

In this section, we will carry out the numerical analysis for parameters satisfying neutrino oscillation data [15]. The neutrino oscillation data at 3σ interval are given in Table 27.2.

Here, we numerically diagonalize the neutrino mass matrix of (27.11) through $U^\dagger M U = diag(m_1^2, m_2^2, m_3^2)$, where U is a unitary matrix. From U, the neutrino mixing angles can be extracted using the standard relations:

$$\sin^2\theta_{13} = |U_{13}|^2,\ \sin^2\theta_{12} = \frac{|U_{12}^2|}{1-|U_{13}^2|},\ \sin^2\theta_{23} = \frac{|U_{23}^2|}{1-|U_{13}^2|} \tag{27.12}$$

Also, the range of solar mass splitting Δm_{21}^2 and atmospheric mass splitting Δm_{31}^2 are taken as the input parameters for the model.

Apart from this, the Jarlskog invariant which controls the size of CP violation in the quark and leptonic sector is given by the equation

$$\begin{aligned} J_{CP} &= \text{Im}[U_{e1} U_{\mu 2} U_{e2}^* U_{\mu 1}^*] \\ &= \sin\theta_{23}\cos\theta_{23}\sin\theta_{12}\cos\theta_{12}\sin\theta_{13}\cos^2\theta_{13}\sin\delta_{CP} \end{aligned} \tag{27.13}$$

To satisfy the current neutrino oscillation data, we can choose the following ranges of the free parameters for the model:

$$\mathrm{Re}[\tau] \in [0.5, 3],\ \mathrm{Im}[\tau] \in [0.1, 2],\ \{\alpha_D, \beta_D\} \in [0.1, 10],$$
$$\{\alpha_{NS}, \beta_{NS}\} \in [0.1, 5],\ v_{NS} \in [10^{13}, 10^{14}]\ \mathrm{eV},\ \mu_0 \in [10^2, 10^3]\ \mathrm{eV}$$

Figures 27.1 and 27.2 show the variation of modular Yukawa couplings with allowed region of $\mathrm{Re}(\tau)$ and $\mathrm{Im}(\tau)$ for NO and IO respectively.

Also, the mixing angles for the model can be obtained from the relations given in (27.12). Figures 27.3, 27.4, and 27.5 show the variation of the sum of neutrino masses (Σm_e) with the mixing angles $\sin\theta_{13}$, $\sin\theta_{12}$, and $\sin\theta_{23}$ for both normal and inverted ordering respectively.

It is evident from the figures that a wide number of parameter space lie within the allowed region for upper bound for sum of neutrino masses ($\Sigma m_\nu \leq 0.12$ eV) and the mixing angles. The lower bound for sum of neutrino masses is found to be around 0.06 eV for both NO and IO.

In Fig. 27.6, we show the correlation of Jarsklog invariant with Dirac CP phase, which is found to be of the order of 10^{-2}. The Dirac CP phase is constrained in the region $0° \leq \delta_{CP} \leq 90°$ and the J_{CP} is found to be in the region -0.05 to 0.05 for NO and -0.06 to 0.06 for IO.

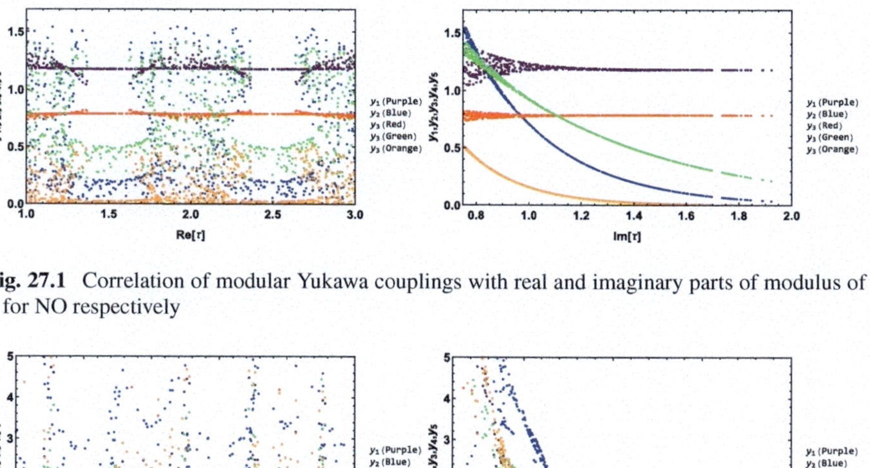

Fig. 27.1 Correlation of modular Yukawa couplings with real and imaginary parts of modulus of τ for NO respectively

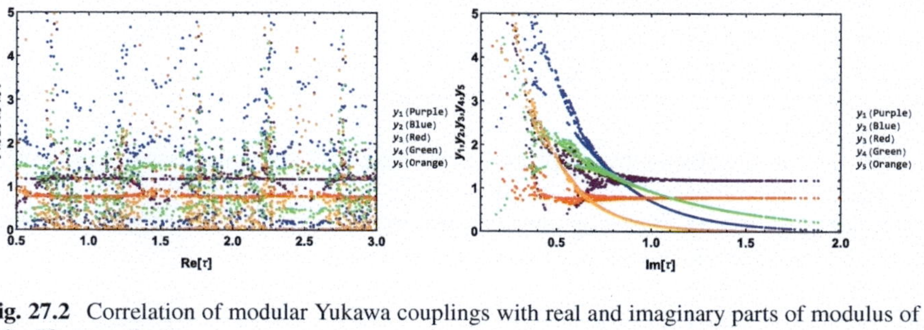

Fig. 27.2 Correlation of modular Yukawa couplings with real and imaginary parts of modulus of τ for IO respectively

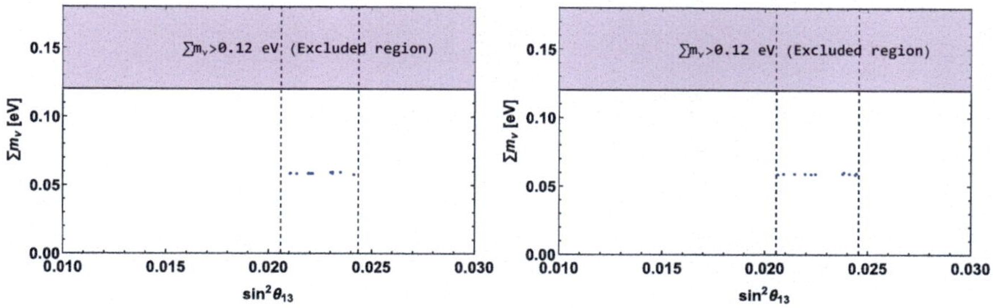

Fig. 27.3 Correlation between sum of neutrino masses (Σm_ν) and $\sin^2\theta_{13}$ for NO (left) and IO (right)

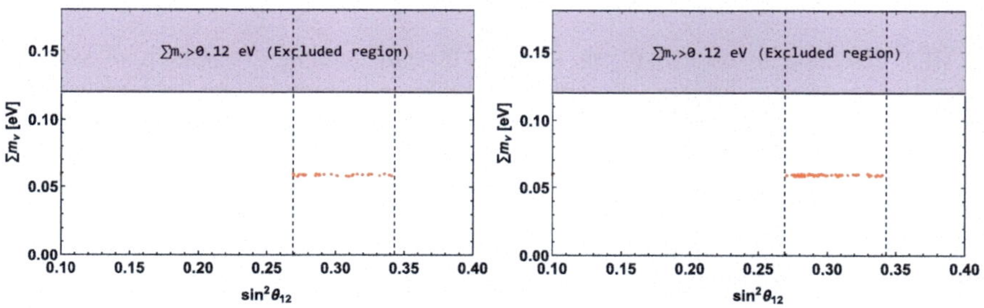

Fig. 27.4 Correlation between sum of neutrino masses (Σm_ν) and $\sin^2\theta_{12}$ for NO (left) and IO (right)

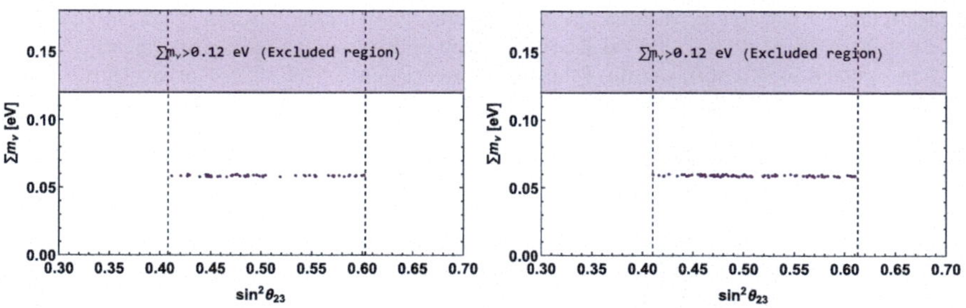

Fig. 27.5 Correlation between sum of neutrino masses (Σm_ν) and $\sin^2\theta_{23}$ for NO (left) and IO (right)

27.4 Conclusion

Here, we have studied an inverse seesaw model imposed with modular S_4 symmetry. This symmetry restricts the use of multiple flavon fields and structures of relevant Yukawa interactions. Further, the model is augmented with Z_4 symmetry to avoid the unnecessary terms in the Lagrangian. The neutrino mass matrix for the model is characterized by the modulus τ and the free parameters of the model.

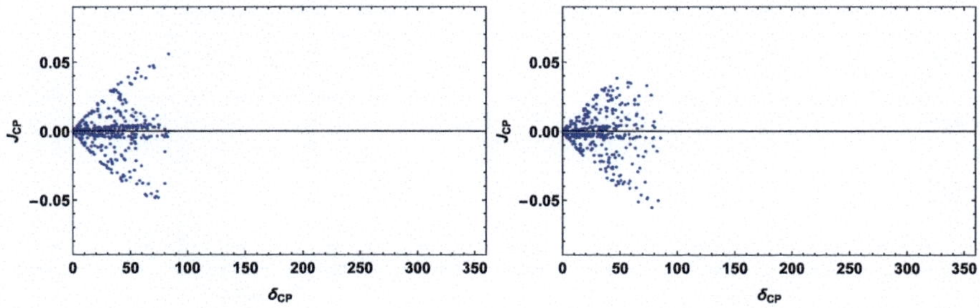

Fig. 27.6 Correlation between Jarsklog invariant (J_{CP}) with Dirac CP phase (δ_{CP}) for NO (left) and IO (right)

After that, we have carried out some numerical analysis in search of various parameter sets to fit into the 3σ ranges of neutrino oscillation data for both normal and inverted orderings. Here, the Dirac CP phase is restricted in the region 0° to 90° for both NO and IO and the lower bound for sum of the neutrino masses is around 0.06 eV. Furthermore, we have found the J_{CP} of the order of 10^{-2} for both NO and IO.

References

1. R. Acciarri, M.A. Acero, M. Adamowski, C. Adams, P. Adamson, S. Adhikari, Z. Ahmad et al., Long-baseline neutrino facility (LBNF) and deep underground neutrino experiment (DUNE) conceptual design report, volume 4 the DUNE detectors at LBNF (2016). arXiv:1601.02984
2. Constraint on the matter–antimatter symmetry-violating phase in neutrino oscillations. Nature 580(7803), 339–344 (2020)
3. M.C. González-Garciá, JW Valle, Fast decaying neutrinos and observable flavour violation in a new class of majoron models. Phys. Lett. B **216**(3–4), 360–366 (1989)
4. AE Cárcamo Hernández, J.M. González, U.J. Saldaña-Salazar, Viable low-scale model with universal and inverse seesaw mechanisms. Phys. Rev. D **100**(3), 035024 (2019)
5. M. Malinský, J.C. Romao, J.W.F. Valle, Supersymmetric SO (10) seesaw mechanism with low $B - L$ scale. Phys. Rev. Lett. **95**(16), 161801 (2005)
6. R.N. Mohapatra, S. Antusch, K.S. Babu, G. Barenboim, M.-C. Chen, A. De Gouvêa, P. De Holanda et al., Theory of neutrinos: a white paper. Rep. Progress Phys. **70**(11), 1757 (2007)
7. B. Brahmachari, S. Choubey, M. Mitra, A 4 flavor symmetry and neutrino phenomenology. Phys. Rev. D **77**(7), 073008 (2008)
8. A. Mukherjee, M.K. Das. Neutrino phenomenology and scalar Dark Matter with A4 flavor symmetry in inverse and type II seesaw. Nuclear Phys. B **913**, 643–663 (2016)
9. A. Di Iura, M.L. López-Ibáñez, D. Meloni, Neutrino masses and lepton mixing from A5 and CP. Nuclear Phys. B **949**, 114794 (2019)
10. E. Ma, Neutrino mass matrix from S4 symmetry. Phys. Lett. B **632**(2–3), 352–356 (2006)
11. M. Chakraborty, R. Krishnan, A. Ghosal, Predictive S4 flavon model with TM1 mixing and baryogenesis through leptogenesis. J. High Energy Phys. **9**, 1–48 (2020)
12. S. Pakvasa, H. Sugawara, Discrete symmetry and Cabibbo angle. Phys. Lett. B **73**(1), 61–64 (1978)
13. F. Feruglio, Are neutrino masses modular forms?, in *From My Vast Repertoire ... Guido Altarelli's Legacy* (2019), pp.227–266

14. Z.-Z. Xing, Flavor structures of charged fermions and massive neutrinos. Phys. Rep. **854**, 1–147 (2020)
15. M.C. Gonzalez-Garcia, M. Maltoni, T. Schwetz, NuFIT: three-flavour global analyses of neutrino oscillation experiments. Universe **7**(12), 459 (2021)
16. F. Deppisch, J.W.F. Valle, Enhanced lepton flavor violation in the supersymmetric inverse seesaw model. Phys. Rev. D **72**(3), 036001 (2005)

Chapter 28
The Dark Hypercharge Symmetry

Hemant Prajapati and Rahul Srivastava

Abstract $U(1)$ extension of the Standard Model (SM) is well motivated, where the charges of SM fermions are fixed by gauge anomaly cancellations and Yukawa interactions. While the literature extensively discusses anomaly cancellation solutions in which SM fermions are vector-like under new $U(1)_X$ symmetry, allowing the Yukawa structure to remain invariant, chiral solutions in which SM fermions are chiral under new symmetry are not well explored. In this work, we venture into these relatively unexplored chiral solutions. We presented a particularly intriguing chiral solution with the introduction of three right-handed fermions (RHFs). We termed this symmetry as Dark-Hyper Charge (DHC) Symmetry. This terminology stems from the resemblance of the $U(1)_X$ charges of the SM particles to hypercharge, with the term 'dark' referring to the relatively large charges of the RHFs. This implies that the branching fraction of Z' to these RHFs should be significantly high. The introduced RHFs could potentially serve as Dark Matter (DM) candidates. We will demonstrate in a model-independent manner using only the Z' interaction channel, that the lightest RHF, denoted as F_1, is a viable DM candidate, and it can meet all current DM constraints with a mass of $M_{F_1} \gtrsim 150$ GeV.

28.1 Introduction

In Gauge theories, gauge symmetry is crucial for establishing unitarity and renormalizability. An anomaly in the gauge symmetry would lead to a gauge anomaly. In a consistent gauge theory, these anomalies must mutually cancel out when considering contributions from multiple chiral fermions [1, 2]. The Standard Model (SM) is a mathematically consistent, gauge anomaly-free theory. Hypercharges of all the SM fermions add up in a way that cancels these anomalies. New gauged symmetries

H. Prajapati (✉) · R. Srivastava
Department of Physics, Indian Institute of Science Education and Research - Bhopal, Bhopal Bypass Road, Bhauri 462066, Bhopal, India
e-mail: hemant19@iiserb.ac.in

R. Srivastava
e-mail: rahul@iiserb.ac.in

J. W. F. Valle et al. (eds.), *Proceedings of the International Conference on Future Prospects in Neutrino and Astroparticle Physics*, Springer Proceedings in Physics 361,
https://doi.org/10.1007/978-981-97-7441-8_29

beyond the Standard Model (BSM) alter these gauge anomaly cancellation conditions. These symmetries are motivated by the desire to explain observations that go beyond the predictions of the SM. The simplest and highly motivated extension among these is an extra $U(1)_X$ gauge symmetry. These new $U(1)$ symmetry is highly inspired by Grand Unified Theories (GUT). Grand unified groups bigger than $SU(5)$ like $SO(10)$ or $E6$ inspired by superstrings, could break to the $SM \otimes U(1)_X$ at a scale lower than the GUT scale. Additionally, scenarios such as Extended Technicolor, top flavour models and Little Higgs Models also involve the presence of extra abelian gauged symmetries.

The transformation of left-handed and right-handed fermionic pair under the given symmetry plays a crucial role in the mass generation of the fermion. Under chiral symmetry, the left-handed and right-handed fermionic fields have distinct charges. However, under vector symmetry, fermionic fields carry the same charges. If the given fermion pair is a vector with respect to all the symmetries then the Dirac mass term is allowed. For a chiral fermion, such a term is not allowed, and their masses are produced through Yukawa couplings following the symmetry breaking. The SM is a chiral theory, and as such, fermionic masses are generated via Yukawa couplings after the electroweak symmetry is broken. In the realm of BSM physics, where new $U(1)_X$ symmetries are introduced, a common conjecture posits that SM fermions exhibit vector-like behaviour under the $U(1)_X$. This assumption results in the preservation of SM Yukawa interactions under $U(1)_X$, enabling the generation of SM fermionic masses without the necessity for additional scalar doublets. As a consequence, two critical conditions, namely anomaly cancellations and Yukawa invariance, play a pivotal role in the incorporation of a new $U(1)$ symmetry. This consideration prompts the categorization of anomaly cancellation solutions into two distinct types: vector and chiral. In vector solutions, SM fermions are treated as vectors under $U(1)_X$, while chiral solutions involve SM fermions exhibiting chiral characteristics. The scientific literature extensively explores and investigates these vector solutions, with notable examples such as $U(1)_{B-L}$, $B - 3L_i$, $B_i - 3L_j$, $L_i - L_j$, among others [3–5].

In this study [6], we explored these chiral solutions. The induced gauge anomalies can be cancelled by adding SM singlets right-handed fermions (RHFs). The $U(1)_X$ charges of all the fermions are fixed by the anomaly conditions and structure of Yukawas in the SM. The introduced RHFs could potentially serve as DM candidates. If the $U(1)_X$ charges of these RHFs are relatively higher than those of SM fermions, then the associated Z' gauge bosons primarily decay to DM, thereby keeping the symmetry concealed. To illustrate this, we will present an intriguing solution, which we will refer to as DHC symmetry. We will treat these RHFs as Majorana particles and show in a model-independent way, using only the Z' interaction channel, that the lightest RHF is a viable DM candidate and can satisfy all current DM constraints. We will also show that under DHC symmetry, the Z' branching fraction in RHFs is significantly large, and hence, collider limits on Z' gauge coupling g_x are less stringent, allowing us to explore low Z' masses at larger gauge coupling.

28.2 Gauge Anomalies

In Quantum Field Theory (QFT), symmetries play a fundamental role, with classical action invariance leading to conserved quantities. However, quantum anomalies can disrupt these conservation laws, resulting in a violation of Ward's identity. This violation signifies a lack of invariance in the quantum effective action, indicating an anomaly. Anomalies can affect both global and gauge symmetries, with the latter having severe consequences, leading to what is known as a gauge anomaly. These anomalies must cancel out in a consistent gauge theory when considering contributions from the multiple chiral fermions.

In perturbation theory, gauge anomalies manifest within triangle loop diagrams featuring fermionic propagation and bosonic emission from vertices, illustrated in Fig. 28.1. These diagrams should vanish to make the theory gauge anomaly-free. In SM hypercharges of fermions add up in a way to cancel these anomalies. Gauging new $U(1)$ symmetry modifies these gauge anomaly cancellation conditions. The anomaly cancellation condition for new gauged $U(1)_X$ symmetry are shown below. It includes all gauge and mixed gauge-gravitational anomalies.

$$[SU(3)_C]^2[U(1)_X] = \sum_q X_{q_L} - \sum_q X_{q_R}. \tag{28.1a}$$

$$[SU(2)_L]^2[U(1)_X] = \sum_l X_{l_L} + 3\sum_q X_{q_L}. \tag{28.1b}$$

$$[U(1)_Y]^2[U(1)_X] = \sum_{l,q}(Y_{l_L}^2 X_{l_L} + 3Y_{q_L}^2 X_{q_L}) - \sum_{l,q}(Y_{l_R}^2 X_{l_R} + 3Y_{q_R}^2 X_{q_R}). \tag{28.1c}$$

$$[U(1)_Y][U(1)_X]^2 = \sum_{l,q}(Y_{l_L} X_{l_L}^2 + 3Y_{q_L} X_{q_L}^2) - \sum_{l,q}(Y_{l_R} X_{l_R}^2 + 3Y_{q_R} X_{q_R}^2). \tag{28.1d}$$

$$[U(1)_X]^3 = \sum_{l,q}(X_{l_L}^3 + 3X_{q_L}^3) - \sum_{l,q}(X_{l_R}^3 + 3X_{q_R}^3). \tag{28.1e}$$

$$[G]^2[U(1)_X] = \sum_{l,q}(X_{l_L} + 3X_{q_L}) - \sum_{l,q}(X_{l_R} + 3X_{q_R}). \tag{28.1f}$$

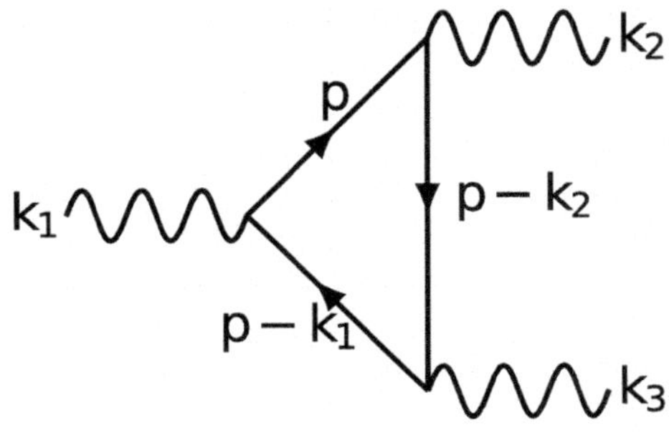

Fig. 28.1 A triangle diagram

Here Y_ψ and X_ψ are the $U(1)_Y$ and $U(1)_X$ charges respectively, and $\psi = \{L, Q, e_R, u_R, d_R\}$. l and q denote leptons and quarks respectively. These charges also get constraints from the SM Yukawa interactions,

$$-\mathscr{L}_{yukawa} \supset Y_e \overline{L}\phi e_R + Y_u \overline{Q}\tilde{\phi} u_R + Y_d \overline{Q}\phi d_R + \text{h.c.}. \tag{28.2}$$

All flavour indices have been suppressed. Here, ϕ denotes the SM Higgs doublet, and $\tilde{\phi}$ corresponds to $i\sigma_2\phi^*$, where σ_2 is the second pauli matrix. The invariance of the Yukawa couplings under hypercharge symmetry gives us the two extra conditions,

$$X_{u_R} = X_Q + X_L - X_{e_R}, \text{ and } X_{d_R} = X_Q - X_L + X_{e_R}. \tag{28.3}$$

To fully preserve the SM Yukawa structure, we will assume equal $U(1)_X$ charges for all generations of SM fermions ($X^i_\psi = X^j_\psi$). In the next section, we pick one particularly fascinating chiral solution and study its properties.

28.3 Dark Hyper Charge Symmetry and Its Properties

Our attention will be primarily directed towards a solution presented in Table 28.1. This solution exhibits a symmetry where the $U(1)_X$ charges of the SM particles bear a striking resemblance to hypercharge. Furthermore, the charges of the right-handed fermions (RHFs) are considerably higher than those of the SM fermions. This suggests that the branching fraction of Z' to these RHFs could be significantly large. As we proceed, our discourse will be centred around this symmetry and its associated properties. We will also argue that the lightest RHF is a good candidate for DM.

The Z' boson has several intriguing properties that can be probed experimentally. One such electroweak observable is the ρ parameter, defined as $\rho = M_W^2/(M_Z \cos\theta_w)^2$. The mixing between SM Z and new Z' boson impacts the ρ parameter [7]. In addition to the SM Higgs, we will also take into account the new scalar field χ_0, which is the SM singlet. The VEV of χ_0 and the Higgs are v_0 and v, respectively, and we will assume $M_{Z'}[g_x, v_0] > M_Z[g_x, v_0]$. In Fig. 28.2 we plotted the ρ parameter as function of v_0 for various values of g_x. The orange band

Table 28.1 Charges of particles under $U(1)_Y$ and DHC symmetry

$U(1)$	Q	u_R	d_R	L	e_R	f_1	f_2	f_3	ϕ	χ_0
$U(1)_Y$	$\frac{1}{3}$	$\frac{4}{3}$	$\frac{-2}{3}$	-1	-2	0	0	0	1	0
$U(1)_X$	$-\frac{1}{3}$	$\frac{5}{3}$	$-\frac{7}{3}$	1	-1	10	-18	17	2	-6

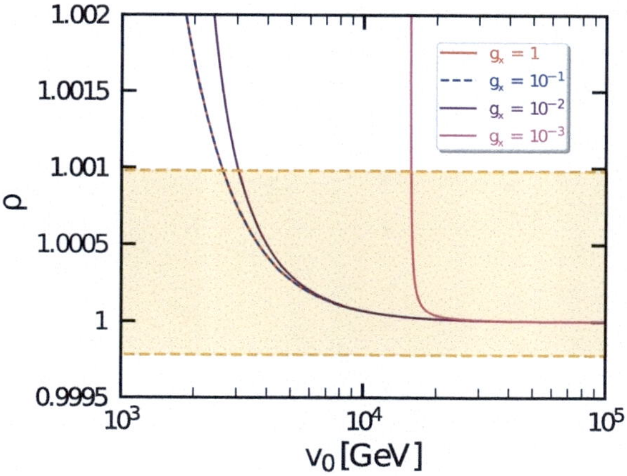

Fig. 28.2 ρ parameter as function of v_0 for various values of g_x

corresponds to the ρ parameter's current experimental value, 1.00038 ± 0.00060 at 3σ [8].

The LHC's, ATLAS and CMS searches for dilepton resonances place the tightest limits on $M_{Z'}$ and its gauge coupling g_x via the process $pp \to Z' \to l^+l^-$, where $l = e, \mu$. However, if the Z' boson primarily couples to Dark Matter (DM) particles with a suppressed coupling to SM particles, its decay remains unobservable in colliders.

Figure 28.3 illustrates the two-body invisible decay channels of Z'. Here, F_i represents the mass eigenstates of RHFs. The vertical dashed lines correspond to $M_{Z'} = 2M_{F_i}$, beyond which the decay channels into that specific RHF begin. The solid (dashed) red line represents the total invisible decay, in DHC (B-L) symmetry, encompassing both RHFs and SM neutrinos. Due to the high $U(1)_X$ charges of RHFs, the invisible branching fraction of Z' is notably high in DHC symmetry, reaching approximately 90% at saturation. Conversely, under $B - L$ symmetry, it is around

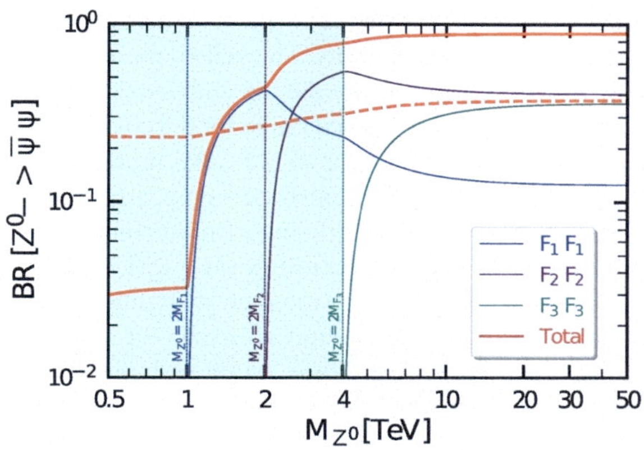

Fig. 28.3 Z' invisible branching fraction versus $M_{Z'}$. The solid (dashed) lines represent DHC ($B - L$) symmetries. RHF masses are set at 0.5 TeV, 1 TeV and 2 TeV. Taken from [6]

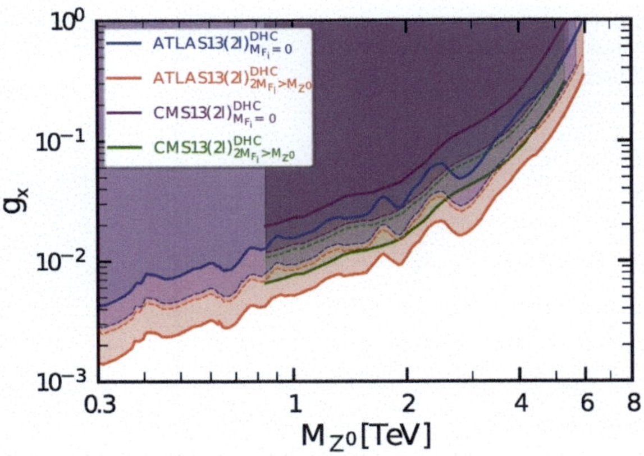

Fig. 28.4 The constraint on $(g_x - M_{Z'})$ plane. Shaded areas indicate regions excluded by ATLAS and CMS dilepton searches. The solid (dashed) lines represent DHC $(B - L)$ symmetries. Taken from [6]

38%. To put the constraints on $(g_x - M_{Z'})$ plane we used the ATLAS [9] and CMS [10] search for Z' in Dilepton resonance at pp collisions, Fig. 28.4 shows these constraints. The constraints significantly depend on whether Z' decays to RHFs. If it does not, the restrictions are considerably stricter compared to $B - L$, due to the higher quark charges in DHC symmetry resulting in a higher Z' production cross-section. In the next section, we will explore the DM constraints on the RHFs.

28.4 Dark Matter Constraints

As discussed earlier, the Lightest RHF could potentially serve as DM candidates. Constraints on $g_x/M_{Z'}$ from ATLAS and CMS, restrict the annihilation channel of RHFs into SM fermions. Nevertheless, we will show that only gauge interactions of F_1 are enough to satisfy the dark matter properties. We will consider the RHFs to be majorana fermions with quasi-degenerate masses, the lightest being the DM. In the left panel of Fig. 28.5, we show the relic density vs DM mass. Cyan points are ruled out by the ρ parameter limits. The yellow points are ruled out by collider constraints. The Blue (Gray) points are the underabundant (overabundant) points. The magenta points satisfy the 3σ allowed range for the cold DM derived from the Planck satellite data, $0.1126 \leq \Omega h^2 \leq 0.1246$ [11]. We found that in the low mass range of dark matter ($\lesssim$ 150 GeV), the constraints on g_x imposed by the ρ parameter and collider data do not permit the fulfilment of relic conditions, even at the Z resonance. Our Dark matter F_1 satisfies the relic at $M_{F_1} \gtrsim 150$ GeV.

Majorana DM F_1 can interact with nucleons through tree-level Z and Z' exchanges. Various direct and indirect DM detection experiments put constraints on the DM mass and its cross-section with nucleons [12–17]. IceCube [12] and the recent results from the LUX-ZEPLIN (LZ) [13] collaborations impose the most strin-

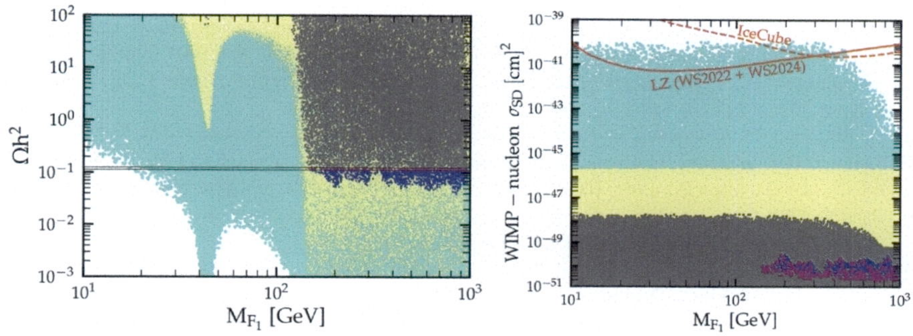

Fig. 28.5 Left panel shows the relic density plotted against DM mass. Cyan (Yellow) points are ruled out by the ρ parameter (collider constraints). Blue (Gray) are the underabundant (overabundant) points. Magenta points meet all criteria and are within 3σ of Planck's cold DM range. The right panel shows the Spin-dependent WIMP-nucleon cross-section averaged over proton and neutron versus DM mass. Points above the dashed red line are excluded by IceCube [12], and those above the solid red line are excluded by LZ collaboration [13]. Taken from [6]

gent constraints. In Fig. 28.5 right panel, we show the DM-nucleon elastic scattering cross-section averaged over proton and neutron versus DM mass. We found that all the points that satisfy the relic also satisfy the constraints from LZ and IceCube.

28.5 Conclusion

In summary, extensions of the SM with $U(1)$ gauge symmetries hold strong motivations. These new symmetries alter the SM's gauge anomaly conditions, placing constraints on charges of new fermions beyond the SM. In this work, we have examined chiral solutions, in which SM fermions are charged non-trivially. Our presented solution involves RHFs with higher $U(1)_X$ charges compared to SM fermions. These RHFs serve as promising DM candidates, with their interactions through Z' being sufficient to fulfil DM properties. We confirm that our dark matter candidate, F_1, with a mass $M_{F_1} \gtrsim 150$ GeV, satisfies all the DM relevant properties and current constraints.

References

1. S.L. Adler, Axial vector vertex in spinor electrodynamics. Phys. Rev. **177**, 2426–2438 (1969). https://doi.org/10.1103/PhysRev.177.2426
2. A. Bilal, Lectures on Anomalies, arXiv:0802.0634 [hep-th]
3. E. Ma, Gauged B—3L(tau) and radiative neutrino masses. Phys. Lett. B **433**, 74–81 (1998). https://doi.org/10.1016/S0370-2693(98)00599-1, arXiv:hep-ph/9709474

4. H.-S. Lee, E. Ma, Gauged $B - x_i L$ origin of R parity and its implications. Phys. Lett. B **688**, 319–322 (2010). https://doi.org/10.1016/j.physletb.2010.04.032, arXiv:1001.0768 [hep-ph]
5. S. Mandal, H. Prajapati, R. Srivastava, $B - L$ model in light of the CDF II result. arXiv:2301.01522 [hep-ph]
6. H. Prajapati, R. Srivastava, in *The Dark HyperCharge Symmetry*. arXiv:2411.02512 [hep-ph]
7. J. Erler, P. Langacker, Constraints on extended neutral gauge structures. Phys. Lett. B **456**, 68–76 (1999). https://doi.org/10.1016/S0370-2693(99)00457-8, arXiv:hep-ph/9903476
8. Particle Data Group Collaboration, P.A. Zyla et al., Review of particle physics. *PTEP***2020**(8), 083C01 (2020). https://doi.org/10.1093/ptep/ptaa104
9. ATLAS Collaboration, G. Aad et al., Search for high-mass dilepton resonances using 139 fb^{-1} of pp collision data collected at $\sqrt{s}$ =13 TeV with the ATLAS detector. Phys. Lett. B **796**, 68–87 (2019). https://doi.org/10.1016/j.physletb.2019.07.016 , arXiv:1903.06248 [hep-ex]
10. CMS Collaboration, A.M. Sirunyan et al., Search for resonant and nonresonant new phenomena in high-mass dilepton final states at $\sqrt{s}$ = 13 TeV. JHEP **07**, 208 (2021). https://doi.org/10.1007/JHEP07(2021)208, arXiv:2103.02708 [hep-ex]
11. Planck Collaboration, N. Aghanim et al., Planck 2018 results. VI. Cosmological parameters. Astron. Astrophys. **641**, A6 (2020). https://doi.org/10.1051/0004-6361/201833910, arXiv:1807.06209 [astro-ph.CO]. [Erratum: Astron.Astrophys. 652, C4 (2021)]
12. IceCube Collaboration, M.G. Aartsen et al., Search for annihilating dark matter in the Sun with 3 years of IceCube data. Eur. Phys. J. C **77**(3), 146 (2017). https://doi.org/10.1140/epjc/s10052-017-4689-9, arXiv:1612.05949 [astro-ph.HE]. [Erratum: Eur.Phys.J.C 79, 214 (2019)]
13. LUX-ZEPLIN Collaboration, J. Aalbers et al., First dark matter search results from the LUX-ZEPLIN (LZ) experiment. Phys. Rev. Lett. **131**(4), 041002 (2023). https://doi.org/10.1103/PhysRevLett.131.041002, arXiv:2207.03764 [hep-ex]
14. ANTARES Collaboration, S. Adrian-Martinez et al., Limits on dark matter annihilation in the sun using the ANTARES neutrino telescope. Phys. Lett. B **759**, 69–74 (2016). https://doi.org/10.1016/j.physletb.2016.05.019, arXiv:1603.02228 [astro-ph.HE]
15. Super-Kamiokande Collaboration, K. Frankiewicz, Searching for dark matter annihilation into neutrinos with super-kamiokande, in *Meeting of the APS Division of Particles and Fields* (2015), p. 10. arXiv:1510.07999 [hep-ex]
16. (XENON Collaboration)âŁ âŁ , XENON Collaboration, E. Aprile et al., First dark matter search with nuclear recoils from the XENONnT experiment. Phys. Rev. Lett. **131**(4), 041003 (2023). https://doi.org/10.1103/PhysRevLett.131.041003, arXiv:2303.14729 [hep-ex]
17. XENON Collaboration, E. Aprile et al., Dark matter search results from a one ton-year exposure of XENON1T. Phys. Rev. Lett. **121**(11), 111302 (2018). https://doi.org/10.1103/PhysRevLett.121.111302, arXiv:1805.12562 [astro-ph.CO]

Chapter 29
Neutrino Mass Sum Rules from Modular A_4 Invariance

Salvador Centelles Chuliá, Ranjeet Kumar, Oleg Popov, and Rahul Srivastava

Abstract Modular symmetries are a novel approach to understanding the flavour structure of leptonic mixing. Using the modular A_4 flavour symmetry integrated into a type-II seesaw, we propose a simple and minimalistic model that restricts the neutrino oscillation parameter space. Most importantly, this setup leads to a sum rule in the physical neutrino masses. When combined with the mass squared differences observed in neutrino oscillations, this sum rule determines the absolute neutrino mass scale. This has significant implications for cosmology, neutrinoless double beta decay experiments and direct neutrino mass measurements. In particular, the model predicts $\sum_i m_i \approx 0.1$ eV for both normal and inverted ordering, and thus can be fully probed by the current generation of cosmological probes in the upcoming years. Furthermore, our model has precise predictions for mixing angles which can be tested in future experiments.

29.1 Introduction

The quest to comprehend the origin of neutrino masses and mixing parameters has persisted for over two decades since the seminal discovery of neutrino oscillations. There are now several theoretical approaches and proposals in the literature, aimed

---

S. C. Chuliá
Max-Planck-Institut für Kernphysik, Saupfercheckweg 1, 69117 Heidelberg, Germany
e-mail: chulia@mpi-hd.mpg.de

R. Kumar (✉) · R. Srivastava
Department of Physics, Indian Institute of Science Education and Research - Bhopal, Bhopal ByPass Road, Bhauri, Bhopal 462066, India
e-mail: ranjeet20@iiserb.ac.in

R. Srivastava
e-mail: rahul@iiserb.ac.in

O. Popov
Department of Biology, Shenzhen MSU-BIT University, 1, International University Park Road, Shenzhen 518172, China
e-mail: opopo001@ucr.edu

J. W. F. Valle et al. (eds.), *Proceedings of the International Conference on Future Prospects in Neutrino and Astroparticle Physics*, Springer Proceedings in Physics 361,
https://doi.org/10.1007/978-981-97-7441-8_30

at understanding the neutrino mass and mixing pattern. One of the most popular approach is to use new symmetries, often taken as non-abelian discrete groups, called "flavour symmetries". The flavour symmetries often lead to predictions for the neutrino mixing angles and phases. These predictions can be tested in various experiments, particularly in neutrino oscillations. Another interesting consequence of flavour symmetries is occurrences of neutrino mass "sum rules". These sum rules typically manifest as relationships among the complex eigenvalues inherent to the neutrino mass matrix, which in turn can be read as a constraint on the physical neutrino masses and the Majorana phases, thus restricting, for example, the neutrinoless double beta decay ($0\nu ee$) parameter space. This approach enables a complementary path to neutrino oscillation experiments as a way to distinguish between given flavour models, as detailed in [1–3] and the references within.

A further development that has recently gained considerable attention is to promote the flavour symmetry to a modular symmetry, first achieved in [4]. In the modular symmetry approach, all the Yukawa couplings are promoted to modular forms that transform non-trivially under the modular group. This approach leads to the prediction of the lepton and/or quark mixing angles.

Motivated by this powerful framework of modular symmetries we aim to obtain a stronger version of the traditional sum rules [5], where a relation is obtained for the singular values of the mass matrix, i.e. the physical neutrino masses, instead of the complex eigenvalues. Once we consider the measured mass squared differences constraints of neutrino oscillations, the absolute neutrino mass scale is fixed. We derive this sum rule using type-II seesaw mass mechanism [6] with the Yukawas transforming as modular forms under A_4 modular symmetry. This set-up is highly predictive where along with the sum rules we also obtain predictions for the leptonic mixing angles and CP phases. The work presented here closely aligns with [7].

29.2 Model Set-Up

To discuss the model framework we have introduced the finite modular group $\Gamma_3 \simeq A_4$. Within this framework, neutrino masses and mixing will be generated from a type-II seesaw mechanism. Here, we assign the leptonic superfields L_i and E_i^c to be triplet ($\mathbf{3}$) and singlets ($\mathbf{1}, \mathbf{1}', \mathbf{1}''$) under A_4 respectively, with weights of –3 and –1. The Higgs doublets H_u and H_d are trivial singlets ($\mathbf{1}$) of A_4 with weight 0. For the type-II seesaw mechanism we introduce a $SU(2)_L$ triplet Δ, which is also a trivial singlet ($\mathbf{1}$) of A_4 with weight 0. The charge assignment of the superfields and their weights are summarized in Table 29.1.

Using these charge assignments and modular weight, the superpotential of our model is given as follows:

$$\begin{aligned}\mathcal{W} &= \alpha_1 \left(\boldsymbol{Y_e} L\right)_1 E_1^c H_d + \alpha_2 \left(\boldsymbol{Y_e} L\right)_{1''} E_2^c H_d + \alpha_3 \left(\boldsymbol{Y_e} L\right)_{1'} E_3^c H_d \\ &\quad + \alpha \left(\boldsymbol{Y_{\nu,1}} \left(LL\right)_{3_S}\right)_1 \Delta + \beta \left(\boldsymbol{Y_{\nu,2}} \left(LL\right)_{3_S}\right)_1 \Delta \\ &\quad + \mu H_u H_d + \mu_\Delta H_d H_d \Delta\end{aligned} \tag{29.1}$$

Table 29.1 The charge assignments of the superfields under $SU(2)_L \otimes U(1)_Y \otimes A_4$, and $-k$ is modular weight

Fields	$SU(2)_L$	$U(1)_Y$	$\Gamma_3 \simeq A_4$	$-k$
L_i	2	$-\frac{1}{2}$	**3**	-3
E_i^c	1	1	**1, 1′, 1″**	-1
H_u	2	$\frac{1}{2}$	**1**	0
H_d	2	$-\frac{1}{2}$	**1**	0
Δ	3	1	**1**	0

Table 29.2 Modular transformations of Yukawa couplings and their weights

Yukawas	$\Gamma_3 \simeq A_4$	k
$\boldsymbol{Y_e = Y_3^{(4)}}$	**3**	4
$\boldsymbol{Y_{\nu,1} = Y_{3a}^{(6)}}$	**3**	6
$\boldsymbol{Y_{\nu,2} = Y_{3b}^{(6)}}$	**3**	6

Here, we are exploiting the fact that for modular A_4, at weight 6, there are two types of triplets denoted as $3a$ and $3b$. Hence, we have two terms, $\alpha \mathbf{Y}_{\nu,1} L L \Delta$ and $\beta \mathbf{Y}_{\nu,2} L L \Delta$ which contribute to the neutrino mass matrix. In the (29.1), charged lepton Yukawa ($\mathbf{Y_e}$) and neutrino sector Yukawas ($\mathbf{Y}_{\nu,1}$, $\mathbf{Y}_{\nu,2}$) are triplets of A_4 with weight 4 and 6 respectively, as given in Table 29.2. The explicit form of the Yukawas in terms of Dedekind eta-function $\eta(\tau)$ of modulus τ and its derivative $\eta'(\tau)$ is given in [4].

Using (29.1), we can derive the charged lepton mass matrix which is given by

$$M_\ell = v_H \begin{pmatrix} Y_{3,1}^{(4)} & Y_{3,2}^{(4)} & Y_{3,3}^{(4)} \\ Y_{3,3}^{(4)} & Y_{3,1}^{(4)} & Y_{3,2}^{(4)} \\ Y_{3,2}^{(4)} & Y_{3,3}^{(4)} & Y_{3,1}^{(4)} \end{pmatrix} \begin{pmatrix} \alpha_1 & 0 & 0 \\ 0 & \alpha_2 & 0 \\ 0 & 0 & \alpha_3 \end{pmatrix}. \tag{29.2}$$

On the other hand, the neutrino mass matrix is given by

$$M_\nu = v_\Delta \begin{pmatrix} 2Y_1 & -Y_3 & -Y_2 \\ * & 2Y_2 & -Y_1 \\ * & * & 2Y_3 \end{pmatrix} \tag{29.3}$$

where $Y_i \equiv \alpha \, \boldsymbol{Y_{3a,i}^{(6)}} + \beta \, \boldsymbol{Y_{3b,i}^{(6)}}$ with $i \in \{1, 2, 3\}$, v_Δ is the VEV of superfield Δ and $(*)$ represents the symmetric part of neutrino mass matrix. We can write the neutrino mass invariants in terms of $r^2 \equiv |Y_1|^2 + |Y_2|^2 + |Y_3|^2$

$$\mathrm{Tr}(M_\nu^\dagger M_\nu) = 6r^2$$
$$\mathrm{Tr}(M_\nu^\dagger M_\nu \, M_\nu^\dagger M_\nu) = 18r^4$$

Therefore, we find $\frac{1}{2}\text{Tr}(M_\nu^\dagger M_\nu)^2 = \text{Tr}(M_\nu^\dagger M_\nu\, M_\nu^\dagger M_\nu)$, which in turn implies

$$\frac{1}{2}(m_1^2 + m_2^2 + m_3^2)^2 = m_1^4 + m_2^4 + m_3^4$$

We can solve one of the masses, for example for m_3

$$m_3^2 = (m_1 \pm m_2)^2 \tag{29.4}$$

After imposing that the masses are positive we find a unique solution for each ordering

$$\begin{aligned} m_3^{NO} &= m_1^{NO} + m_2^{NO}, \qquad m_3^{NO} > m_2^{NO} > m_1^{NO} \\ m_2^{IO} &= m_1^{IO} + m_3^{IO}, \qquad m_2^{IO} > m_1^{IO} > m_3^{IO} \end{aligned}$$

Alternatively, adding the heaviest of the masses on both sides

$$m_{\text{heaviest}} = \frac{1}{2}\sum_i m_i \tag{29.5}$$

Hence, the neutrino mass matrix features the interesting neutrino mass ordering independent sum rule where m_i; $i = 1, 2, 3$ are the three physical masses of the neutrinos and m_{heaviest} is the heaviest out of the three light neutrinos.

29.3 The Sum Rule Implications

The sum rule in (29.5) can be rewritten in an ordering-dependent way and, after imposing the mass squared differences, the neutrino masses become fixed. If we consider the best fit values [8, 9] of Δm_{ij}^2 we get

NO:

$$m_3 = m_1 + m_2$$
$$\Delta m_{21}^2 = 7.5 \times 10^{-5}\,\text{eV}^2, \quad \Delta m_{31}^2 = 2.55 \times 10^{-3}\,\text{eV}^2$$
$$m_1 = 0.0282\,\text{eV}, \quad m_2 = 0.0295\,\text{eV}, \quad m_3 = 0.0578\,\text{eV}$$

IO:

$$m_2 = m_1 + m_3$$
$$\Delta m_{21}^2 = 7.5 \times 10^{-5}\,\text{eV}^2, \quad \Delta m_{31}^2 = -2.45 \times 10^{-3}\,\text{eV}^2$$
$$m_3 = 7.5 \times 10^{-4}\,\text{eV}, \quad m_1 = 0.049\,\text{eV}, \quad m_2 = 0.050\,\text{eV}$$

where NO (IO) refers to the normal (inverted) mass ordering of the neutrinos.

This result has important consequences for cosmological observations. The sum of neutrino masses which if we allow Δm_{21}^2 and Δm_{31}^2 to vary inside their 3σ regions are predicted by our sum rule with great precision

$$\sum_i m_i^{\mathrm{NO}} \in [0.1138, 0.1176]\ \mathrm{eV}, \tag{29.6a}$$

$$\sum_i m_i^{\mathrm{IO}} \in [0.1007, 0.1041]\ \mathrm{eV}. \tag{29.6b}$$

These values are compatible with the Planck 2018 results [10]. Its successor, the Euclid mission, which was launched in July 2023, will probe the sum of neutrino masses [11], and similarly the ground-based microwave background experiments CMB-S4 [12] and SPT-3G [13] will also be able to rule out the sum rule. As such, the sum rule's validity will be under rigorous examination in the upcoming future.

Another important feature of the sum rule appears in neutrinoless double beta decay experiments. The main model contribution to this process comes from the exchange of light neutrinos. In this case, the total rate is proportional to $|m_{ee}|$, defined as

$$|m_{ee}| = \left|\sum_i U_{ei}^2 m_i\right| = |c_{12}^2 c_{13}^2 m_1 + s_{12}^2 c_{13}^2 e^{2i\phi_{12}} m_2 + s_{13}^2 e^{2i\phi_{13}} m_3|\,, \tag{29.7}$$

where ϕ_{12} and ϕ_{13} are Majorana phases, while the combination $\delta_{CP} = \phi_{13} - \phi_{12} - \phi_{23}$ is the usual Dirac CP phase measured in neutrino oscillations.

Not only the neutrino masses are nearly fixed, but additionally the Majorana phases are not free but highly correlated between each other and the other oscillation parameters as shown in the left panel of Fig. 29.1. Hence $|m_{ee}|$ is also tightly predicted

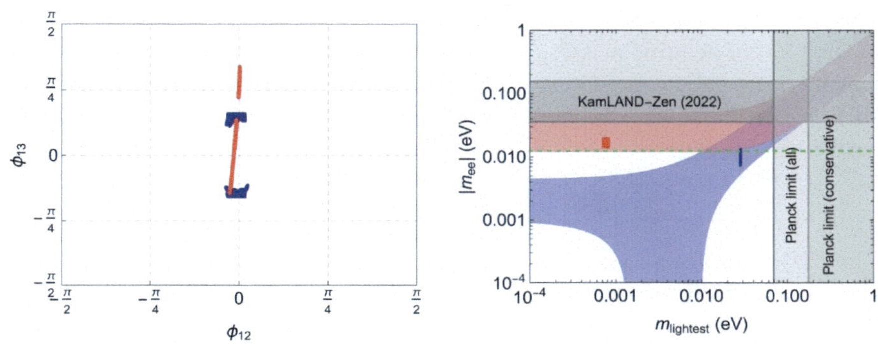

Fig. 29.1 Neutrinoless double beta decay predictions for both NO (blue) and IO (red). **Left panel:** Correlation between Majorana phases. **Right panel:** The combination of the sum rule, the measured mass squared differences and the correlation of the Majorana phases lead to a very precise prediction for $0\nu ee$. The dashed green line is the expected sensitivity of nEXO [14]. The Planck limits shown are for the NO case

in our setup as shown in the right panel of Fig. 29.1. As a visual guide, we have shown the current experimental limits from KamLAND-Zen (2022) [15].

Finally, KATRIN directly measures the effective mass of the electron neutrino which is defined as [16]

$$m_{\nu_e}^{\text{eff}} = \sqrt{\sum_i |U_{ei}|^2 m_i^2} = \sqrt{c_{12}^2 c_{13}^2 m_1^2 + s_{12}^2 c_{13}^2 m_2^2 + s_{13}^2 m_3^2}, \tag{29.8a}$$

$$m_{\nu_e}^{\text{eff}} < 0.8\,\text{eV},\ 90\%\ \text{CL (2023)}, \tag{29.8b}$$

$$m_{\nu_e}^{\text{eff}} < 0.2\,\text{eV},\ 90\%\ \text{CL (2025 expected sensitivity)}. \tag{29.8c}$$

If the mixing parameters are taken to their best fit values, the sum rule predicts

$$\textbf{NO: } m_{\nu_e}^{\text{eff}} = 0.028\ \text{eV}, \tag{29.9a}$$

$$\textbf{IO: } m_{\nu_e}^{\text{eff}} = 0.049\ \text{eV}. \tag{29.9b}$$

Therefore, if KATRIN measures the neutrino mass during its current run the model would be ruled out.

29.4 Neutrino Oscillations Predictions

In addition to the sum rule predictions, our model also can fit the mixing parameters.

Firstly, we emphasize the importance of modular symmetry in constraining the mixing angles. Since the Yukawas now transform as modular forms, their values are controlled by the modulus τ parameter. As a result, the atmospheric angle is tightly correlated with the imaginary part of the modulus τ, see Fig. 29.2.

We now delve into the noteworthy results for the Normal and Inverted Ordering (NO and IO) of neutrino masses. As of the current date, both orderings are experimentally open, but the JUNO experiment is expected to begin collecting data soon and is projected to resolve the hierarchy to the $3-4\sigma$ level over a 6-year period [17]. In what follows, we will use the results of the AHEP global fit [8, 9] by imposing the 2D 1σ or 3σ constraints for the mixing angles θ_{ij}.[1]

29.4.1 Normal Ordering

For the normal ordering of neutrino masses we have $m_3 > m_2 > m_1$, as well as $\Delta m_{31}^2 > 0$. We impose all the mixing angles and mass squared differences to be

[1] 1σ, 2σ and 3σ ranges of AHEP global fit [8, 9] in all 2D contour plots are shown in purple, orange and yellow colour respectively.

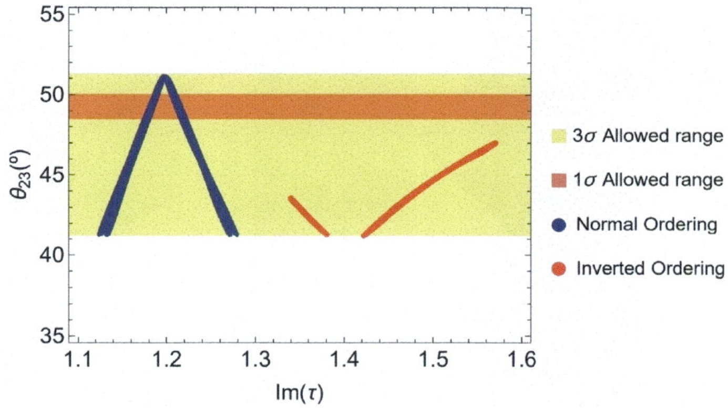

Fig. 29.2 The model predicts a tight correlation between the imaginary part of the modulus τ and the atmospheric angle θ_{23} in both orderings

inside their 2D 3σ regions, but for the sake of representation, we let δ_{CP} in its 5σ regions. For the NO case our model prediction has been shown in Fig. 29.3. The main neutrino oscillation predictions of the model in the NO case are as follows:

(i) $\theta_{13} > 8.36°$, which will be put to test in the next run of T2K [18]
(ii) a correlation between θ_{13} and θ_{23}, which may be resolved by a combination of T2K and Dune [18, 19]
(iii) A correlation between θ_{23} and δ_{CP} which may be falsified by Dune [19].

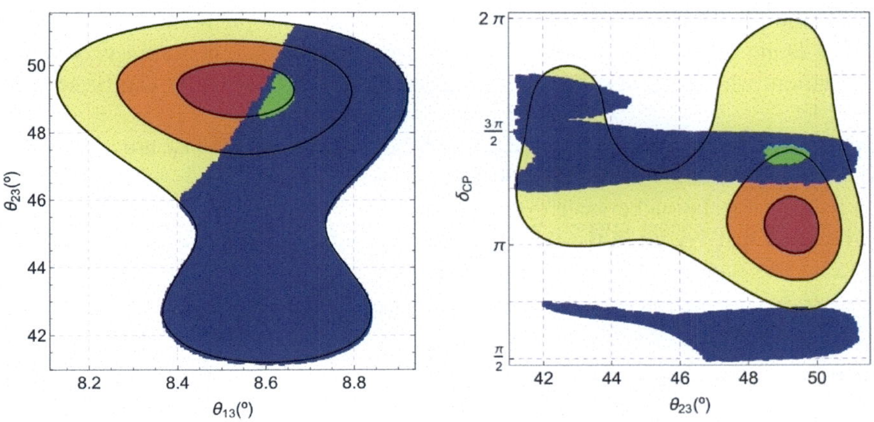

Fig. 29.3 Model predictions in the NO case. In both panels, the blue dots satisfy all the 2D 3σ mixing angles correlations, while the CP violation phase δ_{CP} is in its 5σ region. The green dots instead have the mixing angles inside their 1σ regions and δ_{CP} in their 3σ ones. See text for details. **Left panel:** Correlation between θ_{13} and θ_{23}. **Right panel:** Correlation between θ_{23} and δ_{CP}

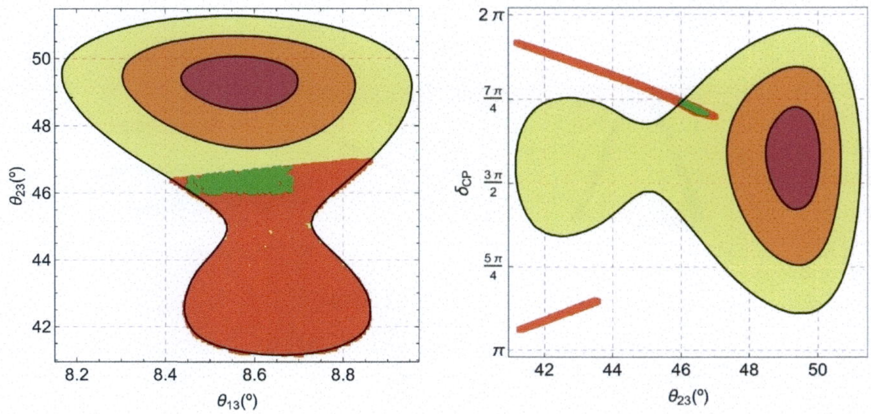

Fig. 29.4 Model predictions in the IO case. In both panels, the red dots satisfy all the 2D 3σ mixing angles correlations, while the CP violation phase δ_{CP} is in its 5σ region. The dark green dots instead have the mixing angles θ_{12} and θ_{13} inside their 1σ regions, while δ_{CP} and θ_{23} are in their 3σ ones. See text for details. **Left panel:** θ_{13} vs θ_{23}. The model predicts an upper bound on $\theta_{23} < 46.8°$. **Right panel:** Sharp correlation between θ_{23} and δ_{CP}

29.4.2 Inverted Ordering

Similarly, for the inverted ordering of neutrino masses, we have $m_2 > m_1 > m_3$, as well as $\Delta m_{31}^2 < 0$. Unlike in the NO case, here the mixing angles θ_{12}, θ_{13} and θ_{23} are nearly uncorrelated. However, θ_{23} has an upper bound of $\theta_{23} < 46.8°$. Since oscillation experiments seem to prefer the lower octant, this prediction may be ruled out by Dune in the near future [19]. Moreover, θ_{23} and δ_{CP} feature a very sharp correlation, which can also be tested by Dune [19]. This behaviour can be seen in Fig. 29.4.

The main neutrino oscillation predictions of the model in the IO case are as follows:
(i) Atmospheric angle has an upper bound $\theta_{23} < 46.8°$
(ii) a sharp correlation between θ_{23} and δ_{CP}.
If the hints that θ_{23} lie in the upper octant get confirmed, the available parameter space of this scenario will be greatly restricted.

29.5 Conclusions

We have introduced a minimal extension of the MSSM based on a modular A_4 symmetry which substantially restricts the number of parameters in the flavour space. The BSM fields and symmetries ingredients of the model are just a $SU(2)_L$ triplet Δ and modulus τ, which gives rise to neutrino masses via a type-II seesaw mechanism. As a consequence, the model is remarkably predictive. In particular, the neutrino mass structure leads to a sum rule for the physical neutrino masses. Combined with neutrino oscillation data this sum rule fixes the absolute neutrino mass scale. The upcoming cosmological probes such as the Euclid mission, the CMB-S4 and SPT-3G experiments, whose first datasets are expected soon, will be able to fully test this sum rule, see (29.6a) and (29.6b). Furthermore, the nEXO experiment will explore part of the relevant parameter space, see Fig. 29.1. On the other hand, the $m_{\nu_e}^{\text{eff}}$ value predicted by the sum rules is below KATRIN's experimental sensitivity, hence any observation in this experiment will rule out the model.

In addition, we also have precise predictions for neutrino oscillation parameters. If neutrino masses follow the normal ordering, the model requires $\theta_{13} > 8.36°$ as well as the correlation between θ_{23} and δ_{CP} shown in Fig. 29.3. If instead, they are arranged in the inverted ordering the predictions are $\theta_{23} < 46.8°$ and an even stronger correlation between θ_{23} and δ_{CP} shown in Fig. 29.4.

References

1. J. Barry and W. Rodejohann, "Neutrino Mass Sum-rules in Flavor Symmetry Models," *Nucl. Phys. B* **842** (2011) 33–50, arXiv:1007.5217 [hep-ph] https://doi.org/10.1016/j.nuclphysb.2010.08.015
2. S.F. King, A. Merle, A.J. Stuart, The Power of Neutrino Mass Sum Rules for Neutrinoless Double Beta Decay Experiments. JHEP series **12**, 005 (2013). https://doi.org/10.1007/JHEP12(2013)005. (arXiv:1307.2901 [hep-ph])
3. J. Gehrlein, A. Merle, and M. Spinrath, "Predictivity of Neutrino Mass Sum Rules," *Phys. Rev. D* **94** no. 9, (2016) 093003, arXiv:1606.04965 [hep-ph] https://doi.org/10.1103/PhysRevD.94.093003
4. F. Feruglio, *Are neutrino masses modular forms?*, pp. 227–266. 2019. arXiv:1706.08749 [hep-ph] https://doi.org/10.1142/9789813238053_0012
5. S. Chuliá Centelles, R. Cepedello, and O. Medina, "Absolute neutrino mass scale and dark matter stability from flavour symmetry," *JHEP* **10** (2022) 080, arXiv:2204.12517 [hep-ph]https://doi.org/10.1007/JHEP10(2022)080
6. J. Schechter, J.W.F. Valle, Neutrino Masses in SU(2) x U(1) Theories. Phys. Rev. series **D22**, 2227 (1980). https://doi.org/10.1103/PhysRevD.22.2227
7. S. Centelles Chuliá, R. Kumar, O. Popov, and R. Srivastava, "Neutrino Mass Sum Rules from Modular $\mathcal{A}_4$ Symmetry," arXiv:2308.08981 [hep-ph]
8. P.F. de Salas, D.V. Forero, S. Gariazzo, P. Martínez-Miravé, O. Mena, C.A. Ternes, M. Tórtola, J.W.F. Valle, 2020 global reassessment of the neutrino oscillation picture. JHEP series **02**, 071 (2021). https://doi.org/10.1007/JHEP02(2021)071. (arXiv:2006.11237 [hep-ph])
9. P. F. De Salas, D. V. Forero, S. Gariazzo, P. Martínez-Miravé, O. Mena, C. A. Ternes, M. Tórtola, and J. W. F. Valle, "Chi2 profiles from Valencia neutrino global fit." http://globalfit.astroparticles.es/, 2021. https://doi.org/10.5281/zenodo.4726908
10. **Planck** Collaboration, N. Aghanim *et al.*, "Planck 2018 results. VI. Cosmological parameters," *Astron. Astrophys.* **641** (2020) A6, arXiv:1807.06209 [astro-ph.CO]. [Erratum: Astron.Astrophys. 652, C4 (2021)] https://doi.org/10.1051/0004-6361/201833910
11. L. Amendola *et al.*, "Cosmology and fundamental physics with the Euclid satellite," *Living Rev. Rel.* **21** no. 1, (2018) 2, arXiv:1606.00180 [astro-ph.CO] https://doi.org/10.1007/s41114-017-0010-3
12. **CMB-S4** Collaboration, K. Abazajian *et al.*, "Snowmass 2021 CMB-S4 White Paper," arXiv:2203.08024 [astro-ph.CO]

13. **SPT-3G** Collaboration, J. S. Avva *et al.*, "Particle Physics with the Cosmic Microwave Background with SPT-3G," *J. Phys. Conf. Ser.* **1468** no. 1, (2020) 012008, arXiv:1911.08047 [astro-ph.CO] https://doi.org/10.1088/1742-6596/1468/1/012008
14. **nEXO** Collaboration, J. B. Albert *et al.*, "Sensitivity and Discovery Potential of nEXO to Neutrinoless Double Beta Decay," *Phys. Rev. C* **97** no. 6, (2018) 065503, arXiv:1710.05075 [nucl-ex] https://doi.org/10.1103/PhysRevC.97.065503
15. **KamLAND-Zen** Collaboration, S. Abe *et al.*, "Search for the Majorana Nature of Neutrinos in the Inverted Mass Ordering Region with KamLAND-Zen," *Phys. Rev. Lett.* **130** no. 5, (2023) 051801, arXiv:2203.02139 [hep-ex] https://doi.org/10.1103/PhysRevLett.130.051801
16. **KATRIN** Collaboration, M. Aker *et al.*, "Direct neutrino-mass measurement with sub-electronvolt sensitivity," *Nature Phys.* **18** no. 2, (2022) 160–166, arXiv:2105.08533 [hep-ex] https://doi.org/10.1038/s41567-021-01463-1
17. **JUNO** Collaboration, A. Abusleme *et al.*, "JUNO physics and detector," *Prog. Part. Nucl. Phys.* **123** (2022) 103927, arXiv:2104.02565 [hep-ex] https://doi.org/10.1016/j.ppnp.2021.103927
18. **T2K** Collaboration, M. A. Ramírez *et al.*, "Measurements of neutrino oscillation parameters from the T2K experiment using 3.6×10^{21} protons on target," arXiv:2303.03222 [hep-ex]
19. **DUNE** Collaboration, B. Abi *et al.*, "Deep Underground Neutrino Experiment (DUNE), Far Detector Technical Design Report, Volume II: DUNE Physics," arXiv:2002.03005 [hep-ex]

Chapter 30
Darko-Lepto-Genesis in Type-III Seesaw

Partha Kumar Paul

Abstract In this study, we propose a concise model in the type-III seesaw framework to address neutrino mass, asymmetric dark matter (DM), and the baryon asymmetry of the Universe. The lepton asymmetry and relic density of DM are generated by the simultaneous CP-violating out-of-equilibrium decay of heavy triplet fermions to the visible sector and dark sector, respectively. Subsequently, the lepton asymmetry transforms into the required baryon asymmetry through electroweak sphalerons, while the asymmetry in the dark sector persists as the observed relic of dark matter. Soft breaking of a Z_2 symmetry, wherein DM is odd, results in an unstable asymmetric DM with a mass ranging from MeV to TeV.

30.1 Introduction

The dark matter (DM), whose existence is well established by various astrophysical and cosmological observations, cannot be accommodated in the Standard Model (SM) of particle physics. The observed relic density of the DM at present epoch is given by $\Omega_{\rm DM}h^2 \sim 0.12$ [1]. The dominance of matter over antimatter in the Universe is one of the unexplained phenomena by the SM. The baryon asymmetry is expressed as $n_B/n_\gamma \sim 6.1 \times 10^{-10}$, where n_B is the baryon number density and n_γ is the photon number density. The current baryon relic density is observed to be $\Omega_{\rm B}h^2 \sim 0.0224$. The DM relic density is $\sim$5.36 times larger than the baryon relic density, which begs for a common origin of both DM and ordinary matter. Apart from this, the non-zero mass of neutrino is one of the most interesting reasons for the search beyond the SM. The mysterious light mass of neutrino, despite evidence, still lacks a known origin. Motivated by these facts, we try to explore the common origin of DM and ordinary matter along with the tiny neutrino mass in the framework of type-III seesaw [2].

P. K. Paul (✉)
Department of Physics, Indian Institute of Technology Hyderabad, Kandi 502285, Telangana, India
e-mail: ph22resch11012@iith.ac.in
URL: https://parthakumarpaul.github.io/pkp/index.html

J. W. F. Valle et al. (eds.), *Proceedings of the International Conference on Future Prospects in Neutrino and Astroparticle Physics*, Springer Proceedings in Physics 361,
https://doi.org/10.1007/978-981-97-7441-8_31

30.2 The Model

We extend the SM with three generations of triplet fermion $\Sigma_i, i = 1, 2, 3$ (1, 3, 0, +), one triplet scalar, Δ (1, 3, 0, –), one vector like singlet fermion, χ (1, 1, 0, –), and a singlet scalar ϕ (1, 1, 0, +), where in brackets the charge assignment of the fields under $SU(3)_c \otimes SU(2)_L \otimes U(1)_Y \otimes Z_2$ is indicated.

The relevant Lagrangian in our model is given as [2]

$$\begin{aligned}\mathcal{L} &= Tr[\overline{\Sigma} i\gamma_\mu D^\mu \Sigma] + Tr[(D_\mu \Delta)^\dagger (D_\mu \Delta)] + (D_\mu H)^\dagger (D_\mu H) + \tfrac{1}{2}\,\partial_\mu \phi\, \partial^\mu \phi \\ &- \Big[Tr[m_\Sigma \overline{\Sigma^c} \Sigma] + y_{_\Sigma} \overline{L} \tilde{H} \Sigma + y_\chi Tr[\overline{\Sigma}\ \Delta\ \chi] + \text{h.c.}\Big] + \overline{\chi} i\gamma_\mu \partial^\mu \chi - M_\chi \overline{\chi}\chi \\ &\qquad -\lambda_{DM} \phi \bar{\chi}\chi - V(H, \Delta, \phi), \end{aligned} \tag{30.1}$$

where the scalar potential, $V(H, \Delta, \phi)$ is given as

$$\begin{aligned} V(H, \Delta, \phi) &= \tfrac{1}{2} m_\phi^2 \phi^2 - \mu_H^2 (H^\dagger H) + \tfrac{1}{2} M_\Delta^2 (\mathbf{\Delta}\cdot\mathbf{\Delta}) + \tfrac{1}{2}\lambda_H (H^\dagger H)^2 + \tfrac{1}{8}\lambda_\Delta (\mathbf{\Delta}\cdot\mathbf{\Delta})^2 \\ &+ \tfrac{1}{2}\lambda_{H\Delta}(H^\dagger H)(\mathbf{\Delta}\cdot\mathbf{\Delta}) + \sqrt{2}\mu H^\dagger (\vec{\sigma}\cdot\mathbf{\Delta}) H + \tfrac{\lambda_\phi}{2}\phi^4 \\ &+ \rho_1 \phi (H^\dagger H) + \lambda_{H\phi}(H^\dagger H)\phi^2\,. \end{aligned} \tag{30.2}$$

In 30.2, due to presence of tri-linear term, $\sqrt{2}\mu H^\dagger(\vec{\sigma}\cdot\mathbf{\Delta})H$, Δ obtains an induced vacuum expectation value (vev), $v_1 \simeq \frac{\mu v_0^2}{2M_\Delta^2}$, and as a result the Z_2 symmetry is softly broken. Here v_0 = 246 GeV. The DM stability can be achieved with $v_1 \sim 1$ eV, i.e. the soft breaking coupling, μ has to be $\mathcal{O}(10^{-8})$ GeV, which results in a negligible mixing between H (SM Higgs) and Δ ($\mathcal{O}(10^{-11})$). After the electroweak symmetry breaking (EWSB) H and ϕ mix and give two mass eigen-states h_1 (identified as SM Higgs) and h_2 (dominantly ϕ). Electroweak corrections induce a mass splitting of approximately $\mathcal{O}(170)$ MeV between the charged ($\delta^\pm$) and neutral (δ^0) constituents of Δ. This mass difference results in a suppression of the phase space for the decay of $\delta^\pm$, potentially leading to displaced vertex signatures [2].

30.3 Neutrino Mass

Neutrino mass can be realized via the type-III seesaw mechanism [3], and the light neutrino mass matrix can be written as

$$(m_\nu)_{\alpha\beta} = (m_D)_{\alpha i} \left(M^{-1}\right)_{ij} (m_D^T)_{j\beta}, \tag{30.3}$$

where m_D represents the 3×3 Dirac mass matrix, while M denotes the 3×3 mass matrix for triplet fermions. Without loss of generality, one can assume M to be diagonal.

30.4 Darko-Lepto-Genesis

To generate successful leptogenesis, three Sakharov's conditions need to be satisfied which are baryon/lepton number violation, C and CP violation, and out-of-thermal equilibrium. In our model, the Lagrangian responsible for neutrino mass generation (i.e. $-Tr[m_\Sigma \overline{\Sigma^c}\Sigma] - y_\Sigma \overline{L}\tilde{H}\Sigma + \text{h.c.}$) breaks the lepton number by 2 units. The expansion of the Universe provides the out-of-equilibrium condition. Now the CP-violating out-of-equilibrium decay of Σ_is to SM leptons (L) and Higgs generates a net lepton asymmetry. We consider the fermion triplet mass spectrum as: $M_{\Sigma_1} < M_{\Sigma_2} < M_{\Sigma_3}$. As the temperature falls below the mass scales of Σ_2 and Σ_3, their interactions go out-of-equilibrium and the asymmetries produced by them are diminished through the lepton number-violating interactions involving Σ_1. So we consider the final lepton asymmetry is produced by only $\Sigma_1 \rightarrow LH$. When the Σ_1 is decaying to visible sector (L, H), at the same time it is also decaying to χ, Δ which we call as dark sector. The simultaneous CP-violating out-of-equilibrium decay of Σ_1 to dark sector (i.e. $\Sigma_1 \rightarrow \chi\Delta$) generates a net dark sector asymmetry provided the χ carries lepton number $+1$. The generated lepton asymmetry gets converted to baryon asymmetry by electroweak sphalerons. The asymmetries generated in H and Δ are washed away once they obtain vev after EWSB. The asymmetry produced in χ remains as the present DM relic which is termed as asymmetric DM. The singlet scalar ϕ helps the symmetric component of χ to annihilate completely. The abundances of Σ_1 and the asymmetries are governed by Boltzmann equations (BE) given as

$$\frac{dY_{\Sigma_1}}{dz} = -\frac{\Gamma_D}{\text{H}'z}(Y_{\Sigma_1} - Y_{\Sigma_1}^{eq}) + (2 \leftrightarrow 2) + \text{Gauge interactions}, \tag{30.4}$$

$$\frac{dY_{\Delta L}}{dz} = \epsilon_L \frac{\Gamma_D}{\text{H}'z}(Y_{\Sigma_1} - Y_{\Sigma_1}^{eq}) + \text{Inverse decay} + (2 \leftrightarrow 2 \text{ washout} + \text{transfer}), \tag{30.5}$$

$$\frac{dY_{\Delta\chi}}{dz} = \epsilon_\chi \frac{\Gamma_D}{\text{H}'z}(Y_{\Sigma_1} - Y_{\Sigma_1}^{eq}) + \text{Inverse decay} + (2 \leftrightarrow 2 \text{ washout} + \text{transfer}), \tag{30.6}$$

where $Y_x = n_x/n_\gamma$; $x = \Sigma_1, \Delta_L, \Delta_\chi$, $z = M_{\Sigma_1}/T$, H' is the Hubble parameter, ϵ_L, and ϵ_χ are the CP asymmetry parameters in visible and dark sectors, respectively. The first term in RHS of 30.4 is the decay term, $(2 \leftrightarrow 2)$ accounts for all $\Delta L = 1, 2$ scattering processes, and the $\Delta L = 0$ gauge scatterings contribute to the last term. In 30.5 and 30.6, the first terms in the RHS are the source terms for the asymmetries whereas the other terms account for the washout which involves inverse decay, $\Delta L = 1, 2$ scatterings, and $\Delta L = 0$ transfer processes which transfer the asymmetries from one sector to other. The details of the BE can be found in [2]. It has to be noted that the gauge scatterings, which are absent in leptogenesis via type-I seesaw, play an important role in determining the number density of the triplet fermion. Since they do not violate the lepton number, they come only in the evolution of the fermion triplet. Even if we start with an initial abundance of triplet fermions set to zero, the presence of these gauge interactions drives their number density towards equilibrium, particularly when the mass, $M_{\Sigma_1} < 10^{14}$ GeV, leading to rapid thermal equilibrium. As a result, the asymmetry generated remains independent of the initial conditions related to the triplet number density. Based on the strength of the interaction rates, the washout regime is categorized into two : strong washout (sw) and weak washout (ww). When the interaction rate is greater than the Hubble expansion rate of the Universe, the regime is called the strong washout regime and the opposite one is called the weak washout regime. The efficiency of the washout, η is defined as

$$\eta_i = \frac{Y_{\Delta i}}{\epsilon_i \, Y^{eq}_{\Sigma_1}|_{T \gg M_{\Sigma_1}}} \; ; \quad i = L, \, \chi. \tag{30.7}$$

In the ww regime $\eta \to 1$, and in sw regime $\eta \ll 1$.

The asymmetries can be expressed in terms of the efficiency factor and CP asymmetry parameter as

$$Y_{\Delta i} \equiv \frac{n_{\Delta i}}{n_\gamma} = \epsilon_i \eta_i \frac{n_{\Sigma_1}}{n_\gamma} \; ; \quad i = L, \, \chi. \tag{30.8}$$

The generated lepton asymmetry gets converted to baryon asymmetry by sphalerons and is given as

$$Y_B = -0.55 Y_{\Delta L}. \tag{30.9}$$

Using (30.8) and (30.9), we get the DM to baryon relic density ratio as

$$\frac{\Omega_\chi h^2}{\Omega_B h^2} = \frac{1}{0.55} \frac{M_\chi}{M_p} \frac{\epsilon_\chi}{\epsilon_L} \frac{\eta_\chi}{\eta_L} = \frac{1}{0.55} \frac{M_\chi}{M_p} \frac{Y_{\Delta\chi}}{Y_{\Delta L}}, \tag{30.10}$$

where M_p is the proton mass. We have performed a numerical scan over wide range of parameter space satisfying the observed lepton asymmetry ($1.05455 \times 10^{-9} \leq Y_{\Delta L} = \frac{Y_B}{S} \leq 1.16364 \times 10^{-9}$) and the ratio of DM to baryon relic density ($5.27472 \leq \Omega_\chi h^2 / \Omega_B h^2 \leq 5.45394$). In the scan we have fixed the mass of the triplet scalar, Δ at M_Δ=1 TeV and used the Davidson-Ibarra bound [4] on the triplet fermion mass which is $M_{\Sigma_1} \gtrsim 3 \times 10^9$.

From (30.10) we can see that for a fixed value of DM mass, M_χ, as the ratio of dark sector efficiency factor to visible sector efficiency factor increases, the ratio of the CP asymmetries in dark sector to visible sector has to be decreased to explain the observed DM to baryon relic density ratio. To illustrate this we have shown the parameter space in $\eta_\chi/\eta_L - \epsilon_\chi/\epsilon_L$ plane as a function of M_χ in Fig. 30.1. We see that for a fixed value of $M_\chi = 100$ GeV : $\epsilon_\chi/\epsilon_L \sim 1$ corresponds to $\eta_\chi/\eta_L \sim 10^{-2}$ and $\epsilon_\chi/\epsilon_L \sim 10^{-4}$ corresponds to $\eta_\chi/\eta_L \sim 10^2$. The details of the scan can be found in [2]. In left panel of Fig. 30.2, we have shown the cosmological evolution of Y_{Σ_1}, $Y_{\Delta L}$, $Y_{\Delta\chi}$, and $Y^{eq}_{\Sigma_1}$ w.r.t. z in the strong washout regime for a benchmark point (BP1) from our scan. The BP1 is : $\{\epsilon_\chi = 1.55 \times 10^{-9}, \; \epsilon_L = 1.25 \times 10^{-7}, \; y_\Sigma = 3.79 \times 10^{-3}, \; y_\chi = 3.97 \times 10^{-3}, \; M_\chi = 112.83 \text{ GeV}, \; M_{\Sigma_1} = 8.84 \times 10^9 \text{ GeV}, \; M_\Delta = 1 \text{ TeV}\}$. As can be seen from the plot, the abundance of Σ_1 closely traces its equilib-

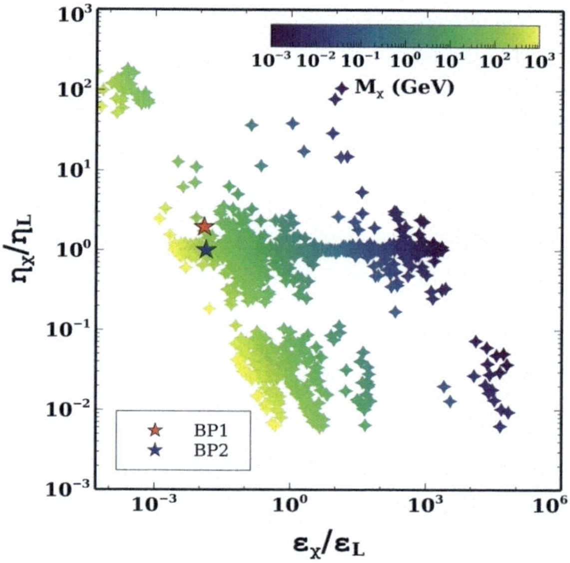

Fig. 30.1 Parameter space in $\eta_{\chi}/\eta_{L} - \epsilon_{\chi}/\epsilon_{L}$ plane as a function of DM mass, M_{χ}

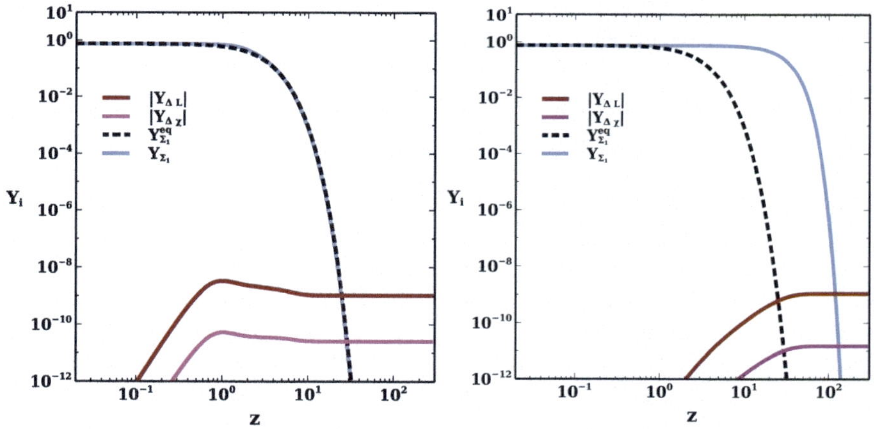

Fig. 30.2 Cosmological evolution of Y_{Σ_1}, $Y_{\Delta L}$, $Y_{\Delta\chi}$, and $Y^{eq}_{\Sigma_1}$ w.r.t. z in the strong [left] and weak [right] washout regimes

rium abundance. In this case the branching ratio to the visible sector is 64.66% and its to the dark sector is 35.34% which results in more suppression in visible sector asymmetry than dark sector asymmetry due to the strong washout effect. We have shown a benchmark point (BP2) in weak washout regime in the right panel of Fig. 30.2. The BP2 is : $\{\epsilon_{\chi} = 2.01 \times 10^{-11},\ \epsilon_L = 1.48 \times 10^{-9},\ y_{\Sigma} = 5.92 \times 10^{-4},\ y_{\chi} = 4.29 \times 10^{-4},\ M_{\chi} = 204.99\,\text{GeV},\ M_{\Sigma_1} = 2.17 \times 10^{12}\,\text{GeV},\ M_{\Delta} = 1\,\text{TeV}\}$. In this regime, the fermion triplets cannot follow the equilibrium number density throughout its evolution. Even though the branching to the visible sector is 79.26% and that to the dark sector is 20.74%, there is no observed suppression in the asymmetry within the visible sector which is a manifestation of the weak washout effects.

Now the generated asymmetry in the χ remains as the DM relic. To achieve this, it is necessary for the symmetric component of χ to undergo complete annihilation. For this purpose, we introduce a singlet scalar ϕ and the process contributing to the complete annihilation of the symmetric component of dark matter is $\bar{\chi}\chi \to \phi\phi$. The decay width of ϕ to SM fermions after mixing with the SM Higgs is given as

$$\Gamma_{\phi\to f\bar{f}} = \sin^2\gamma\left(\frac{m_f}{v_0}\right)^2 \frac{M_{h_2}}{8\pi}\left(1-4\frac{m_f^2}{M_{h_2}^2}\right)^{\frac{3}{2}}, \tag{30.11}$$

where $\sin\gamma \sim \frac{\rho_1 v_0}{M_{h_1}^2}$ is the mixing between H and ϕ. For the analysis of the later part of this work, we ensure that ϕ decays to the SM fermions well before the Big Bang Nucleosynthesis (BBN) by appropriately choosing the values of ϕ mass, M_{h_2} and its mixing with SM Higgs, $\sin\gamma$ so that it does not alter the predictions of BBN.

30.5 Dark Matter Phenomenology

For a particle to be considered a dark matter candidate, its lifetime must exceed 10^{25} s [5]. However, the Z_2 symmetry, ensuring the stability of dark matter, experiences a soft breaking when Δ obtains an induced vev following the EWSB and makes the DM unstable. The DM then decays to the SM particle and ϕ after mixing with Σ_1. The different decay modes are $\chi \to W^+e^-$, $\chi \to Z\nu$, $\chi \to \phi\nu$, $\chi \to H\nu$, $\chi \to \nu f\bar{f}$. We have shown the lifetime of the DM as function of its mass in Fig. 30.3. Here we have used the points that satisfy correct DM relic density and correct lepton asymmetry from the previous section as input in the decay rates and taken $M_{h_2} = 50$ MeV, $\sin\gamma = 10^{-5}$, $\lambda_{DM} = 0.02$, and $v_1 = 1$ eV. The red shaded region is excluded by the requirement of DM lifetime which is $\tau_{DM} > 10^{25}$ s. We see that the DM in a wide mass range (MeV to TeV) is allowed by the DM lifetime constraint as well as these points also satisfy the required DM relic density and lepton asymmetry of the Universe simultaneously.

The singlet scalar, ϕ, which was introduced for the complete annihilation of the symmetric component of the dark matter, mixes with Higgs and opens up the window for DM direct detection. We have computed the DM nucleon spin-independent cross-section and shown the parameter space in the plane of $\lambda_{DM} - \sin\gamma$ for a DM mass of 10 GeV (1 TeV) and ϕ mass of 5 GeV in left (right) panel of Fig. 30.4. We have imposed direct detection experiment bounds such as XENONnT [6], LZ [7], perturbative bound which is $\lambda_{DM} < \sqrt{4\pi}$. For the full annihilation of the sym-

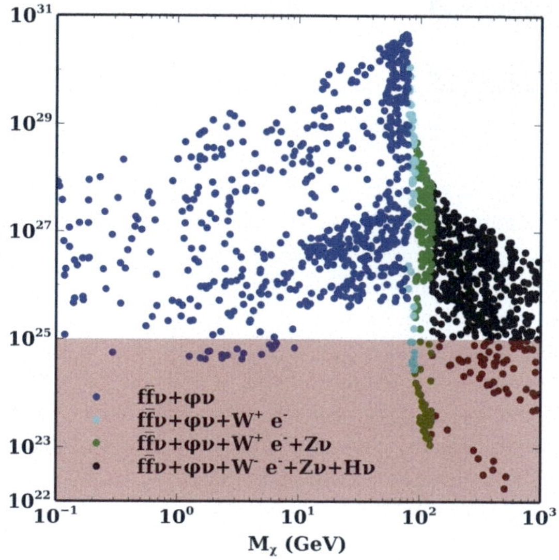

Fig. 30.3 DM lifetime as a function of mass of the DM

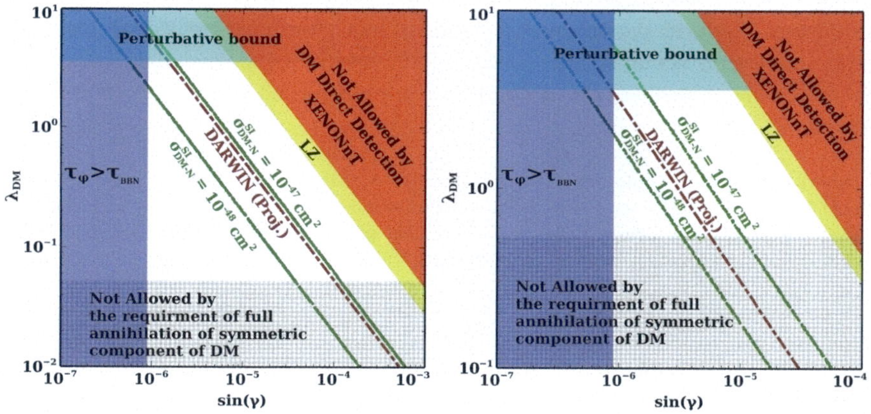

Fig. 30.4 The Higgs portal mediated spin-independent scattering between DM and nucleons (N) in the plane $\lambda_{DM} - \sin\gamma$. [left]: $M_\chi = 10\,\text{GeV}$, $M_{h_2} = 5\,\text{GeV}$, [right]: $M_\chi = 1\,\text{TeV}$, $M_{h_2} = 5\,\text{GeV}$

metric component of the DM, its annihilation cross-section has to be $> 2.6 \times 10^{-9}$ GeV $^{-2}$. The region shaded in grey in Fig. 30.4 is excluded by the requirement of complete annihilation of symmetric component of dark matter. We have also shown the projected sensitivity of DARWIN [8]. Certainly, the future direct search experiments featuring enhanced sensitivities will have the capability to explore this parameter space.

30.6 Conclusion

We investigate an economical model in type-III seesaw framework to elucidate phenomena such as non-zero neutrino mass, dark matter, and baryon asymmetry of the Universe. The SM of particle physics is extended by three hypercharge zero fermion triplets (Σ_1, Σ_2, Σ_3), one vector like singlet fermion (χ), one hypercharge zero triplet scalar (Δ), and one singlet scalar (ϕ). The CP-violating out-of-equilibrium decay of the lightest triplet, Σ_1, to both the visible and dark sectors, facilitates the simultaneous generation of lepton and DM asymmetries. It has to be noted that, due to additional gauge interactions in the type-III seesaw scenario, the fermion triplets undergo thermalization in the early universe, and for $M_{\Sigma_1} < 10^{14}$ GeV, the final lepton asymmetry becomes independent of the initial condition on the triplet number density. The generated lepton asymmetry gets converted to baryon asymmetry by sphalerons. The singlet scalar, ϕ helps in full annihilation of the symmetric component of the DM. The asymmetry in the χ remains as the DM relic which is called as asymmetric DM. The asymmetry generated in the Δ vanishes once it obtains a vev and Z_2 symmetry gets softly broken. However, breaking of the Z_2 symmetry makes the DM unstable. A wide range of DM mass is allowed after imposing the DM lifetime constraint by appropriately tuning the vev of the scalar triplet, Δ. The parameter space consistent with DM phenomenology and successful leptogenesis is explored. We have also discussed the direct detection prospect of DM considering various experimental bounds. The small mass splitting between the charged and the neutral components of the Δ can give rise to displaced vertex signatures at the collider [2].

Acknowledgements I would like to thank Satyabrata Mahapatra, Narendra Sahu, and Prashant Shukla for useful discussions. I would like to acknowledge the Department of Atomic Energy-Board of Research in Nuclear Sciences, Government of India, (Ref. Number: 58/14/15/2021-BRNS/37220) for providing financial support during the work.

References

1. Planck Collaboration, N. Aghanim et al., Planck 2018 results. VI. Cosmological parameters, https://doi.org/10.48550/arXiv.1807.06209
2. S. Mahapatra, P.K. Paul, N. Sahu, P. Shukla, Cogenesis of matter and dark matter from triplet fermion seesaw (2023). arXiv:2305.11138
3. R. Foot, H. Lew, X. He, G.C. Joshi, Seesaw neutrino masses induced by a triplet of leptons. Z. Phys. C **44**, 441 (1989)
4. S. Davidson, A. Ibarra, A lower bound on the right-handed neutrino mass from leptogenesis. Phys. Lett. **B535**, 25–32 (2002). https://doi.org/10.48550/arXiv.hep-ph/0202239
5. M.G. Baring, T. Ghosh, F.S. Queiroz, K. Sinha, new limits on the dark matter lifetime from dwarf spheroidal galaxies using fermi-LAT. Phys. Rev. D **93**(10), 103009 (2016). https://doi.org/10.48550/arXiv.1510.00389

6. XENON Collaboration, E. Aprile et al., First dark matter search with nuclear recoils from the XENONnT experiment. Phys. Rev. Lett. **131**, 041003 (2023). https://doi.org/10.48550/arXiv.2303.14729
7. LZ Collaboration, J. Aalbers et al., First dark matter search results from the LUX-ZEPLIN (LZ) experiment. Phys. Rev. Lett. **131**(4), 041002 (2023). https://doi.org/10.48550/arXiv.2207.03764
8. DARWIN Collaboration, J. Aalbers et al., DARWIN: towards the ultimate dark matter detector. JCAP **11**, 017 (2016). https://doi.org/10.48550/arXiv.1606.07001

Chapter 31
Neutrino Phenomenology in Texture Zeros of Minimal Inverse Seesaw

Nayana Gautam and Mrinal Kumar Das

Abstract We study texture zeros of Dirac mass matrix in a minimal inverse seesaw (2, 3) model with S_4 flavor symmetry. The seesaw model is extended with the addition of two right-handed neutrinos and three sterile states in intermediate mass range. The structure of the mass matrices is obtained from S_4 symmetry. There are six possible two zero textures of the Dirac mass matrix. We study the neutrino mass and mixing within the six different textures emphasizing Dirac CP phase and the mixing angle θ_{23} that may shed light on octant problem. Based on the allowed ranges of the mass and mixing with the active neutrinos, we study the viability of the different textures.

31.1 Introduction

Standard Model (SM) of particle physics is successful in incorporating our current knowledge of elementary particles along with many observed phenomena in particle physics and cosmology as well. However, many important issues like the origin of neutrino mass, mass ordering, flavor puzzle, matter-antimatter asymmetry, dark matter remain unexplained within the framework of SM. Consequently, different BSM frameworks have been introduced in the literature that can account for such issues. Inverse seesaw is one of the most promising mechanisms as the neutrino mass is generated at a lower scale than the conventional seesaw. In this paper, we study an inverse seesaw framework where two right-handed neutrinos and three sterile fermions are added to the standard model particle content. The number of free parameters in the model can be significantly reduced by implementing specific structures of the mass matrices with some zero entries at certain positions [1–4]. In inverse seesaw ISS (2, 3), light neutrino mass matrix is generated by three mass matrices namely Dirac neutrino mass matrix M_D, heavy neutrino mass matrix M_N,

N. Gautam (✉)
Department of Physics, Lakhimpur Girls' College, Khelmati, North Lakhimpur 787031, India
e-mail: nayanagtm72@gmail.com

M. K. Das
Department of Physics, Tezpur University, Tezpur 784028, India
e-mail: mkdas@tezu.ernet.in

J. W. F. Valle et al. (eds.), *Proceedings of the International Conference on Future Prospects in Neutrino and Astroparticle Physics*, Springer Proceedings in Physics 361,
https://doi.org/10.1007/978-981-97-7441-8_32

and sterile neutrino mass matrix μ [5]. In our study, we have taken maximal possible zeros in M_N as well as in μ, and the vanishing elements are postulated in the matrix M_D. We also address the explicit origin of such textures of matrices, assuming S_4 flavor symmetry [6]. The maximal allowed zeros in M_D is two. After constructing the mass matrices, we study the predictions of the textures on neutrino phenomenology like mass and mixing and Dirac CP phase.

The paper is structured as follows. In Sect. 31.2, we describe the neutrino mass production within the framework of ISS (2, 3). Section 31.3 is the brief discussion on different zero textures of Dirac mass matrix along with the S_4 flavor symmetric origin. In Sect. 31.4, we discuss the results of the numerical analysis. Finally, we present our conclusion in Sect. 31.5.

31.2 The Framework

In minimal inverse seesaw, the standard model of particle physics is extended by the sequential addition of two RH neutrinos and three SM singlet fermions s_i [5, 7]. The Lagrangian in this model can be written as,

$$L = -\frac{1}{2} n_L^T C M n_L + h.c \tag{31.1}$$

The mass matrix in the basis $n_L = (\nu_L, N_R, s)$ can be written as,

$$M = \begin{pmatrix} 0 & M_d & 0 \\ M_d^T & 0 & M_N \\ 0 & M_N^T & \mu \end{pmatrix} \tag{31.2}$$

The light active neutrino mass in this framework can be obtained as,

$$m_\nu \approx M_d d M_d^T \tag{31.3}$$

where d is a 2×2 matrix given by,

$$M_H^{-1} = \begin{pmatrix} d_{2\times 2} & \cdots \\ \cdots\cdot & \cdots\cdot \end{pmatrix} with\ M_H = \begin{pmatrix} 0 & M_N \\ M_N^T & \mu \end{pmatrix} \tag{31.4}$$

31.3 Texture Zeros of Dirac Mass Matrix with S_4 Flavor Symmetry

In ISS (2, 3) model, the light neutrino mass matrix can be obtained from three mass matrices M_D, M_N, and μ. Therefore, the allowed number of zeros in the light neutrino

Table 31.1 Fields and their respective transformations under the symmetry group of the model

Field	L	l_R	$\bar{N_R}$	H	s	ϕ	ϕ'	ϕ_s	ϕ_l	χ	χ'
S_4	3_1	3_1	2	1_1	1_1	3_2	3_1	1_1	1_1	3_2	3_1
$SU(2)_L$	2	1	1	2	1	1	1	1	1	1	1
Z_4	1	1	i	1	$+i$	$-i$	$-i$	-1	$-i$	$-i$	$-i$
Z_3	ω^2	1	1	1	ω	ω	ω	ω	ω	ω^2	ω^2

mass matrix further constrains the structure of these three matrices [8]. In our study, maximal possible zeros in M_N and μ are considered [9]. The charge assignments of the particles in the model under the symmetry group S_4 are shown in Table 31.1 [7].

The Yukawa Lagrangian for the charged leptons and also for the neutrinos can be expressed as:

$$-\mathcal{L} = \mathcal{L}_{\mathcal{M}_\mathcal{L}} + \mathcal{L}_{\mathcal{M}_\mathcal{D}} + \mathcal{L}_{\mathcal{M}_\mathcal{N}} + \mathcal{L}_{\mathcal{M}_\mathcal{S}} + h.c \tag{31.5}$$

As mentioned in our work [7], the matrices M_N and μ and the charged lepton mass matrix m_l are,

$$M_N = \frac{1}{\Lambda}\begin{pmatrix} y_{1r}v_r v_r' & 0 & 0 \\ 0 & y_{2r}v_r' v_r' & 0 \end{pmatrix}. \tag{31.6}$$

$$\mu = y_s \begin{pmatrix} 1 & 0 & 0 \\ 0 & 1 & 0 \\ 0 & 0 & 1 \end{pmatrix} v_s \tag{31.7}$$

$$m_l = \frac{v_h}{\Lambda}\begin{pmatrix} y_l v_l & 0 & 0 \\ 0 & y_l v_l & 0 \\ 0 & 0 & y_l v_l \end{pmatrix}, \tag{31.8}$$

Here, $p = y_s v_s$, $f = y_{1r} v_r v_r'$ and $g = y_{2r} v_r' v_r'$. With these structures of M_N and μ, the allowed two zero texture zeros of M_D have already been shown in [10].

The Dirac Lagrangian in this model can be written as [10],

$$\mathcal{L}_{\mathcal{M}_\mathcal{D}} = \frac{y}{\Lambda}\bar{N_R}LH\phi + \frac{y\prime}{\Lambda}\bar{N_R}LH\phi' \tag{31.9}$$

The cut-off scale Λ lowers the mass dimension to 4. Here, we have also taken extra scalar ϕ and ϕ' with SM Higgs. Different VEV alignments of ϕ and ϕ' lead to different six classes of texture zeros of Dirac mass matrix in ISS (2, 3) as shown in our previous work [10]. These are

Class A1:

$$M_D^1 = \begin{bmatrix} 0 & c & e \\ b & 0 & h \end{bmatrix} \tag{31.10}$$

Class A2:

$$M_D^2 = \begin{bmatrix} a & 0 & e \\ b & d & 0 \end{bmatrix} \quad (31.11)$$

Class A3:

$$M_D^3 = \begin{bmatrix} a & c & 0 \\ 0 & d & h \end{bmatrix} \quad (31.12)$$

Class B1:

$$M_D^4 = \begin{bmatrix} a & c & e \\ 0 & 0 & h \end{bmatrix} \quad (31.13)$$

Class B2:

$$M_D^5 = \begin{bmatrix} a & c & e \\ 0 & d & 0 \end{bmatrix} \quad (31.14)$$

Class B3:

$$M_D^6 = \begin{bmatrix} a & c & e \\ b & 0 & 0 \end{bmatrix} \quad (31.15)$$

31.4 Numerical Analysis and Results

The light neutrino mass matrix in ISS (2, 3) can be diagonalized with the unitary PMNS matrix [11] as follows,

$$m_\nu = U_{PMNS} m_\nu^{diag} U_{PMNS}^T \quad (31.16)$$

In the framework of ISS (2, 3), the lightest neutrino mass is found to be zero. The model parameters in all the textures can be evaluated by comparing the neutrino mass matrix arising from the model with the one which is parametrized by the available 3σ global fit data [12]. Comparing the two matrices we evaluate the masses, mixing angles, and Dirac CP phase. The results obtained are shown below.

From Figs. 31.1 and 31.2, it is evident that a wide range of parameter space leads to the allowed range of $\sin^2\theta_{23}$. Figure 31.1 shows that maximum data lead to the upper octant of $\sin^2\theta_{23}$. However, there are data that satisfy lower octant too. The range of δ_{CP} obtained from the model lies within 120°–400° as shown in Figs. 31.3 and 31.4. The range obtained lies within the allowed in 3σ range.

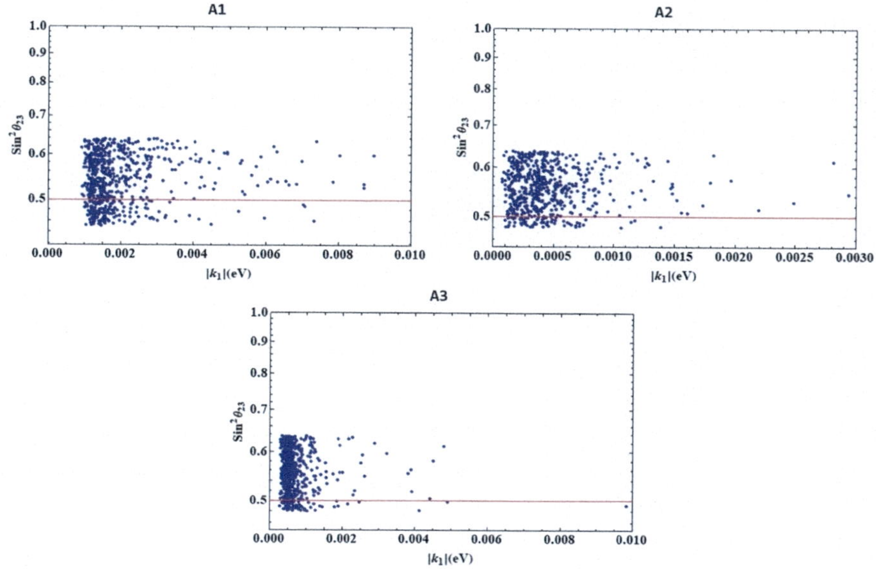

Fig. 31.1 $\sin^2\theta_{23}$ as a function of the model parameter k_1 in textures A1, A2, A3

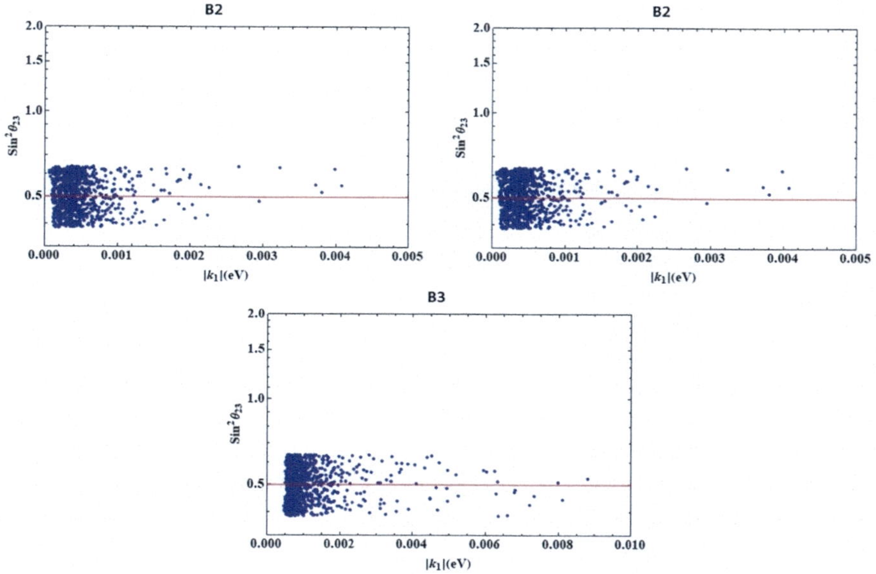

Fig. 31.2 $\sin^2\theta_{23}$ as a function of the model parameter k_1 in textures B1, B2, B3

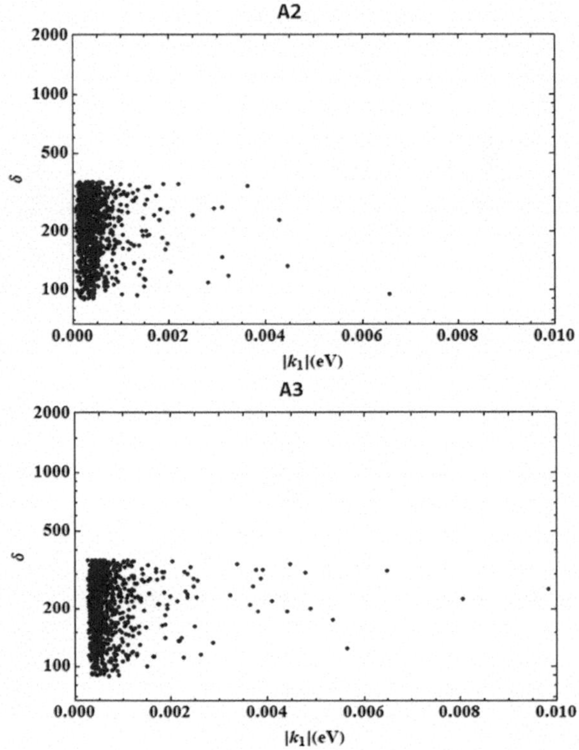

Fig. 31.3 Predictions of different one zero textures of neutrino mass matrix on Dirac CP phase

31.5 Summary and Conclusion

We have carried out the study of neutrino phenomenology in the framework of ISS (2, 3) that leads to three light active neutrino states, two pseudo-Dirac states, and an additional sterile state in keV range. We have explored different two zero textures of Dirac mass matrix and their impacts on neutrino phenomenology. There are six possible allowed two zero textures in the framework. We have calculated neutrino mass, mixing angles, and Dirac CP phase using the evaluated model parameters in all the textures. The parameter space predicted by the model lies within the range of neutrino oscillation data. Three of the textures lead to range of θ_{23} that lie in upper octant. Other three textures support both the upper and lower octant of θ_{23}. The range of δ_{CP} obtained from the model lies within 120°–400°. Thus, we may conclude that the two zero textures in the framework of ISS (2, 3) are suitable for explaining neutrino phenomenology.

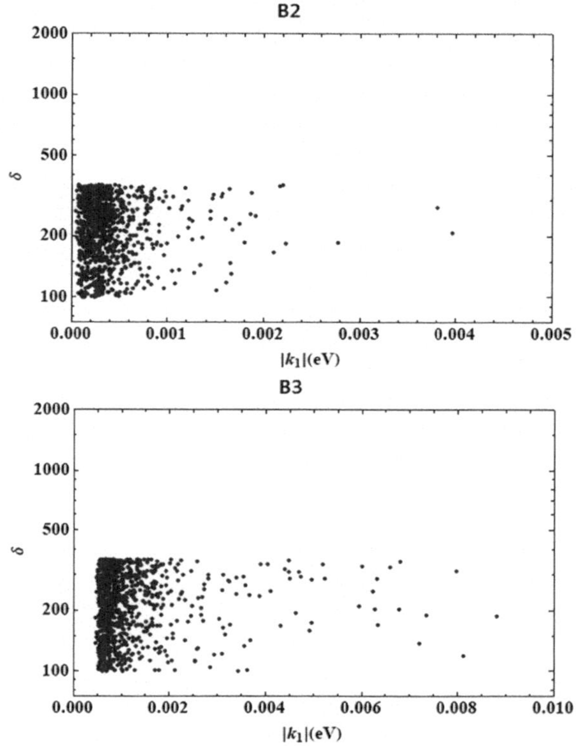

Fig. 31.4 Predictions of different two zero textures of Dirac mass matrix on Dirac CP phase

References

1. A. Ghosal, R. Samanta, Probing texture zeros with scaling ansatz in inverse seesaw. JHEP **05**, 077 (2015)
2. A. Kageyama, S. Kaneko, N. Shimoyama, M. Tanimoto, Seesaw realization of the texture zeros in the neutrino mass matrix. Phys. Lett. B **538**, 96–106 (2002)
3. G.C. Branco, D. Emmanuel-Costa, M.N. Rebelo, P. Roy, Four zero neutrino Yukawa textures in the minimal seesaw framework. Phys. Rev. D **77**, 053011 (2008)
4. S. Choubey, W. Rodejohann, P. Roy, Phenomenological consequences of four zero neutrino Yukawa textures. Nucl. Phys. B **808**, 272–291 (2009). [Erratum: Nucl. Phys. B 818, 136–136 (2009)]
5. A. Abada, M. Lucente, Looking for the minimal inverse seesaw realisation. Nucl. Phys. B **885**, 651–678 (2014)
6. W. Grimus, A.S. Joshipura, L. Lavoura, M. Tanimoto, Symmetry realization of texture zeros. Eur. Phys. J. C **36**, 227–232 (2004)
7. N. Gautam, M.K. Das, Phenomenology of keV scale sterile neutrino dark matter with S_4 flavor symmetry. JHEP **01**, 098 (2020)
8. D. Meloni, A. Meroni, E. Peinado, Two-zero Majorana textures in the light of the Planck results. Phys. Rev. D **89**(5), 053009 (2014)
9. H.B. Camara, R.G. Felipe, F.R. Joaquim, Minimal inverse-seesaw mechanism with Abelian flavour symmetries. JHEP **05**, 021 (2021)

10. N. Gautam, M.K. Das, Impact of texture zeros on dark matter and neutrinoless double beta decay in inverse seesaw. Nucl. Phys. B **971**, 115519 (2021)
11. C. Giganti, S. Lavignac, M. Zito, Neutrino oscillations: the rise of the PMNS paradigm. Prog. Part. Nucl. Phys. (2017)
12. P.F. de Salas, D.V. Forero, S. Gariazzo, P. Martínez-Miravé, O. Mena et al., 2020 global reassessment of the neutrino oscillation picture. JHEP **02**, 071 (2021)

Chapter 32
Mass Varying Neutrino and Quintessence: A Comparative Study of Scaling Solution

Hemanshi Bundeliya and S. D. Pathak

Abstract We investigate the comparative study of mass varying neutrino and canonical scalar field (quintessence) as a source of dark energy in the spatially flat ($k = 0$) universe. We obtained the scaling solution of dominating components of the universe in the absence of interaction between them.

32.1 Introduction

The accelerated expansion of the universe is supported by a number of astrophysical and cosmological observations in the recent past [1–10]. The next step is to explain this observed accelerated expansion, so far substantial efforts have been put in by several authors [11–14] by considering exotic components of energy having negative pressure which provides repulsive gravity characteristics called dark energy. The cosmic energy budgets constrained by Planck-2018 observations of CMB-data [7–10], confirmed that 70% of the energy budget in the universe is in the form of dark energy while the remaining 30% of the energy budget of the universe is attributed to non-relativistic baryonic and non-baryonic pressure-less ($p_m = 0$) dust matter with the equation of state (EoS) $w_m = 0$. In the development of the fundamental theory of dark energy, there are several candidates which have been proposed in the recent past [15–35, 37, 37–57]. Among all possible candidates of dark energy, the cosmological constant is considered as one of the simplest candidates and used in the model of the constant dark energy, with an EoS of $w_\lambda = -1$. The constant dark energy model suffers a serious cosmological issues addressed in the literature [15–19]. These two issues become a motivation to introduce the dynamical dark energy models that have gained prominence [20–25, 29]. The scalar field serves as a

H. Bundeliya (✉) · S. D. Pathak
Department of Physics, Lovely Professional University, Phagwara, Punjab 144411, India
e-mail: bundeliyahemanshi.20.phy@ishls.indusuni.ac.in

S. D. Pathak
e-mail: prince.pathak19@gmail.com

J. W. F. Valle et al. (eds.), *Proceedings of the International Conference on Future Prospects in Neutrino and Astroparticle Physics*, Springer Proceedings in Physics 361,
https://doi.org/10.1007/978-981-97-7441-8_33

candidate of dynamical dark energy dubbed as ϕCDM having dynamical EoS w_ϕ in which quintessence [20, 27, 28] is deemed to be its simplest candidate. Apart from quintessence, the cosmological dynamics of different scalar fields (namely Phantom and Tachyon) have been documented in the literature [30–35, 37]. In the modification of Einstein's theory of gravity, the repulsive source of gravity can be obtained from the modification of the geometrical part of field equations. Apart from the constant and dynamical dark energy model, another candidate for dark energy is studied by authors [58–69] in cosmology dubbed as Mass Varying Neutrino (MVN).

In Sect. 32.2, we discuss the coupled quintessence model while Sect. 32.3, is devoted to mass varying neutrino and its basic mathematical theory. In Sect. 32.4, we conclude our work.

32.1.1 Background Mathematics

Let us consider the following Friedmann–Lemaitre–Robertson–Walker (FRLW) metric

$$ds^2 = g_{\mu\nu}dx^\mu dx^\nu \equiv -dt^2 + a^2(t)\frac{dr^2}{1-kr^2} + r^2d\theta^2 + r^2\sin^2\theta d\phi^2 \tag{32.1}$$

where we take $c = 1$, $a(t)$ as the scale factor, and k as the curvature parameter having values $k = 1, 0, 1$ corresponding to spatially open, flat, and closed geometry of the universe, respectively. Einstein Hilbert's action is given as

$$S = \int d^4x\sqrt{-g}\left(\frac{R}{2} + \mathcal{L}_m + \mathcal{L}_\phi\right) \tag{32.2}$$

and energy-momentum tensor $T_{\mu\nu}$ for Lagrangian L is defined as

$$T^{\mu\nu} = \frac{\partial\mathcal{L}}{\partial(\partial_\mu\phi)}\partial^\nu\phi - g^{\mu\nu}\mathcal{L} \tag{32.3}$$

Here $\mathcal{L}_m, \mathcal{L}_\phi$ are Lagrangian for matter and scalar field, respectively. Field equation in general relativity is written as

$$R_{\mu\nu} - \frac{1}{2}Rg_{\mu\nu} = 8\pi GT_{\mu\nu} \tag{32.4}$$

where $R_{\mu\nu}$, R are Ricci tensor and Ricci scalar. Solving Eq. (32.4) for a spatially flat universe ($k = 0$) leads to the Friedmann equations as

$$H^2 = \left(\frac{\dot{a}}{a}\right)^2 = \frac{8\pi G}{3}\rho \tag{32.5}$$

and

$$\frac{\ddot{a}}{a} = -\frac{4\pi G}{3}(\rho + 3\rho) \tag{32.6}$$

where $a(t)$ is the expansion factor of the universe and $H(t)$ is the Hubble parameter. For cosmic acceleration, we have $\rho + 3p < 0$ as equation of state $w = \frac{p}{\rho} < -\frac{1}{3}$ and for cosmic deceleration we should have $w > -\frac{1}{3}$. In the Brane-World model, the attractive gravitational force dominates in the early universe, and in late-time the repulsive gravity term dominates and provides cosmic acceleration. The continuity equation is given as

$$\dot{\rho} + 3H(1 + w)\rho = 0 \tag{32.7}$$

For constant equation of state (EoS) w one can obtain from Eq. (32.7) the following scaling solution of ρ as given by

$$\rho = \rho_0 \left(\frac{a}{a_0}\right)^{-3(w+1)} \tag{32.8}$$

where ρ_0, a_0 are present values of ρ and a, respectively. One can obtain the following form of expansion factor with the constant EoS w as

$$a(t) = a_0 \left(\frac{t}{t_0}\right)^{\frac{2}{3(1+w)}} \tag{32.9}$$

where t_0 is the present epoch. The scaling solution discussed in Eqs. (32.8) with (32.9) is valid for a single perfect fluid. If there are multiple components (for example two components one is pressure-less matter and the other one is scalar field as a source of dynamical dark energy) interacting with each other then the scaling behavior of components will be changed.

32.2 Coupled Quintessence Model

The interacting dark energy model proposed by several authors [70–76] motivated to solve the fine-tuning and cosmic coincidence problem by considering the transfer of energy between the two components of the universe and thus conservation of energy locally violated while globally it is conserved. Thus, one can write [77] energy-momentum tensor $T^{(1)}_{\mu\nu}$ and $T^{(2)}_{\mu\nu}$ for the two respective components as

$$\nabla^{\mu} T^{(1)}_{\mu\nu} = Q_{\nu} \tag{33.10a}$$

and

$$\nabla^{\mu} T^{(2)}_{\mu\nu} = -Q_{\nu} \quad (33.10b)$$

where $Q_{\nu} \neq 0$ and denotes the energy–momentum exchanged between two components.

The non-canonical Lagrangian of scalar field ϕ is given [78–84] by

$$\mathcal{L}_{\phi} = Y\left(\frac{Y}{M^4}\right)^{\gamma-1} - V(\phi) \quad (32.11)$$

where $Y = \frac{1}{2}\dot{\phi}^2$ here dot represents the time derivative, γ is a dimensionless parameter, and M is constant with the dimension of mass. In this case, one can obtain the energy density and pressure of the non-canonical scalar field from Eq. (32.3) as

$$\rho_{\phi} = \left(\frac{\partial \mathcal{L}}{\partial Y}\right)(2Y) - \mathcal{L} \quad (33.12a)$$

$$p_{\phi} = \mathcal{L}. \quad (33.12b)$$

The non-canonical Lagrangian Eq. (32.11) leads to its canonical (quintessence) form for $\gamma = 1$, and thus from Eqs. (33.10a) and (33.10b) we have the following form of energy density and the pressure for spatially homogeneous ($\nabla\phi = 0$) quintessence field

$$\rho_{\phi} = \frac{1}{2}\dot{\phi}^2 + V(\phi) \quad (33.13a)$$

$$p_{\phi} = \frac{1}{2}\dot{\phi}^2 - V(\phi) \quad (33.13b)$$

with EoS

$$w_{\phi} = \frac{p_{\phi}}{\rho_{\phi}} = \frac{\frac{1}{2}\dot{\phi}^2 - V(\phi)}{\frac{1}{2}\dot{\phi}^2 + V(\phi)}. \quad (32.14)$$

In the coupled quintessence model total global energy is conserved but locally there is a violation of the conservation of energy. Thus, expressions Eqs. (33.10a) and (33.10b) can be taken as

$$\dot{\rho_{\phi}} + 3H\left(1 + w_{\phi}\right)\rho_{\phi} = -Q, \quad (33.15a)$$

$$\dot{\rho_m} + 3H(1 + w_m)\rho_m = Q \quad (33.15b)$$

where Q is the time component of Q_ν dubbed as the coupling parameter between matter and quintessence. One can obtain the following equation of motion of quintessence from Eqs. (33.13a) and (33.15a) with coupling parameter Q as

$$\ddot{\phi} + 3H\dot{\phi} + \frac{dV}{d\phi} - \frac{Q}{\dot{\phi}} = 0. \tag{32.16}$$

Recently, in interacting dark energy, the different forms of coupling parameter Q have been proposed and analyzed by different authors [85–90]. They consider the linear and nonlinear forms of coupling strengths. In [91] author considers the two stable interacting dark energy models by considering interaction strength as a linear combination of energy densities of components. In the absence of coupling strength, Q one can obtain a scaling solution of dark energy (as quintessence scalar field) and matter from Eqs. (33.15a) and (33.15b). If dark energy does not interact with background matter then it is one of the most difficult tasks to test the fundamental origin of dark energy. Dark energy as a scalar field has been contemplated by a number of authors [93–100] in the development of cosmic inflation theory.

32.3 Mass Varying Neutrinos as Dark Energy

The concept of mass varying neutrinos suggests that the masses of neutrinos could change over time, and this variation could potentially have cosmological implications, including explaining the observed acceleration of the universe's expansion. In cosmology, the fluid description of neutrinos is not possible [62]. One can solve the distribution function $f\ (q^i,\ p^i,\ t)$ in phase space. The general Boltzmann equation [101] is given as

$$\frac{Df}{Dt} = \frac{\partial f}{\partial t} + \frac{dx^i}{dt}\frac{\partial f}{\partial x^i} + \frac{dq}{dt}\frac{\partial f}{\partial q} + \frac{dn^i}{dt}\frac{\partial f}{\partial n^i} = \left(\frac{\partial f}{\partial t}\right)_c. \tag{32.17}$$

From the solution of Eq. (32.17), one can obtain the energy density of neutrinos as

$$\rho_\nu = a^{-4}\int \frac{d^3q}{(2\pi)^3}\sqrt{q^2 + a^2 m_\nu^2} f_0(q) \tag{32.18}$$

where $f_0(q)$ is the background neutrino distribution function, $q^i = ap^i$ is the co-moving momentum, and $m_\nu(\phi) = m_i e^{\beta\phi}$(say) is the mass of neutrino coupled with quintessence. The pressure of neutrinos can be obtained as

$$P_\nu = \frac{1}{3a^4}\int \frac{d^3q}{(2\pi)^3}\frac{q^2}{\sqrt{q^2 + a^2 m_\nu^2}} f_0(q). \tag{32.19}$$

The equation of motion of MVN as the candidate of dark energy can be obtained from Eqs. (32.18) and (32.19) as

$$\dot{\rho_\nu} + 3H(1 + w_\nu)\rho_\nu = \frac{\partial \ln m_\nu}{\partial \phi}\dot{\phi}(1 - 3w_\nu)\rho_\nu \tag{32.20}$$

where $w_\nu = \frac{P_\nu}{\rho_\nu}$ be the equation of the state of MVN dark energy. The dot is taken as a derivative concerning time. The equation of motion for the coupled scalar field can be obtained from the continuity equation of the combined fluid of neutrinos and quintessence given as

$$\ddot{\phi} + 3H\dot{\phi} + \frac{dV_{ef}(\phi)}{\phi} = \frac{\partial \ln m_\nu}{\partial \phi}(1 - 3w_\nu)\rho_\nu \tag{32.21}$$

and continuity equation of combined fluid can be written as

$$\dot{\rho_\nu} + \dot{\rho_\phi} + 3H\left(\rho_\nu + \rho_\phi + P_\nu + P_\phi\right) = 0. \tag{32.22}$$

The choice of $m_\nu(\phi)$ of field can be understood by potential $V_{ef}(\phi)$ as

$$V_{ef}(\phi) = V(\phi) + V_{co}(\phi) \tag{32.23}$$

here $V(\phi)$ is quintessence potential and $V_{co}(\phi)$ arises due to coupling of neutrinos given as

$$V_{co}(\phi) = a^{-4}\int \frac{d^3q}{(2\pi)^3}\sqrt{q^2 + a^2m_\nu^2}f_0(q) = \left(\tilde{\rho_\nu} - 3\tilde{P_\nu}\right)e^{\beta\phi} \tag{32.24}$$

where $\rho\tilde{}_\nu = \rho_\nu e^{\beta\phi}$ and $P\tilde{}\nu = P_\nu e^{\beta\phi}$ are taken as independent of field ϕ. For a given potential one can obtain the scaling solution of MVN from the following continuity equation:

$$\dot{n_\nu} = -3Hn_\nu \tag{32.25}$$

where n_ν is the total neutrinos and antineutrinos density under the non-relativistic limit.

32.4 Conclusion

Homogeneous quintessence is considered a potential candidate of dynamical dark energy. The concept of mass varying neutrinos suggests that the masses of neutrinos could change over time, and this variation could potentially have cosmological implications, including explaining the observed acceleration of the universe's expansion.

The masses of neutrinos change over time and could potentially explain the observed acceleration of the universe's expansion. The scaling solution of quintessence can be obtained from the continuity Eq. (32.7) in the absence of coupling strength. Under the non-relativistic limit, the scaling solution of neutrino can be obtained from Eq. (32.25). Energy density and pressure of neutrino can be obtained from the Boltzmann equation (32.17) given by Eqs. (32.18) and (32.19). The equation of motion is obtained as given in Eq. (32.20).

References

1. A.G. Riess et al., Astron. J. **116**, 1009 (1998)
2. S. Perlmutter et al., Astrophys. J. **517**, 565 (1999)
3. D. Spergel et al., Astrophys. J. S. **148**, 175 (2003)
4. H. Seo, D.J. Eisenstein, Astrophys. J. **598**, 720 (2003)
5. C. Blake, K. Glazebrook, Astrophys. J. **594**, 665 (2003)
6. E. Komatsu et al., Astrophys. J. S. **192**, 18 (2011)
7. N. Aghanim et al., [Planck Collaboration], Planck 2018 results VI. Cosmol Parameters: A & A **641**, A6 (2020)
8. N. Aghanim et al., [Planck Collaboration], Planck 2018 results I. Overview Cosmol. Legacy Planck: A & A **641**, A1 (2020)
9. R.A. Knop et al., Ap. J. **598**, 102 (2003)
10. A.G. Riess et al., Ap. J. **607**, 665 (2004)
11. F.R.E. George, E. Platts, D. Sloan, A. Weltman, JCAP **04**, 026 (2016)
12. M.S. lowski, A. Stachowski, JCAP **10**, 066 (2015)
13. Y. Bisabr, H. Salehi, Class. Quantum Grav. **19**, 2369 (2002)
14. Y. Fujii, T. Nishioka, Phys. Rev. D **42**, 361 (1990)
15. P. Bull et al., Phys. Dark Universe **12**, 56 (2016)
16. L. Perivolaropoulos, F. Skara, New Astron. Rev. **95**, 101659 (2022)
17. A. Del Popolo, M. Le Delliou, Galaxies **5**, 17 (2017)
18. S. Weinberg, Rev. Mod. Phys. **61**, 1 (1989)
19. A.V. Astashenok, A. Del Popolo, Class. Quantum Gravity **29**, 085014 (2012)
20. B. Ratra, P.J.E. Peebles, Phys. Rev. D, **37**, 3406 (1988)
21. C. Escamilla-Rivera, A. Najera, JCAP **03**, 060 (2022)
22. G.B. Zhao, M. Raveri, L. Pogosian et al., Nat Astron **1**, 627 (2017)
23. R. Mainini et al., ApJ **599**, 24 (2003)
24. P. Brax, C. Burrage, C. Englert, M. Spannowsky, Phys. Rev. D **94**, 084054 (2016)
25. E.J. Copeland, M. Sami, S. Tsujikawa, Int. J. Mod. Phys. D **15**, 1753 (2006)
26. L. Amendola, Phys. Rev. D **64**, 043511 (2000)
27. H.Y. Chang, R.J. Scherrer, Phys. Rev. D **88**, 083003 (2013)
28. S.D. Storm, R.J. Scherrer, Phys. Lett. B **829**, 137126 (2022)
29. J. Kaur, S.D. Pathak, V.K. Ojha, M. Khlopov, Astropart. Phys. https://doi.org/10.1016/j.astropartphys.2024.102926
30. M.M. Verma, S.D. Pathak, Astrophys. Space Sci. **350**, 381 (2014)
31. M.M. Verma, S.D. Pathak, Int. J. Mod. Phys. D **23**(9), 1450075 (2014)
32. M.M. Verma, S.D. Pathak, Astrophys. Space Sci. **344**, 505 (2013)
33. D. Rajeeb Kumar, S.D. Pathak, V.K. Ojha, Chinese Phys. C **47**(5), 055102 (2023)
34. V.B. Johri, Phys. Rev. D **70**, 041303R (2004)
35. V.B. Johri, Class. Quantum Gravity Quantum Gravity **19**, 5959 (2002)
36. V.B. Johri, Phys. Rev. D **63**, 103504 (2001)
37. S. Choudhury, A. Mazumdar, Nucl. Phys. B **9**, 386 (2014)

38. R. Allahverdi, B. Dutta, A. Mazumdar, Phys. Rev. D. **78**, 063507 (2008)
39. Y. Bai, D. Stolarski, JPAC, **03**, 091 (2021)
40. S. Choudhury, A. Mazumdar, E. Pukartas, JHEP **4**, 1–27 (2014)
41. K. Enqvist, A. Mazumdar, Phys. Rep. **380**, 99–234 (2003)
42. K. Enqvist, A. Jokinen, A. Mazumdar, JCAP **01**, 008 (2004)
43. K. Enqvist, A. Mazumdar, P. Stephens, JCAP **2010**, 020 (2010)
44. R. Allahverdi, K. Enqvist, J. Garcia-Bellido, A. Jokinen, A. Mazumdar, JCAP **2007**, 019 (2007)
45. A. Mazumdar, J. Rocher, Phys. Reps. **497**, 85–215 (2011)
46. S. Choudhury, A. Mazumdar, Phys. Lett. B **733**, 270–275 (2014)
47. S. Choudhury, A. Mazumdar (2014). arXiv:1403.5549
48. R. Allahverdi, K. Enqvist, J. Garcia-Bellido, A. Mazumdar, Phys. Rev. Lett. **97**, 191304 (2006)
49. R. Allahverdi, B. Dutta, A. Mazumdar, Phys. Rev. D, **78**, 063507 (2008)
50. S.M. Choi, H.M. Lee, EPJ C **76**, 1–15 (2016)
51. K. Dimopoulos, C. Owen, A. Racioppi, Astropart. Phys. **103**, 16–20 (2018)
52. L. Iacconi et al., JCAP **06**, 007 (2022)
53. J.J. Blanco-Pillado et al., JCAP **02**, 034 (2013)
54. K. Kefala et al., Phys. Rev. D **104**, 023506 (2021)
55. S. DavidStorm, R.J. Scherrer, Phys. Lett. B, **829**, 137126 (2022)
56. R. Cerezo, J.G. Rosa, J. High Energ. Phys. **2013**, 24 (2013)
57. S. Hotchkiss et al., JCAP **06**, 002 (2011)
58. S. Mandal et al., JCAP **05**, 018 (2021)
59. R. Fardon, A.E. Nelson, N. Weiner, JCAP **10**, 005 (2004)
60. R.D. Peccei, Phys. Rev. D **71**, 023527 (2005)
61. N. Afshordi, M. Zaldarriaga, K. Kohri, Phys. Rev. D **72**, 065024 (2005)
62. A.W. Brookfield, C. van de Bruck, D.F. Mota, D. Tocchini-Valentini, Phys. Rev. Lett. **96**, 061301 (2006)
63. G.Y. Chitov, T. August, A. Natarajan, T. Kahniashvili, Phys. Rev. D **83**, 045033 (2011)
64. L.S. Kisslinger, Mod. Phys. Lett. A **34**, 1950327 (2019)
65. H. Mohseni Sadjadi, V. Anari, Phys. Rev. D, **95**, 123521 (2017)
66. J. Sakstein, M. Trodden, Phys. Rev. Lett. **124**, 161301 (2020)
67. G.Y. Chitov, PoS QFTHEP-2011, 074 (2013)
68. P. Gu, X. Wang, X. Zhang, Phys. Rev. D **68**, 087301 (2003)
69. M. Costanzi et al., JCAP **12**, 012 (2013)
70. L.P. Chimento, Phys. Rev. D **81**, 043525 (2010)
71. J. H. He, B. Wang, P. Zhang, Phys. Rev. D, **80**, 063530 (2009)
72. B. Wang, Y.G. Gong, E. Abdalla, Phys. Lett. B **624**, 141 (2005)
73. S.D. Campo, R. Herrera, D. Pavon, Int. J. Mod. Phys. D **20**(4), 561 (2011)
74. K.R. Mishra et al., Phys. Dark Universe **40**, 101211 (2023)
75. J. Kang, Phys. Dark Universe **31**, 100784 (2021)
76. M.B. Gavela et al., JCAP **2009**, 034 (2009)
77. S. Bahamonde et al., Phys. Rep. **775**, 1–122 (2018)
78. A. Diez-Tejedor, A. Feinstein, Phys. Rev. D **74**, 023530 (2006)
79. V. Mukhanov, A. Vikman, JCAP **0602**, 004 (2006)
80. S. Unnikrishnan, Phys. Rev. D **78**, 063007 (2008)
81. S. Unnikrishnan, V. Sahni, A. Toporensky, JCAP **1208**, 018 (2012)
82. S. Unnikrishnan, V. Sahni, JCAP **1310**, 063 (2013)
83. D. Wands, J. De-Santiago, Y. Wang, Class. Quantum Gravity **29**, 145017 (2012). arXiv:1203.6776[astro-ph.CO]
84. V. Sahni, A.A. Sen, Eur. Phys. J. C **77**, 225 (2017)
85. W. Zimdahl, D. Pavon, Phys. Lett. B **521**, 133 (2011)
86. A. Kundu, S.D. Pathak, V.K. Ojha, Commun. Theor. Phys. **73**, 025402 (2021)
87. M.M. Verma, S.D. Pathak, Int. J. Theor. Phys. **53**, 2370 (2012)
88. M. Shahalam et al., Eur. Phys. J. C **75**, 395 (2015)

89. M. Shahalam et al., Eur. Phys. J. C **77**, 686 (2017)
90. E. Di Valentino et al., Phys. Rev. D **101**, 063502 (2020)
91. W. Yang et al., JCAP **09**, 019 (2018)
92. K. Akash et al., J. Phys. Conf. Ser. **1531**, 012086 (2020)
93. R. Allahverdi et al., Phys. Rev. Lett. **97**, 191304 (2006)
94. A.A. Sen, R.J. Scherrer, Phys. Rev. D. **72**, 063511 (2005)
95. F.C. Carvalho et al., Phys. Rev. Lett. **97**, 081301 (2006)
96. S. Downes, B. Dutta, Phys. Rev. D. **87**, 083518 (2013)
97. S. Panda, M. Sami, S. Tsujikawa, Phys. Rev. D **76**, 103512 (2007)
98. S. Choudhury, A. Mazumdar, S. Pal, JCAP **7**, 041 (2013)
99. K. Enqvist, A. Mazumdar, P. Stephens, JCAP **6**, 020 (2010)
100. S. Unnikrishnan, V. Sahni, JCAP **10**, 063 (2013)
101. Y.Y. Keum, Modern Phys. Lett. A, **22**(25–28), 2131 (2007)

Chapter 33
GeV Scale Dark Matter and $B \rightarrow K+$ Missing Energy Phenomenology in the $U(1)_{B-L}$ Model

Ajay Kumar Yadav, Shivaramakrishna Singirala, Manas Kumar Mohapatra, and Suchismita Sahoo

Abstract We elucidate the recently observed discrepancy in the branching ratio of $B^+ \rightarrow K^+$ +inv decay mode using a GeV scale scalar dark matter in an anomaly-free $U(1)_{B-L}$ gauge extension of the Standard Model. We constrain the new parameters by using consistency with the existing bounds on Dark matter relic density, direct detection, collider and BR($B \rightarrow K^* \nu\bar{\nu}$) measured at the Belle II experiment. We investigate couplings between the mediator and the SM fermions as well as the dark matter particle. We then estimate the branching ratios of $b \rightarrow s\nu\bar{\nu}$ decay processes such as $B \rightarrow (K^+, K^*)\nu\bar{\nu}$, $B_s \rightarrow (\eta, \eta')\nu\bar{\nu}$,$B_s \rightarrow \phi\nu\bar{\nu}$ and $B_c \rightarrow (D_s, D_s^*)\nu\bar{\nu}$ in a common parameter space, meeting the current experimental bounds of both sectors simultaneously.

33.1 Introduction

The extensively established foundational theory in particle physics Standard Model (SM) falls short in elucidating the asymmetry between matter and antimatter, the existence of dark matter (DM), and the observation of minute neutrino mass, which distinctly point towards the existence of new physics (NP) beyond the limitations of the SM. Over the recent years, noteworthy deviations of $(2-3)\sigma$ [1–4] from the Standard Model have surfaced in a variety of angular observables linked to flavor changing neutral current (FCNC) $b \rightarrow sl^+l^-$ and flavor changing charge current (FCCC) $b \rightarrow c\tau\nu_l$ decay modes. The notable anomalies involve the lepton flavor universality violating (LFUV) ratios such as $R_{K^{(*)}}$, with disparities of 0.2σ [2], $R_{D^{(*)}}$, with disparities of 2σ (2.2σ) [4] and $R_{J/\psi}$, with 2σ discrepancy [3], in which the hadronic uncertainties canceled out significantly. In addition, the Belle II experiment [5] has recently announced the initial measurement of $B^+ \rightarrow K^+\nu\bar{\nu}$ branching ratio

A. K. Yadav (✉) · S. Sahoo
Department of Physics, Central University of Karnataka, Kalaburagi 585367, India
e-mail: yadavajaykumar286@gmail.com

S. Singirala · M. K. Mohapatra
University of Hyderabad, Hyderabad 500046, Telangana, India

J. W. F. Valle et al. (eds.), *Proceedings of the International Conference on Future Prospects in Neutrino and Astroparticle Physics*, Springer Proceedings in Physics 361,
https://doi.org/10.1007/978-981-97-7441-8_34

$$\mathcal{BR}(B^+ \to K^+ \nu\bar{\nu})|^{\text{Expt}} = (2.4 \pm 0.7) \times 10^{-5} \tag{33.1}$$

which deviates from the SM prediction

$$\mathcal{BR}(B^+ \to K^+ \nu\bar{\nu})|^{\text{SM}} = (5.06 \pm 0.14 \pm 0.28) \times 10^{-6} \tag{33.2}$$

by 2.8σ [6]. This scenario requires a thorough examination of these discrepancies in both the SM and scenarios beyond the SM, with the goal of investigating the framework of New Physics.

In $B \to K + \nu_l \bar{\nu}_l$ decay, the undetected neutrinos carry away energy and momentum, contributing to the observed missing energy. Neutrinos escape detection due to their weak interaction, and any process with missing energy E in the final state, such as $B^+ \to K^+ + \not{E}$, contributes to the measured branching ratio of the decay $B^+ \to K^+ \nu\bar{\nu}$. The Standard Model requires supplementation with extra symmetries or particles to address this anomaly in missing energy. The missing energy can be thought of as dark matter pair, a pair of sterile neutrino or any other light particle participating in this $b \to s + \not{E}$ transition with a weak enough interaction strength. Therefore, precision investigations into B-meson decay are vital to identify or constrain Dark Matter or other light Beyond SM elements. The combined study of Dark matter and NP suggestion from B-meson decay is an exciting window towards the building of new physics. In this work, the SM is extended with $U(1)_{B-L}$ gauge extensions to fix the missing energy anomaly with a light dark matter. Furthermore, we compute the branching ratios of other $b \to s\nu\bar{\nu}$ decay modes such as $B \to K^{(*)}\nu\bar{\nu}$, $B_s \to \eta^{(\prime)}\nu\bar{\nu}, B_s \to \phi\nu\bar{\nu}$ and $B_c \to D_s^{(*)}\nu\bar{\nu}$ to recheck the real existence of NP.

The paper is organized as follows. Section 33.2 describes the particle content and the interaction Lagrangian of the gauge extended model. We present the constraints on new parameters from both the dark and flavor sectors in Sect. 33.3. Section 33.3 also includes the general effective Hamiltonian for $b \to s\nu_l\bar{\nu}_l$ channels. The impacts of new parameters on $B \to P(V)\nu_l\bar{\nu}_l$ decay modes are discussed in Sect. 33.4. Section 33.5 outlines the concluding remarks.

33.2 Model Description

Within the framework of $U(1)_{B-L}$ gauge extensions of the Standard Model, we illustrate a fundamental case involving singlet scalar dark matter to explain the absence of energy as dark matter. For anomaly cancelation, we strengthen the model by incorporating the three Majorana right-handed neutrinos N_{iR} with $i = 1, 2, 3$, each carrying a charge -1 under the extended $U(1)_{B-L}$ gauge group. Additionally, we include two singlet scalars, ϕ (spontaneously breaks the new $U(1)$) and ϕ_{DM} (possible dark matter candidate) with $B - L$ charges 2 and $1/3$, respectively, as given in Table 33.1.

The pertinent interaction Lagrangian is given as follows:

Table 33.1 Fields and their charges of the proposed $U(1)_{B-L}$ model

	Field	$SU(2)_L \times U(1)_Y$	$U(1)_{B-L}$
Fermions	$Q_L \equiv (u, d)_L^T$	$(\mathbf{2},\ 1/6)$	$1/3$
	u_R	$(\mathbf{1},\ 2/3)$	$1/3$
	d_R	$(\mathbf{1},\ -1/3)$	$1/3$
	$\ell_L \equiv (\nu,\ e)_L^T$	$(\mathbf{2},\ -1/2)$	-1
	e_R	$(\mathbf{1},\ -1)$	-1
	N_{1R}, N_{2R}, N_{3R}	$(\mathbf{1},\ 0)$	-1
Scalars	H	$(\mathbf{2},\ 1/2)$	0
	ϕ	$(\mathbf{1},\ 0)$	2
	$\phi_{\rm DM}$	$(\mathbf{1},\ 0)$	$1/3$

$$
\begin{aligned}
\mathcal{L}_{\rm BL} &= -\frac{1}{3}\, g_{\rm BL}\overline{Q}_L\, Z'_\mu \gamma^\mu Q_L - \frac{1}{3}\, g_{\rm BL}\overline{u}_R\, Z'_\mu \gamma^\mu u_R - \frac{1}{3}\, g_{\rm BL}\overline{d}_R\, Z'_\mu \gamma^\mu d_R + g_{\rm BL}\overline{\ell}_L\, Z'_\mu \gamma^\mu \ell_L \\
&\quad + g_{\rm BL}\overline{e}_R\, Z'_\mu \gamma^\mu e_R + i\,\overline{N}_{iR}\left(\partial\!\!\!/ + i\, g_{\rm BL}\, Z'_\mu \gamma^\mu\right) N_{iR} - \frac{y_{\alpha\beta}}{2}\left(\sum_{\alpha,\beta=1}^{3} \overline{N^c_{\alpha R}}\, N_{\beta R}\, \phi + h.c.\right) \\
&\quad + \left|(\partial_\mu - 2 i g_{\rm BL} Z'_\mu)\,\phi\right|^2 + \left|\left(\partial_\mu - \frac{1}{2} i g_{\rm BL}\, Z'_\mu\right)\phi_{\rm DM}\right|^2 - V(H, \phi, \phi_{\rm DM}) + \mathcal{L}_{\rm SM}, \qquad (33.3)
\end{aligned}
$$

where the scalar potential $V(H, \phi, \phi_{\rm DM})$ of the model is given by

$$
\begin{aligned}
V(H, \phi, \phi_{\rm DM}) &= \mu_H^2 H^\dagger H + \lambda_H (H^\dagger H)^2 + \mu_\phi^2 \phi^\dagger \phi + \lambda_\phi (\phi^\dagger \phi)^2 + \mu_{\rm DM}^2 \phi_{\rm DM}^\dagger \phi_{\rm DM} \\
&\quad + \lambda_{\rm DM} (\phi_{\rm DM}^\dagger \phi_{\rm DM})^2 + \lambda_{H\phi} (H^\dagger H)(\phi^\dagger \phi) + \lambda_{\rm HD} (H^\dagger H)(\phi_{\rm DM}^\dagger \phi_{\rm DM}) \\
&\quad + \lambda_{\rm D\phi} (\phi^\dagger \phi)(\phi_{\rm DM}^\dagger \phi_{\rm DM}) \qquad (33.4)
\end{aligned}
$$

with $\phi_{\rm DM} = \frac{S+iA}{\sqrt{2}}$, $\langle H \rangle = (0, v/\sqrt{2})^T$, $\langle \phi_2 \rangle = v_2/\sqrt{2}$. After spontaneous symmetry breakdown, ϕ attains a vev v_2 and the new gauge field Z' receives a mass $M_{Z'} = 2 g_{\rm BL} v_2$, where $g_{\rm BL}$ is new gauge coupling. The mixing of the two CP-even scalars, H_1 and ϕ, results in two mass eigenstates: H_1 (the observed 125 GeV Higgs at LHC) and a lighter scalar H_2 (with a mass below 5 GeV).

33.3 Constraints of New Parameters

After acquiring an understanding of the new particles and their interactions, we now proceed to constrain the new parameters from both the flavor phenomenology and the dark matter sectors.

33.3.1 Dark Matter Phenomenology

33.3.1.1 Relic Density

Both components of the singlet scalar have the potential to contribute to the relic density of the Universe. The dark matter relic abundance is calculated using

$$\Omega \mathrm{h}^2 = \frac{1.09 \times 10^9 \ \mathrm{GeV}^{-1}}{{g_*}^{1/2} M_{\mathrm{pl}}} \frac{1}{J(x_f)}, \tag{33.5}$$

where $M_{\mathrm{pl}} = 1.22 \times 10^{19}$ GeV is the Planck mass, $g_* = 106.75$ denotes the total number of effective relativistic degrees of freedom, and $J(x_f)$ is constructed as

$$J(x_f) = \int_{x_f}^{\infty} \frac{\langle \sigma v \rangle (x)}{x^2} dx. \tag{33.6}$$

with $\langle \sigma v \rangle$ is the thermally averaged annihilation cross-section [7]. The scalar dark matter (DM) can undergo annihilation through scalar bosons $H_{1,2}$, or Z'. We focus on the scalar portal to achieve a light DM ($M_{DM} < 2.5$ GeV), aiming to account for the missing energy signal observed in B-meson decays. The annihilation channels encompass $SS(AA) \to f\bar{f}$, where f represents quark-antiquark and lepton-antilepton pairs, constituting s-channel processes (left panel of Fig. 33.1). Consequently, a resonance is evident when the DM mass is approximately half of the propagator mass, $M_{\mathrm{DM}} \approx \frac{M_{H_2}}{2}$, in the plot (Fig. 33.2) of Ωh^2 versus DM mass. We have implemented the model in LanHEP [8] and then micrOMEGAs [9] packages for dark matter study.

33.3.1.2 Direct Detection

The dark matter can disperse off the nucleus through the effective interaction of the scalars H_1 and H_2. The DM-nucleon effective Lagrangian is given as

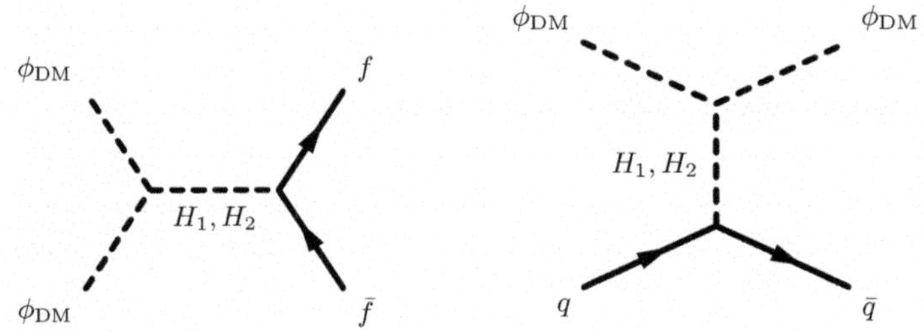

Fig. 33.1 Feynman diagrams contributing to relic density (left panel) and direct detection (right panel)

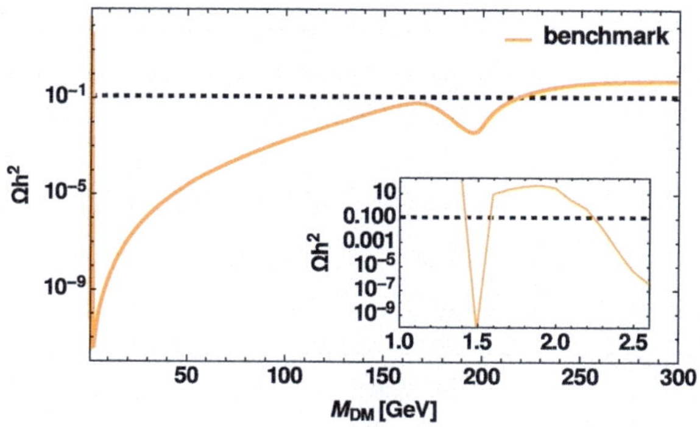

Fig. 33.2 Relic density versus DM mass with $M_{H_2} = 3$ GeV. Black horizontal dotted lines denote the 3σ range of Planck limit [10]

$$\mathcal{L}_{\text{eff}} = \frac{\lambda}{M_{H_i}^2} S S \bar{q} q + \frac{\lambda}{M_{H_i}^2} A A \bar{q} q. \tag{33.7}$$

The spin-independent (SI) cross-section versus DM mass is projected in Fig. 33.3 where the cross-section is discovered to be significantly less than DarkSide-50's experimental upper limit [11]. In case of Z'-portal, the effective Lagrangian is as follows:

$$\mathcal{L}'_{\text{eff}} \supset -\frac{n_{\text{DM}} g_{\text{BL}}^2}{3M_{Z'}^2} \left(S\partial^\mu A - A\partial^\mu S\right) \bar{u}\gamma_\mu u - \frac{n_{\text{DM}} g_{\text{BL}}^2}{3M_{Z'}^2} \left(S\partial^\mu A - A\partial^\mu S\right) \bar{d}\gamma_\mu d, \tag{33.8}$$

where n_{DM} is the charge of DM under new $U(1)$. The expression for SI WIMP-nucleon cross-section is given by [7, 12]

$$\sigma_{Z'} = \frac{\mu^2}{16\pi} \frac{n_{\text{DM}}^2 g_{\text{BL}}^4}{M_{Z'}^4}, \tag{33.9}$$

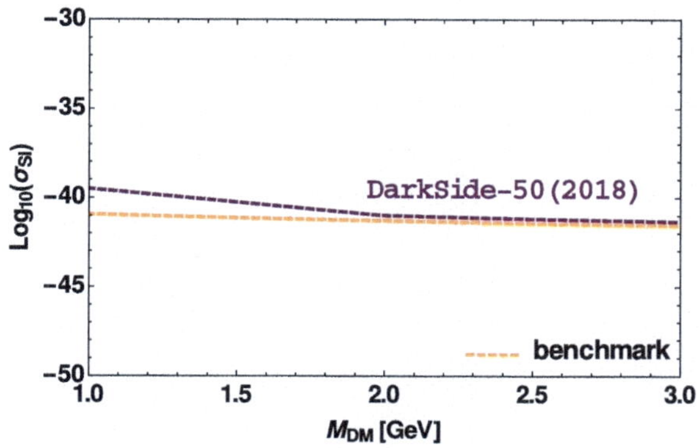

Fig. 33.3 WIMP-nucleon spin-independent cross-section versus DM mass. Here the dashed line corresponds to DarkSise-50 [11]

where $\mu = \left(\frac{M_n M_{\mathrm{DM}}}{M_n + M_{\mathrm{DM}}}\right)$ is the reduced mass of DM-nucleon with M_n being the nucleon mass. It is convenient to write (in cm^2) as

$$\sigma_{Z'} = 7.75 \times 10^{-42} \times n_{\mathrm{DM}}{}^2 \times \left(\frac{\mu}{1\ \mathrm{GeV}}\right)^2 \times \left(\frac{1\ \mathrm{TeV}}{\left(\frac{M_{Z'}}{g_{\mathrm{BL}}}\right)}\right)^4 . \tag{33.10}$$

From the above expression, one can see that the model can produce SI contribution below DarkSide-50 limit as well. We truncate discussion in gauge portal to this point as our emphasis is on scalar-portal dark matter for this work.

33.3.1.3 Higgs Invisible Width

Since Higgs can decay to light scalars (S, A, H_2), constraints on Higgs invisible width may be able to limit the scalar couplings. The effective Lagrangian with scalar H_2 is given as

$$\Gamma_{\mathrm{inv}} = \Gamma_{H_2 H_2} + \Gamma_{SS} + \Gamma_{AA}. \tag{33.11}$$

We utilize the upper limit on Higgs invisible width, i.e., 20% of total width in Fig. 33.4.

33.3.2 *Flavor Phenomenology: Constraints from $B \to K + \not{E}$*

The general effective Hamiltonian of $b \to s\nu\bar{\nu}$ quark level transitions at the scale of the bottom quark mass is stated as follows [13, 14]:

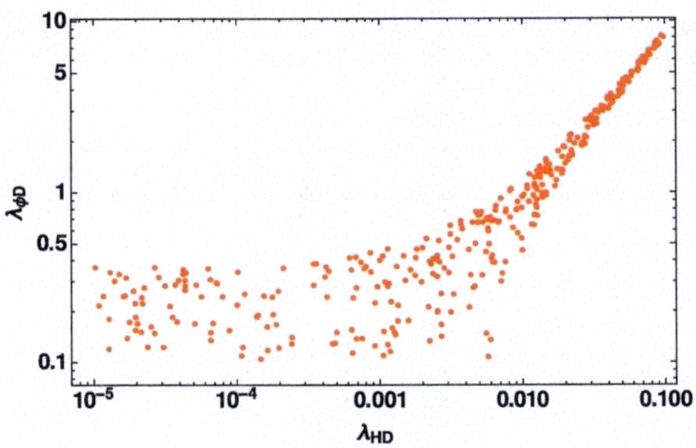

Fig. 33.4 Constraint on scalar couplings from Higgs invisible width

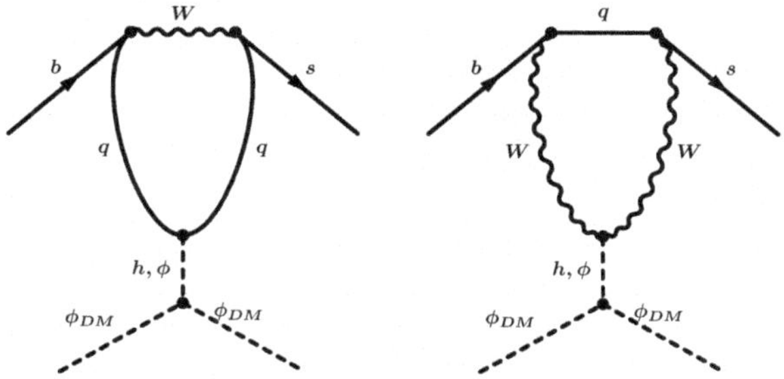

Fig. 33.5 Feynman diagrams contributing to missing energy

$$\mathcal{H}_{eff} = \frac{-4G_F}{\sqrt{2}} V_{tb} V_{ts}^* (C_L^\nu O_L^\nu + C_R^\nu O_R^\nu) + h.c., \tag{33.12}$$

where G_F is the Fermi constant, $V_{qq'}$ are the CKM elements, $\mathcal{O}_{L,R}^\nu$ are the six-dimension operators

$$O_L^\nu = \frac{\alpha^2}{4\pi}(\bar{s}\gamma_\mu P_L b)(\bar{\nu}\gamma^\mu(1-\gamma_5)\nu), \quad O_R^\nu = \frac{\alpha^2}{4\pi}(\bar{s}\gamma_\mu P_R b)(\bar{\nu}\gamma^\mu(1-\gamma_5)\nu) \tag{33.13}$$

and $C_{L,R}^\nu$ are the corresponding Wilson coefficients. Here, $C_L^\nu = -6.38 \pm 0.06$ and $C_R^\nu = 0$ in the SM, which can be nonzero in the presence of new physics.

In $B \to K+$ Missing Energy, the disagreement between theoretical and experimental branching ratio $\Longrightarrow$ The presence of a new source of missing energy $\Longrightarrow$ can be a pair of DM.

$$\mathcal{BR}(B \to K + \not{E}) = \mathcal{BR}(B \to K\nu\bar{\nu}) + \mathcal{BR}(B \to K\phi_{DM}\phi_{DM}). \tag{33.14}$$

Here the branching fraction for $B \to K\nu\bar{\nu}$ decays with respect to q^2 is given by [15]

$$\frac{d\mathcal{BR}(B \to K\nu\bar{\nu})}{dq^2} = \frac{G_F^2\alpha^2}{256\pi^5 m_B^3}\tau_B \mid V_{tb}V_{ts}^* \mid^2 \lambda^{3/2}(m_B^2, m_K^2, q^2)[f_+(q^2)]^2 \mid C_L^\nu + C_R^\nu \mid^2, \tag{33.15}$$

with $m_B(m_K)$ are the masses of $B(K)$ meson and $f_+(q^2)$ form factors are taken from the Light Cone Sum Rules (LCSRs) method [16]. All possible one loop penguin diagrams contributing to the $B \to K\phi_{DM}\phi_{DM}$ decay modes are presented in Fig. 33.5.

33.4 Impact on $B \to P(V)\nu\bar{\nu}$ Decay Modes

This section discusses the implication of the constrained new parameters on the other $b \to s\nu_l\bar{\nu}_l$ decay modes. The branching ratio expression for $B \to P\nu\bar{\nu}_l$ decay modes, where P is a pseudo scalar meson, is given in the previous section.

For the study of $B \to V\nu\bar{\nu}$ decay modes, where V denotes for vector meson, the branching fraction is given as [14]:

$$\frac{d\mathcal{BR}(B \to V\nu\bar{\nu})}{dq^2} = 3\tau_B[|A_\perp|^2 + |A_\parallel|^2 + |A_0|^2]. \tag{33.16}$$

Here, $A_{\perp,\parallel,0}$ are the $B \to V$ transversity amplitudes, which are expressed in terms of WCs and form factors as

$$A_\perp(q^2) = \frac{2N\sqrt{2\lambda(m_B^2, m_P^2, q^2)}}{m_B}[C_L^\nu + C_R^\nu]\frac{V(q^2)}{[m_V + m_B]} \tag{33.17}$$

$$A_\parallel(q^2) = \frac{-2N\sqrt{2}}{m_B}[m_V + m_B][C_L^\nu - C_R^\nu]A_1(q^2) \tag{33.18}$$

$$A_0(q^2) = \frac{-Nm_B}{m_V\sqrt{q^2}}[C_L^\nu - C_R^\nu]\Big(\Big[1 - \frac{m_V^2}{m_B^2} - \frac{q^2}{m_B^2}\Big][m_V + m_B]A_1(q^2) \tag{33.19}$$

$$- \lambda(m_B^2, m_P^2, q^2)\frac{A_2(q^2)}{[m_B + m_V]}\Big) \tag{33.20}$$

where N is the normalization factor

$$N = |V_{tb}V_{ts}^*|\Big[\frac{G_F^2\alpha^2 q^2\sqrt{\lambda(m_B^2, m_P^2, q^2)}}{3.2^10\pi^5 m_B}\Big]^{1/2}$$

with $\lambda(m_B^2, m_V^2, q^2) = m_B^4 + m_V^4 + q^4 - 2(m_B^2 m_V^2 + m_V^2 q^2 + m_B^2 q^2)$.

For numerical analysis, we use the particle masses, lifetime of $B_{(s)}$ meson, CKM matrix elements and Fermi constant values from the Particle Data Group [17]. We use the form factor results that either permit comparison with the experiment or enable us to make predictions. For $B^+ \to K^+\nu\bar{\nu}$, we used form factor fit result of LCSRs with LQCD [16, 18, 19]. The $B \to K^*, \phi$ form factors, computed in the LCSR method, which is used from [20]. The form factor for $B \to \eta, \eta'$ are taken from [21, 22] which are obtained in the LCSRs approach and the form factors for $B_c \to (D_s, D_s^*)\nu\bar{\nu}$ in the QCD and RQM method, respectively, are used from [23, 24]. The predicted branching ratios of $B \to P(V)\nu_l\bar{\nu}_l$ decay modes in the SM and in the extended $U(1)_{B-L}$ model are presented in Table 33.2.

33.5 Conclusion

In this paper, we have discussed the missing energy anomaly associated with $b \to s\nu_l\bar{\nu}_l$ channel in the context of an anomaly-free $U(1)_{B-L}$ gauge extended model. We addressed the $B \to K+$ missing energy anomaly as the presence of GeV scale scalar dark matter pair. We constrained the new parameters by using the existing

Table 33.2 Predicted branching ratios of $b \to s\nu\bar{\nu}$ channels in the proposed $U(1)_{B-L}$ model

Decay modes	Values in SM	Values in $U(1)_{B-L}$ model	Experimental limit
$B^0 \to K^{*0}\nu\bar{\nu}$	1.6285×10^{-6}	3.91×10^{-6}	$<1.8 \times 10^{-5}$
$B_s \to \phi\nu\bar{\nu}$	2.262×10^{-6}	5.423×10^{-6}	$<5.4 \times 10^{-3}$
$B \to \eta\nu\bar{\nu}$	0.61×10^{-6}	1.47×10^{-6}	...
$B \to \eta'\nu\bar{\nu}$	0.37×10^{-6}	0.97×10^{-6}	...
$B_c^+ \to D_s^+\nu\bar{\nu}$	0.185×10^{-6}	0.52×10^{-6}	...
$B_c^+ \to D_s^{*+}\nu\bar{\nu}$	0.086×10^{-6}	0.215×10^{-6}	...

experimental bounds on the dark matter and flavor physics phenomenology. We then predicted the branching ratios of other $b \to s\nu_l\bar{\nu}_l$ decay modes by using the constrained new parameters, which are found to be within reach of the experimental limit.

Acknowledgements AKY would like to thank the Department of Science and Technology (DST)-Inspire Fellowship division, Govt of India, for the financial support through ID No. IF210687.

References

1. B. Capdevila, A. Crivellin, J. Matias, Review of semileptonic B anomalies. Eur. Phys. J. ST **1**, 20 (2023)
2. R. Aaij et al., Test of lepton universality in $b \to s\ell^+\ell^-$ decays. Phys. Rev. Lett. **131**(5), 051803 (2023)
3. R. Aaij et al., Measurement of the ratio of branching fractions $\mathcal{B}(B_c^+ \to J/\psi\tau^+\nu_\tau)/\mathcal{B}(B_c^+ \to J/\psi\mu^+\nu_\mu)$. Phys. Rev. Lett. **120**(12), 121801 (2018)
4. Y.S. Amhis et al., Averages of b-hadron, c-hadron, and τ-lepton properties as of 2021. Phys. Rev. D **107**(5), 052008 (2023)
5. Eps conference. EPS conference 2023
6. D. Bečirević, G. Piazza, O. Sumensari, Revisiting $B \to K^{(*)}\nu\bar{\nu}$ decays in the standard model and beyond. Eur. Phys. J. C **83**(3), 252 (2023)
7. S. Singirala, R. Mohanta, S. Patra, Singlet scalar dark matter in $U(1)_{B-L}$ models without right-handed neutrinos. Eur. Phys. J. Plus **133**(11), 477 (2018)
8. A.V. Semenov, LanHEP: a package for automatic generation of Feynman rules in gauge models (1996)
9. G. Belanger, F. Boudjema, A. Pukhov, A. Semenov, MicrOMEGAs 2.0: a program to calculate the relic density of dark matter in a generic model. Comput. Phys. Commun. **176**, 367–382 (2007)
10. P.A.R. Ade et al., Planck 2015 results. XIII. Cosmological parameters. Astron. Astrophys. **594**, A13 (2016)
11. P. Agnes et al., DarkSide-50 532-day dark matter search with low-radioactivity Argon. Phys. Rev. D **98**(10), 102006 (2018)
12. S. Khalil, H. Okada, T. Toma, Right-handed Sneutrino dark matter in supersymmetric B-L model. JHEP **07**, 026 (2011)

13. W. Altmannshofer, A.J. Buras, D.M. Straub, M. Wick, New strategies for New Physics search in $B \to K^* \nu\bar{\nu}$, $B \to K\nu\bar{\nu}$ and $B \to X_s \nu\bar{\nu}$ decays. JHEP **04**, 022 (2009)
14. A.J. Buras, J. Girrbach-Noe, C. Niehoff, D.M. Straub, $B \to K^{(*)}\nu\overline{\nu}$ decays in the Standard Model and beyond. JHEP **02**, 184 (2015)
15. P. Colangelo, F. De Fazio, P. Santorelli, E. Scrimieri, Rare $B \to K^{(*)}$ neutrino anti-neutrino decays at B factories. Phys. Lett. B **395**, 339–344 (1997)
16. N. Gubernari, A. Kokulu, D. van Dyk, $B \to P$ and $B \to V$ form factors from B-meson light-cone sum rules beyond leading twist. JHEP **01**, 150 (2019)
17. R.L. Workman, et al., Review of particle physics. *PTEP* **2022**, 083C01 (2022)
18. W.G. Parrott, C. Bouchard, C.T.H. Davies, B → K and D→ K form factors from fully relativistic lattice QCD. Phys. Rev. D **107**(1), 014510 (2023)
19. C. Grunwald, G. Hiller, K. Kröninger, L. Nollen, More synergies from beauty, top, Z and Drell-Yan measurements in SMEFT. JHEP **11**, 110 (2023)
20. A. Bharucha, D.M. Straub, R. Zwicky, $B \to V\ell^+\ell^-$ in the standard model from light-cone sum rules. JHEP **08**, 098 (2016)
21. G. Duplancic, B. Melic, Form factors of B, $B_s \to \eta^{(\prime)}$ and D, $D_s \to \eta^{(\prime)}$ transitions from QCD light-cone sum rules. JHEP **11**, 138 (2015)
22. P. Ball, R. Zwicky, New results on $B \to \pi, K, \eta$ decay formfactors from light-cone sum rules. Phys. Rev. D **71**, 014015 (2005)
23. L.J. Cooper, C.T.H. Davies, M. Wingate, Form factors for the processes $B_c^+ \to D^0\ell^+\nu_\ell$ and $B_c^+ \to D_s^+\ell^+\ell^+(\nu\bar{\nu})$ from lattice QCD. Phys. Rev. D **105**(1), 014503 (2022)
24. M.K. Mohapatra, N. Rajeev, R. Dutta, Combined analysis of Bc→ Ds(*)μ+μ- and Bc→ Ds(*)$\nu\nu$ decays within Z' and leptoquark new physics models. Phys. Rev. D **105**(11), 115022 (2022)

The manufacturer's authorised representative in the EU is Springer Nature Customer Service Centre GmbH, Europaplatz 3, 69115 Heidelberg, Germany. If you have any concerns regarding our products, please contact ProductSafety@springernature.com

Printed and bound by CPI Group (UK) Ltd, Croydon, CR0 4YY
30/09/2026
02225298-0011